W0112286

Marshall Harvey Stone

Functional Analysis and Related Fields

Proceedings of a Conference in honor of Professor Marshall Stone, held at the University of Chicago, May 1968

Edited by Felix E. Browder

Springer-Verlag Berlin Heidelberg New York 1970

Felix E. Browder
Department of Mathematics
The University of Chicago
Chicago, Illinois 60637, U.S.A.

ISBN-13: 978-3-642-48274-8 e-ISBN-13: 978-3-642-48272-4
DOI: 10.1007/978-3-642-48272-4

This work is subject to copyright. All rights are reserved, whether the whole or part of the material is concerned, specifically those of translation, reprinting, re-use of illustrations, broadcasting, reproduction by photocopying machine or similar means, and storage in data banks. Under § 54 of the German Copyright Law where copies are made for other than private use, a fee is payable to the publisher, the amount of the fee to be determined by agreement with the publisher.
© by Springer-Verlag Berlin Heidelberg 1970. Library of Congress Catalog Card Number 74-79552.
Softcover reprint of the hardcover 1st edition 1970
Typesetting, printing, and binding: Universitätsdruckerei H. Stürtz AG, Würzburg
Title No. 1582

Preface

On May 20—24, 1968, a Conference on Functional Analysis and Related Fields was held at the Center for Continuing Education of the University of Chicago in honor of Professor MARSHALL HARVEY STONE on the occasion of his retirement from active service at the University. The Conference received support from the Air Force Office of Scientific Research under the Grant AFOSR 68-1497. The Organizing committee for this Conference consisted of ALBERTO P. CALDERON, SAUNDERS MACLANE, ROBERT G. POHRER, and FELIX E. BROWDER (Chairman).

The present volume contains some of the papers presented at the Conference. Other talks which were presented at the Conference for which papers are not included here are:

K. CHANDRASEKHARAN, "Zeta functions of quadratic fields";

J. L. DOOB, "An application of probability theory to the Choquet boundary";

P. HALMOS, "Irreducible operators";

R. KADISON, "Strong continuity of operator functions";

L. NIRENBERG, "Intrinsic norms on complex manifolds";

D. SCOTT, "Some problems and recent results in Boolean algebras";

I. M. SINGER, "A conjecture relating the Reidemeister torsion and the zeta function of the Laplacian".

A banquet in honor of Professor STONE was held during the Conference, with brief talks by S. S. CHERN, A. A. ALBERT, S. MACLANE, E. HEWITT, K. CHANDRASEKHARAN, and F. E. BROWDER (as Toastmaster), as well as a response by Professor STONE.

F. E. BROWDER

Contents

List of Contributors

BROWDER, FELIX E. — Department of Mathematics, The University of Chicago, Chicago, Ill. 60637, U. S. A.

CHERN, SHIING S. — Department of Mathematics, University of California, Berkeley, Calif. 94720, U. S. A.

DO CARMO, M. — Department of Mathematics, University of California, Berkeley, Calif. 94720, U. S. A.

HARISH-CHANDRA — The Institute for Advanced Study, School of Mathematics, Princeton, N. J. 08540, U. S. A.

HEWITT, EDWIN — Department of Mathematics, University of Washington, Seattle 5, Washington 98105, U. S. A.

KATO, TOSIO — Department of Mathematics, University of California, Berkeley, Calif. 94720, U. S. A.

KOBAYASHI, S. — Department of Mathematics, University of California, Berkeley, Calif. 94720, U. S. A.

KURODA, S. T. — University of Tokyo, Hongo, Tokyo, Japan.

MACKEY, GEORGE W. — Department of Mathematics, Harvard University, Cambridge, Mass. 02138, U. S. A.

MACLANE, SAUNDERS — Department of Mathematics, The University of Chicago, Chicago, Ill. 60637, U. S. A.

NACHBIN, LEOPOLDO — Department of Mathematics, University of Rochester, Rochester, N. Y. 14627, U. S. A.

NELSON, EDWARD — Princeton University, Fine Hall, Princeton, N. J. 08540, U. S. A.

SEGAL, IRVING — Massachusetts Institute of Technology, Cambridge, Mass. 02139, U. S. A.

WEIL, ANDRÉ — The Institute for Advanced Study, School of Mathematics, Princeton, N. J. 08540, U. S. A.

ZYGMUND, ANTONI — Department of Mathematics, The University of Chicago, Chicago, Ill. 60637, U. S. A.

Nonlinear Eigenvalue Problems and Group Invariance

By Felix E. Browder

Department of Mathematics, The University of Chicago

Introduction

In the transition from linear to nonlinear functional analysis, a very special position is occupied by the spectral theory of self-adjoint operators in Hilbert space. Its most important nonlinear generalization occurs in the framework of the topological theory of variational problems, the Morse theory and the theory of Lusternik-Schnirelmann, two of the many topics that have been gathered together in recent literature under the general title of *global analysis* or analysis on manifolds. Although the Morse theory has played an exceptionally important role in the development of topology and geometry in the past two decades, from the point of view of analytical problems and particularly problems on the existence of solutions of nonlinear partial differential equations, it seems clear that the Lusternik-Schnirelmann theory must play a more important role. The decisive reason for this is that the Lusternik-Schnirelmann theory yields existence results for critical points of *given* variational problems while in most analytic contexts, the Morse theory yields such results only for *generic* problems; and while in the context of differential topology, the usual problem is one of a class of essentially equivalent ones, for the proof of existence of eigenfunctions of a given partial differential operator it is not satisfactory to obtain results for similar or nearby problems.

It is our purpose in the present discussion to present a complete development of the Lusternik-Schnirelmann theory and of its application to nonlinear eigenvalue problems in a form which yields a generalized Sturm-Liouville theory for eigenfunctions of nonlinear elliptic boundary value problems of the form

$$\sum_{|\alpha|\leqq m} (-1)^{|\alpha|} D^{\alpha} A_{\alpha}(x, u, \ldots, D^m u) = \lambda \sum_{|\beta|\leqq k} (-1)^{|\beta|} D^{\beta} B_{\beta}(x, u, \ldots, D^k u)$$

where $k < m$, and the two operators are Euler-Lagrange operators of two multiple integral functionals:

$$A_\alpha = \frac{\partial G}{\partial \xi_\alpha}, \ g(u) = \int G(x, u, \ldots, D^m u)\, dx,$$

$$B_\beta = \frac{\partial K_\beta}{\partial \xi_\beta}, \ k(u) = \int K(x, u, \ldots, D^k u)\, dx.$$

Our discussion differs in a number of respects from earlier treatments both of the abstract Lusternik-Schnirelmann theory and of its application to eigenvalue problems. The first and most important respect is our emphasis in the applications upon the role of group invariance of the given problems as the feature which makes non-trivial results possible from the theory. The conventional applications in the past have been to even functionals and the writer has been aware for several years that the use of this hypothesis was only a special case of the application of the invariance of the problem under a group of transformations π having at least one element of finite order. We develop the theory in detail from this point of view, both because it makes the theory in its usual form more transparent and because of its significance as part of a general program of relating significant results in nonlinear functional analysis to hypotheses of group invariance (whose importance was emphasized to the writer by V. BARGMANN).

In the second place, we place a great emphasis in the general theory upon obtaining results under the weakest possible regularity assumptions upon the functionals and manifolds involved. There is a very strong motive behind this emphasis, namely the character of the applications. In the nonlinear elliptic eigenvalue problems of the form $A(u) = \lambda B(u)$ of the form which we have written above, hypotheses of differentiability upon the functionals involved beyond the C^1 framework involve burdensome hypotheses in terms of differentiability and growth conditions on the derivatives of the functions A_α and B_β. Ideally, however, the generalized eigenfunction theory for the nonlinear case ought to be valid under hypotheses of the same type as the existence theory of solutions under elliptic boundary conditions of the equation $A(u) = f$. That is in fact possible as shown below (and was shown earlier in the writer's previous papers [3], [4]) on the basis of two somewhat different techniques, transplantation and Galerkin approximations. The results of the first technique in its most general form is embodied in Theorem 9 of Section 6, while the study of the convergence of finite-dimensional Galerkin or Rayleigh-Ritz approximations to the higher order eigenfunctions for the nonlinear case is carried through in detail in Section 8.

In presenting the proof of the abstract Lusternik-Schnirelmann theorems for a C^1 function h on a C^{2-} Finsler manifold X (or on a C^1

Finsler manifold X which has a C^1 diffeomorphism with a C^{2-} manifold), we have followed a variant of the previous arguments whose purpose should be made clear. There are three basic elements to the construction of a LUSTERNIK-SCHNIRELMANN theory. The first of these elements is of a strictly topological character, relating the structure of a topological space X with a given continuous function h from X to the reals, a given set K of points playing the role of generalized "critical points" of the function h, and a deformation F of the space X over itself which eventually lowers the value of h except in neighborhoods of K. The results of the theory on this level (developed in detail in Sections 1 and 2 below) consist of the Lusternik generalized minimax principle for deformation invariant families S of subsets of X, i.e. the existence of a point x_0 such that x_0 lies in K, and

$$h(x_0) = \inf_{A \in S} \sup_{x \in A} h(x),$$

as well as of properties of K and of $K \cap h^{-1}(r)$ in terms of the category of X and minimax values of h over the families S_k of subsets of X having Lusternik-Schnirelmann category $\geqq k$ in X. The theory on this level is of an elementary topological character.

In the second step of the Lusternik-Schnirelmann argument, one specializes X to be a manifold of class C^1 and h to be a C^1 function on X with K being the set of actual critical points of h on X (i.e. points x in X such that $dh_x = 0$). The problem at this level of the argument is to construct the appropriate deformations F to which the topological theory can be applied. The problem has two principal facets: First, to obtain a group of transformations T_t of X into itself for $t \geqq 0$ obtained from a flow in X along the trajectories of a vector field related to the gradient field dh_x and second to apply a condition which guarantees that the transformation semigroup $\{T_t\colon t \geqq 0\}$ moves X in a such a way as to make sure that h decreases substantially except near the critical points of h. In the treatment of the abstract Lusternik-Schnirelmann theory on C^{2-} manifolds (and originally for smoother manifolds) Palais [12] constructed a *pseudo-gradient* vector field for h on X, i.e. a locally Lipschitzian vector field P such that for each x in $X - K$,

$$\|P_x\|_x \leqq 2\,\|dh_x\|_x, \; dh_x(P_x) \leqq -\|dh_x\|_x^2,$$

in terms of a given Finsler structure on the tangent bundle of X. The vector field P can be integrated in the neighborhood of each point x_0 of $X - K$, and the study of flows on vector fields of this forms was combined with the assumption of the condition (C) of PALAIS and SMALE [16] which essentially asserts that $K \cap h^{-1}([a, b])$ is compact for each finite interval $[a, b]$ and that $\|dh_x\|$ is bounded from below on $h^{-1}([a, b])$

on the complement of each neighborhood of K. Such a construction is only possible on C^{2-} manifolds for which an invariant concept of a locally Lipschitz vector field is possible, and its does not yield a global deformation even on $X-K$ since the trajectories are only defined up to the point that they enter K. In addition, the flows generated in this fashion are not continuous, in general, at the points of K.

To simplify this treatment (and thereby to make it possible to use the simple formulation of the topological part of the Lusternik-Schnirelmann theory), we have constructed global deformations by constructing a *quasi-gradient* vector field for h, i.e. a locally integrable vector field P on $X-K$ such that

$$\|dP_x\| \leqq 2q(x)\|dh_x\|_x; \; dh_x(P_x) \leqq -q(x)\|dh_x\|^2,$$

where

$$q(x) = \inf_{z\in K}\{d(x,z) + |h(x) - h(z)|\}.$$

Such fields always generate a continuous one-parameter semigroup of mappings T_t of X into X, a fact which is proved in Section 4, and the resulting deformation is of the appropriate type for the topological results if the functional h satisfies condition (C). The derivation of the abstract Lusternik-Schnirelmann theorems on FINSLER manifolds is given in Section 5 for C^1 functions h satisfying condition (C) on C^{2-} manifolds X, and then extended to manifolds X which are C^1 but diffeomorphic in the C^1 sense to C^{2-} manifolds. We remark that the condition (C) essentially pertains to the case of infinite dimensional manifolds and is trivially satisfied if the manifold X is compact.

To apply the abstract thery to the proof of existence of solutions of eigenvalue problems, we apply the method of Lagrange multipliers to the pair of C^1 functions h and g on the infinite dimensional Banach space B. The critical points of the restriction h_c of h to the level surface M_c of g on which $g(x)=c$ yield solutions of the eigenvalue problem $h'(u)=\xi g'(u)$ with real ξ. The existence of such eigenfunctions in infinite number is established in Section 6 under appropriate hypotheses on the functionals h and g under the crucial hypotheses that there exists a diffeomorphism φ of the level surface M_c on itself such that φ is periodic of prime period q and φ^j for all j with $1\leqq j\leqq q-1$ is fixed-point free on M_c, while $h(\varphi(x))=h(x)$. Such a transformation φ exists, obviously, whenever there exists a group of transformations π acting without fixed points (except for the identity) on M_c such that π includes an element of finite order and h is invariant under π. The crucial topological fact in this proof is based upon the results given in Section 3 that for an infinite dimensional Banach space B, $S_1(B)/Z_p$ has infinite Lusternik-Schnirelmann category.

In Section 7, these results are applied to the case of the nonlinear elliptic eigenvalue problem $A(u) = \lambda B(u)$ in the variational sense under appropriate hypotheses on the coefficient functions A_α and B_β. The hypotheses applied are generalizations of the following: There exists a real number p with $2 \leqq p < \infty$ such that for all α and β

$$|A_\alpha(x, \xi)| \leqq c(1 + |\xi|^{p-1}), \tag{1}$$

$$|B_\beta(x, \xi)| \leqq c(1 + |\xi|^{p-1}), \tag{2}$$

$$\sum_{|\alpha|=m} [A_\alpha(x, \zeta, \eta) - A_\alpha(x, \zeta', \eta)](\zeta_\alpha - \zeta'_\alpha) > 0 \tag{3}$$

for each η of lower order and $\zeta \neq \zeta'$.

$$\sum_{|\alpha| \leqq m} A_\alpha(x, \xi)\, \xi_\alpha \geqq c_0 |\xi|^p - c_1, \; (c_0 > 0). \tag{4}$$

A consequence of these conditions, or the much more general conditions assumed in Section 7, is that for the corresponding functional $g(u)$ written above, the derivative $g' \colon V \to V^*$, (where V is the closed subspace of $W^{m,p}(\Omega)$ on which the variational problem is taken), satisfies the following basic condition (S) of which the condition (C) of the problem is a consequence:

(S) *If v_j converges weakly to v in V and if*

$$(g'(v_j) - g'(v), v_j - v) \to 0$$

then v_j converges strongly to v in V.

Finally in Section 8, condition (S) for g' together with a direct argument using Galerkin approximations by finite dimensional problems yields an existence theorem for nonlinear eigenvalue problems independent of the abstract Lusternik-Schnirelmann theorem on infinite dimensional manifolds for problems invariant under a periodic linear mapping φ of prime period. In particular, this applies to the nonlinear elliptic problems with $1 < p < 2$.

The basic results of LUSTERNIK-SCHNIRELMANN were first published in an extended way in their monograph [11] in 1934, and subsequent results of various members of the Russian school are described in the books of LUSTERNIK [12], KRASNOSELSKI [10], and VAINBERG [19]. An abstract Lusternik-Schnirelmann theorem on Hilbert manifolds for smooth functions satisfying condition (C) was first proved by J. T. SCHWARTZ [17] along lines which were influenced by the earlier work of PALAIS and SMALE on the corresponding generalization of the Morse theory [16]. This theorem was extended to smooth functions on relatively smooth Finsler manifolds by PALAIS [13] who introduced the important concept of the pseudo-gradient field. The Lusternik-Schnirelmann theory was first applied to nonlinear elliptic eigenvalue problems in BROWDER

[2]. More recent improvements of the abstract theory on Finsler manifolds have been given by PALAIS in [14] and BROWDER in [4]. The application of Galerkin approximations to the Lusternik-Schnirelmann theory of eigenvalue problems was first treated in Browder [3], (see also [4]). The study of nonlinear operators from a Banach space V to V^* which satisfy condition (S) is given in Chapter 17 of BROWDER [5]. The proof that nonlinear elliptic operators under our general assumptions have realizations in the appropriate sense which satisfy condition (S) is given in the Appendix to Chapter 1 of [4].

There is an extensive literature on the relation of the Lusternik-Schnirelmann category to various homological and homotopy properties, the most extensive surveys of which were given in the Theses of R. H. FOX [8] and of GANEA [9]. (See also BERSTEIN-GANEA [1], and EILENBERG-GANEA [7]).

An interesting discussion of Lusternik-Schnirelmann theory on C^1 manifolds under minimal hypotheses is given in the Ph. D. Dissertation of STANLEY WEISS [20] at the University of Chicago in 1969, and in particular, an application of the Galerkin approximation argument under somewhat variant hypotheses. The discussion of [20] also contains a perturbation theorem on critical points of non-even functionals differing by small amounts from even functionals, extending the results along these lines of the final Chapter in KRASNOSELSKI [10].

A technically inadequate treatment of nonlinear eigenvalue problems by Lusternik-Schnirelmann theory was published by M. S. BERGER in Ann. Scuola Norm. Sup. Pisa, **20** (1966), 545—582, which contains basic errors both in the abstract theory and in the application to elliptic problems. (The corrections published in 1967 do not suffice, although they abandon the major point of the abstract theory.)

1. The topological foundation of the minimax principle

It is our purpose in the present Section to display the topological basis of the generalized minimax principle of LUSTERNIK and SCHNIRELMANN under minimal hypotheses.

Let X be a topological space, with X always assumed to be a Hausdorff space. The basic data for the minimax principle consists of three parts:

(1) A continuous real-valued function h on X;

(2) A closed subset K of X;

(3) A family S of subsets of X.

(In the fundamental application of the principle, X is a differentiable manifold, h is a differentiable function on X, and K is the set of critical point of h on X, i.e. the set of x in X at which $dh_x = 0$.)

In terms of the above data, the minimax of h over S is defined by

$$m(h, S) = \inf_{A \in S} \sup_{x \in A} h(x).$$

The minimax principle in its most general form asserts that under appropriate hypotheses, there exists a point x_0 in K such that

$$h(x_0) = m(h, S).$$

To obtain hypotheses of this type, we consider hypotheses on the class S of subsets of X and postulate in addition the existence of a suitable deformation of X over itself.

Definition (1.1). *The family S of subsets of X is said to be deformation invariant if for each A in S and each homotopy $F: X \times [0, 1] \to X$ where $F(x, 0) = x$, $f_1(x) = F(x, 1)$ for all x in X, it follows that $f_1(A)$ also lies in the class S.*

Definition (1.2). *If h is a real-valued continuous function on X, K a closed subset of X ($R^+ = \{t \mid t \in R^1, t \geqq\}$), and if F is a continuous mapping of $X \times R^+$ into X with $F(x, 0) = x$ for all x in X, then F is said to be a deformation of X satisfying the condition $c_{h,K}$ if the following holds:*

($c_{h,K}$): If $[a, b]$ is a closed finite interval of the reals, ($a < b$), such that $h^{-1}([a, b])$ contains no points of K, then there exists $t > 0$ such that if $f_t(x) = F(x, t)$ for each x in X, then

$$h(f_t(x)) \leqq a, \quad \text{whenever} \quad h(x) \leqq b.$$

The condition $c_{h,K}$ can be relativized to the class S by replacing it by the apparently weaker condition:

($c_{h,K,S}$) If $[a, b]$ is a closed finite interval, $a < b$, such that $h^{-1}([a, b])$ contains no points of K, and if A is a set in the class S, then there exists $t = t(A) > 0$ such that if x lies in A and $h(x) \leqq b$, then $h(f_t(x)) \leqq a$.

Theorem 1. *Let X be a topological space, h a continuous real-valued function on X, K a closed subset of X such that $h(K)$ is closed in the reals. Let S be a deformation invariant class of subsets of X, and suppose that*

$$m(h, S) = \inf_{A \in S} \sup_{x \in A} h(x)$$

is finite. Suppose that there exists a deformation F of X through itself satisfying the condition ($c_{h,K}$) of Definition (1.2), (or more generally, the condition ($c_{h,K,S}$) given above).

Then there exists at least one element x_0 in K such that

$$h(x_0) = m(h, S).$$

Remark. We note for later reference that the condition that $h(K)$ be closed in the reals will hold if we assume for each finite a and b that $K \cap h^{-1}([a, b])$ is compact.

Proof of Theorem 1. We suppose that the conclusion of Theorem 1 is false and deduce a contradiction. If there exists no point x_0 in K such that $h(x_0) = m(h, S)$, then $m(h, S)$ lies in the complement in the reals of the set $h(K)$, which is closed by hypothesis. Hence we may find two real numbers a and b with $a < m(h, S) < b$ such that $h^{-1}([a, b])$ does not intersect K. Moreover, since $m(h, S)$ is the infinum of the set of numbers $m_{h,A} = \sup_{x \in A} h(x)$, there exists a set A in the class S such that

$$m(h, S) \leqq \sup_{x \in A} h(x) \leqq b.$$

We now apply the hypothesis that we have a deformation F of X through itself which satisfies the condition $(c_{h,K,S})$ or the stronger condition $(c_{h,K})$. By this condition, there exists $t = t(A)$, depending on the set A from the family S constructed above, such that f_t maps $A \cap h^{-1}((-\infty, b])$ into a subset of $\{x \mid h(x) \leqq a\}$. For this value of t, let $A_1 = f_t(A)$. Since S is deformation invariant, it follows that A_1 also belongs to the class S. On the other hand,

$$\sup_{x \in A_1} h(x) \leqq a < m(h, S)$$

contradicting the definition of $m(h, S)$ as $\inf_{B \in S} \sup_{x \in B} h(x)$. This contradiction proves the Theorem. q.e.d.

Remark. The hypothesis that the family S is deformation invariant (or in a slightly more formal terminology, *ambient homotopy invariant*) seems much more natural to the present writer than the weaker hypothesis that S is *ambient isotopy invariant* introduced by PALAIS in [14]. Aside from the most important examples of such classes, those defined by the Lusternik-Schnirelmann category (which we treat in the following Sections), we note the following examples of such classes which are essentially the same as examples cited in [14]:

(a) S consists of the single set X. Then $m(h, S) = \sup_{x \in X} h(x)$.

(b) S consists of all one point subsets of X. Then

$$m(h, S) = \inf_{x \in X} h(x).$$

(c) Let Z be any topological space and let $[Z, X]$ denote the set of homotopy classes of maps of Z into X. Given ξ in $[Z, X]$, let $S = S(\xi) = \{A \mid A = g(Z)$ for some g in the homotopy class $\xi\}$. In particular, if $Z = S^k$, the k-dimensional sphere, this process associates a deformation invariant class $S(\xi)$ to each element ξ of the k-dimensional homotopy group $\pi_k(X)$.

(d) Let H_k denote the k-dimensional homology functor (with arbitrary coefficients and with respect to some given homology theory over the appropriate class of spaces). Given ζ in $H_k(X)$ with $\zeta \neq 0$, let $S = S(\zeta)$ be the class of subsets A of X such that ζ lies in the image of the injection homomorphism $H_k(i_A): H_k(A) \to H_k(X)$.

(e) Let H^k denote a k-dimensional cohomology functor, γ a non-zero element of $H^k(X)$, and let $S = S(\gamma) = \{A \mid \gamma$ is not annihilated by the restriction map $H^k(i_A): H^k(X) \to H^k(A)\}$.

2. The Lusternik-Schnirelmann category

By specializing and strengthening the results of Section 1 in terms of the LUSTERNIK-SCHNIRELMANN category, we now derive an analogue of the principle of Lusternik-Schnirelmann giving the lower bound for the number of critical points of a function on a manifold in terms of the category of the manifold.

As before, X is a topological space. If A is a subset of X, A is said to have Lusternik-Schnirelmann category $\leqq k$ in X (which we write in the form $\operatorname{cat}(A; X) \leqq k$) if A can be covered by k closed subsets of X, each of which is contractible to a point in X. The category of A in X is defined to be the least k such that $\operatorname{cat}(A; X) \leqq k$, if any such k exist. Otherwise, we set $\operatorname{cat}(A; X) = +\infty$. We also set $\operatorname{cat}(M) = \operatorname{cat}(M; M)$.

Lemma (2.1). (a) *$\operatorname{cat}(A; X) = 1$ if and only if A is contractible to a point in X.*

(b) *If $A \subset B$, then* $\operatorname{cat}(A; X) \leqq \operatorname{cat}(B; X)$. *If $B = cl(A)$, the closure of A in X, then* $\operatorname{cat}(A; X) = \operatorname{cat}(B; X)$.

(c) $\operatorname{cat}(A \cup B; X) \leqq \operatorname{cat}(A; X) + \operatorname{cat}(B; X)$.

(d) *If $F: X \times [0, 1] \to X$ is a homotopy of the identity mapping $F(x, 0) = x$ for all x in X, and if we set $f_1(x) = F(x, 1)$, $(x \in X)$, then* $\operatorname{cat}(f_1(A), X) \geqq \operatorname{cat}(A; X)$ *for each subset A of X.*

The proof of Lemma (2.1) is elementary and left to the reader.

From property (d) of Lemma (2.1), it follows that if for an integer $k \geqq 1$, we set

$$S_k = \{A \mid \operatorname{cat}(A; X) \geqq k\},$$

then each S_k is a deformation invariant class of subsets of X in the sense of Definition (1.1).

The k-th LUSTERNIK-SCHNIRELMANN minimax value of h is defined in terms of the class S_k by

$$m_k(h) = m(h, S_k).$$

We now consider the following sharpening of the condition $(c_{h,K})$:

Definition (2.1). *Let X be a Hausdorff space, K a closed subset of X, h a continuous real-valued function on X. If F is a continuous mapping of $X \times R^+$ into X with $F(x, 0) = x$ for all x in X, then F is said to be a deformation of X satisfying condition $(c'_{h,K})$ if the following is true:*

$(c'_{h,K})$ (1) *For any closed finite subinterval $[a, b]$ of the real numbers for which $h^{-1}([a, b]) \cap K = \emptyset$, there exists a constant $T > 0$ such that for any $t \geqq T$ and any x in X with $h(x) \leqq b$, we have*

$$h(f_t(x)) \leqq a,$$

(where $f_t(x) = F(x, t)$ for each x in X and $t \geqq 0$).

(2) *Let r be a real number, V a neighborhood of $K \cap h^{-1}(r)$ in X. Then there exists an interval $[a, b]$ of the reals and a real number $t > 0$ such that $a < b < r$, and for any x in X with $h(x) \leqq b$,*

$$f_t(x) \subset V \cup h^{-1}([-\infty, a]).$$

Since condition $(c'_{h,K})$ holds when both parts (1) and (2) hold and since part (1) of $(c'_{h,K})$ is a uniform version of $(c_{h,K})$ for all $t \geqq T$ rather than a single value of t, condition $(c'_{h,K})$ implies condition $(c_{h,K})$.

Theorem 2. *Let X be a paracompact Hausdorff space with the property that each point x has a neighborhood contractible in X. Let h be a continuous function from X to the reals, bounded from below on X, K a closed subset of X. Suppose that there exists a deformation F of X which satisfies the condition $(c'_{h,K})$, $(F: X \times R^+ \to X)$.*

Then the number of points in K is at least as large as $\operatorname{cat}(X)$ *if* $\operatorname{cat}(X)$ *is finite, and K is infinite if* $\operatorname{cat}(X) = +\infty$. *If h is bounded on K and there exists a largest integer k_0 for which $m_k(h)$ is finite, then* $k_0 = \operatorname{cat}(X) < +\infty$. *If for a real number r, there exists a positive integer s such that $r = m_s(h) = m_{s+1}(h)$, then there exists an infinite number of points in $K \cap h^{-1}(r)$.*

Proof of Theorem 2. Since h is bounded from below on X, $m_1(h)$, which is identical with the infimum of h on X, is finite. Since each class S_k contains its successor S_{k+1}, the sequence of numbers $\{m_k(h)\}$ does not decrease with increasing k and may become infinite for some finite integer. Let k_0 be the largest integer for which $m_k(h) < +\infty$, and if $m_k(h)$ is finite for every k, we set $k_0 = +\infty$.

Since the hypothesis $(c'_{h,K})$ is a strengthening of the condition $(c_{h,K})$, it follows from Theorem 1 that for any k for which $m_k(h)$ is finite, there exists an element x_k of K such that $h(x_k)=m_k(h)$. To prove the various assertions of Theorem 2, if therefore suffices to prove the following:

(A) *If for a real number* $r=m_s(h)=m_{s+1}(h)$, *then* $K\cap h^{-1}(r)$ *is infinite.*

(B) $\operatorname{cat}(X)=k_0$, *if* h *is bounded on* K.

Indeed, it follows from (A) that if two minimax values coincide, then K is infinite. If no two minimax values coincide, there exist as least as many points in K as there are minimax levels, i.e. at least k_0 points in K. Since by (B), $k_0=\operatorname{cat}(X)$, it follows that there are at least $\operatorname{cat}(X)$ distinct points in K. The other assertions of Theorem 2 are contained in Assertions (A) and (B).

Proof of Assertion (A). Suppose that $K\cap h^{-1}(r)$ is a finite set $\{x_1,\ldots,x_q\}$. By hypothesis, each point x_j has a neighborhood V_j in X which is contractible to a point in X. Since each V_j can be shrunk without destroying its contractibility in X, we may assume that the neighborhoods V_j are pairwise disjoint, and moreover that they are closed in X. Let V be the neighborhood of $K\cap h^{-1}(R)$ given by $V=\bigcup_{j=1}^{q} V_j$. Then V is contractible to a point in X by combining the deformations defined on each separate V_j. In particular, $\operatorname{cat}(V;X)=1$.

We now apply the property (2) of the condition $(c'_{h,K})$ to the neighborhood V of $K\cap h^{-1}(r)$. Then there exist real numbers a, b with $a<r<b$ and $t>0$ such that if $f_t(x)=F(x,t)$ for the assumed deformation F, then

$$f_t(h^{-1}((-\infty,b])\subset V\cup h^{-1}((-\infty,a]).$$

Since $r=m_{s+1}(h)=\inf_{A\in S_{s+1}}\ \sup_{x\in A} h(x)$, there exists a subset A of X with $\operatorname{cat}(A;X)\geqq s+1$ such that

$$\sup_{x\in A} h(x)\leqq b.$$

Since

$$A\subset h^{-1}((-\infty,b]),$$

it follows that

$$f_t(A)\subset V\cup h^{-1}((-\infty,a]).$$

Let $B=h^{-1}((-\infty,a])$. Since $f_t(A)$ is the image of A under a continuous deformation of X, it follows from property (d) of the category in Lemma (2.1) that

$$\operatorname{cat}(f_t(A);X)\geqq\operatorname{cat}(A;X)\geqq s+1.$$

On the other hand, by property (c) of the category in Lemma (2.1)

$$\operatorname{cat}(f_t(A); X) \leqq \operatorname{cat}(V; X) + \operatorname{cat}(B; X) = 1 + \operatorname{cat}(B, X).$$

Since $\sup_{x\in B} h(x) \leqq a$, and $\inf_{A\in S_s} \sup_{x\in A} h(x) = m_s(h) = r > a$, $\operatorname{cat}(B; X) \leqq s-1$. Hence

$$s+1 \leqq \operatorname{cat}(f_t(A); X) \leqq 1 + \operatorname{cat}(B; X) \leqq s,$$

a contradiction proving Assertion (A). q.e.d.

Proof of Assertion (B). Let k be any positive integer with $k \leqq \operatorname{cat}(X)$, ($k$ arbitrary if $\operatorname{cat}(X) = +\infty$). To prove that Assertion (B) holds, it suffices to show for each such k that S_k contains a set A with

$$\sup_{x\in A} h(x) < +\infty.$$

Since $k \leqq \operatorname{cat}(X)$, S_k is not a vacuous family since it includes X itself.

Let r be a real number such that $h(x) < r$ for all x in K. For each point x in X and $b(x) > h(x)$, there exists a real number $t(x)$ and a neighborhood $W(x)$ of x in X by condition ($c'_{h,K}$) such that for $t \geqq t(x)$ and all u in $W(x)$, $h(f_t(u)) \leqq r$. Since X is assumed to be paracompact and HAUSDORFF, there exists a partition of unity corresponding to the covering of X by the neighborhoods $W(x)$, i.e. a family of real valued functions p_β from X to the interval $[0, 1]$ such that each p_β has its support in one of the neighborhoods $W(x)$ for some point x in X, only a finite number of the p_β differ from zero on some neighborhood of each point v of X, and for every u in X, $\sum_\beta p_\beta(u) = 1$. For each β, choose a corresponding point $x = x(\beta)$ such that p_β has its support in $W(x)$, and with this choice fixed, set

$$\delta(u) = \sum_\beta p_\beta(u)\, t(x(\beta)).$$

For any β for which $p_\beta(u) \neq 0$, we know that u lies in $W(x(\beta))$. Hence for any $t \geqq t(x(\beta))$, $f_t(u)$ satisfies the inequality $h(f_t(u)) \leqq r$. Since $\delta(u)$ is a convex linear combination of the numbers $t(x_\beta)$, it is at least as large as the least of them. Hence for $t \geqq \delta(u)$ and any u in X, $h(f_t(u)) \leqq r$. Moreover, δ is a continuous function from X to the positive real numbers.

Let $f: X \times [0, 1] \to K$ be defined by

$$f(u, t) = F(u, t\delta(u)).$$

Then f is a deformation of X into a subset of $h^{-1}((-\infty, r])$, and the final image of X under this deformation is a subset A of S_k on which the supremum of h is not greater than r. q.e.d.

Since both Assertions (A) and (B) have been established, the proof of Theorem 2 is complete. q.e.d.

If we strengthen our hypotheses on X and K, we can obtain the following much stronger version of Assertion (A) in the proof of Theorem 2:

Theorem 3. *Let X be a metrizable absolute neighborhood retract, h a continuous real-valued function on X, K a closed subset of X such that $K \cap h^{-1}([a, b])$ is compact for each finite interval $[a, b]$. Suppose that there exists a deformation F of X which satisfies condition $(c'_{h,K})$. Suppose further that for a finite real number r,*

$$r = m_s(h) = m_{s+1}(h) = \cdots = m_{s+n}(h)$$

for positive integers s and n.

Then

$$\operatorname{cat}(K \cap h^{-1}(r)) \geqq n+1 .$$

The proof of Theorem 3 is based upon the following elementary lemma:

Lemma (2.1). *Let X be a metrizable absolute neighborhood retract, K_0 a compact subset of X. Then there exists a neighborhood U of K_0 in X such that*

$$\operatorname{cat}(U; X) = \operatorname{cat}(K_0; X) .$$

Proof of Lemma (2.1). Since for any neighborhood U of K_0, $\operatorname{cat}(U; X) \geqq \operatorname{cat}(K_0, X)$, it suffices to show that if $\operatorname{cat}(K_0; X) \leqq k$, we can choose a neighborhood U of K_0 such that $\operatorname{cat}(U; X) \leqq k$. If $\operatorname{cat}(K_0; X) \leqq k$, K_0 may be written as the union of k_1 subsets K_j with $k_1 \leqq k$, with each K_j a non-empty compact subset of K_0 which is contractible to a point in X. If we knew that there exists a closed neighborhood U_j of K_j in X which is also contractible to a point in X, then the union U of these sets U_j would be a neighborhood of K_0 with $\operatorname{cat}(U; X) \leqq k_1 \leqq k$. Hence it suffices to prove the existence of such a neighborhood U_j for each j.

Since K_j is contractible to a point in X, there exists a deformation $F_j: K_j \times [0, 1] \to X$ which deforms K_j into a point. We extend F_j to a mapping G_j of $(K_j \times [0, 1] \cup X \times \{0\}) \to X$ by setting

$$G_j(x, 0) = x, \ (x \in X) .$$

Then G_j is continuous, and since X is an absolute neighborhood retract and $X \times [0, 1]$ is normal, G_j may be extended to a continuous mapping H_j of a neighborhood W of $(K_j \times [0, 1] \cup X \times \{0\})$ in $X \times [0, 1]$ into X.

Since K_j is compact, there exists a closed neighborhood U_j of K_j in X such that $U_j \times [0, 1]$ is contained in W. The restriction of H_j to $U_j \times [0, 1]$ is obviously a deformation of U_j in X. The final image of U_j under this deformation can be considered as a subset of a prescribed neighborhood V_0 of the point which is the final image of K_j under the deformation F_j, provided that U_j is taken as still a smaller neighborhood of K_j. Since a metrizable absolute neighborhood retract is locally contractible, it follows that the final image of U_j under H_j is contractible in X. Hence U_j is contractible in X. q.e.d.

Proof of Theorem 3. We let $K_0 = K \cap h^{-1}(r)$. Then K_0 is compact by the hypothesis on K, and if we apply Lemma (2.1), there exists a neighborhood V of K_0 in X such that

$$\operatorname{cat}(V; X) = \operatorname{cat}(K_0; X) = q.$$

To prove Theorem 3, it suffices to show that $q \geqq (n+1)$.

We apply the property (2) of the condition $(c'_{h,K})$ with respect to the given neighborhood V of $K \cap h^{-1}(r)$. By that property, there exist real numbers a and b with $a < r < b$, and $t > 0$, such that

$$f_t(h^{-1}(-\infty, b])) \subset V \cup h^{-1}((-\infty, a]).$$

Since $r < b$, while

$$m_{s+n}(h) = \inf_{A \in S_{s+n}} \sup_{x \in A} h(x) = r,$$

there exists a subset A of X with

$$\operatorname{cat}(A; X) \geqq s + n$$

such that

$$\sup_{x \in A} h(x) \leqq b,$$

i.e. $A \subset h^{-1}((-\infty, b])$. Thus

$$f_t(A) \subset V \cup h^{-1}((-\infty, a]).$$

Let $B = h^{-1}(-\infty, a])$. Since $f_t(A)$ is the image of A under a continuous deformation of X, it follows that

$$\operatorname{cat}(f_t(A); X) \geqq \operatorname{cat}(A; X) \geqq s + n.$$

On the other hand,

$$\operatorname{cat}(f_t(A); X) \leqq \operatorname{cat}(V; X) + \operatorname{cat}(B; X).$$

Since

$$\sup_{x \in B} h(x) = a < r,$$

while

$$m_s(h) = \inf_{A \in S_s} \sup_{x \in A} h(x) = r,$$

it follows that $\operatorname{cat}(B; X) \leqq s - 1$. Finally, we obtain

$$s + n \leqq \operatorname{cat}(f_t(A); X) \leqq \operatorname{cat}(V; X) + \operatorname{cat}(B; X) \leqq q + (s-1),$$

which implies that

$$q \geqq (n+1). \quad \text{q.e.d.}$$

Since $\operatorname{cat}(K \cap h^{-1}(r); X) \leqq \operatorname{cat}(K \cap h^{-1}(r)) \leqq \dim(K \cap h^{-1}(r)) + 1$, it follows from Theorem 3 that $\dim(K \cap h^{-1}(r)) \geqq n$.

3. The Lusternik-Schnirelmann category of infinite dimensional lens spaces

The results of the Lusternik-Schnirelmann theory become useful when we can apply them to spaces with a large category, and in particular to spaces X with $\operatorname{cat}(X) = +\infty$. A particularly simple class of such spaces is described by the following theorem:

Theorem 4. *Let B be an infinite dimensional Banach space, $S_1(B)$ its unit sphere about the origin. Let T be a continuous mapping of $S_1(B)$ into itself such that $T^p =$ the identity map of $S_1(B)$ for a given prime $p \geqq 2$. Suppose that for $1 \leqq j \leqq p-1$, T^j has not fixed points in $S_1(B)$. Let X be the quotient space obtained from $S_1(B)$ by identifying any point u with $T^j(u)$ for any j.*

Then $\operatorname{cat}(X) = +\infty$.

Remark. In the particular case, $p = 2$, and the mapping $T(u) = -u$, we get the standard infinite-dimensional projective space obtained from B.

Proof of Theorem 4. The quotient map q of $S_1(B)$ onto X is a covering mapping so that $\pi_1(X)$ is the cyclic group Z_p. Since B is of infinite-dimension, $S_1(B)$ is a retract of the closed unit ball in B and hence is contractible. (This follows from the wellknown theorem of Dugundji which asserts that the closed unit ball of B has the fixed point property if and only if B is of finite dimension.) Since the higher homotopy groups $\pi_k(X)$ coincide with the corresponding groups for $S_1(B)$ for $k \geqq 2$, it follows that X is aspherical in all dimensions $\geqq 2$. Hence X is an Eilenberg-MacLane space $K(Z_p, 1)$.

The category of the spaces $K(\Pi, 1)$ was determined by Eilenberg-Ganea [7] to be the $1+$ algebraic dimension of Π if $\operatorname{alg\,dim}(\Pi) \geqq 2$. In particular, $\operatorname{cat}(K(Z_p, 1)) = +\infty$. (See also Ganea [9], page 203.)

A more direct argument can be derived using the fact that cat (X) is bounded from below by cup-long (X), where the latter denotes the maximal number of cohomology classes of positive dimension (over any coefficient domain) whose product is non-null. (For this fact, see BERSTEIN-GANEA [1], FOX [8], or GANEA [9]). The cohomology ring of $K(Z_p, 1)$ with coefficients in Z_p is a polynomial algebra on a single element of dimension 1 for $p=2$, and the tensor product of a polynomial algebra on an element of dimension 2 and an exterior algebra on an element of dimension 1 for $p>2$. (See STEENROD-EPSTEIN [18], Corollary (5.3), page 68.) In all cases, the cup-long is infinite. q.e.d.

4. The construction of deformations and quasi-gradient vector fields

We now begin the process of specializing the topological results of Sections 1 and 2 to the case of principal interest, that in which X is a differentiable manifold of class C^1 modelled on some Banach space B, h is a differentiable function of class C^1 on X, and K is the set of critical points of h. In order to apply the preceding results to this case, it is necessary (under appropriate hypotheses) to find ways of constructing deformations $F: X \times R^+ \to K$ satisfying the condition $(c'_{h,K})$.

To formulate the appropriate hypotheses, we assume the existence of a Finsler structure on the manifold X, or more precisely on the tangent bundle $T(X)$, i.e. for each point x in X, we are given a norm $\|\cdot\|_x$ on the tangent space $T_x(X)$ to X at x which is equivalent to the B-norm and with the property that in a coordinate neighborhood N_0 at x in which $T(M)$ can be trivialized as $B \times N_0$, then for each $k>1$, there exists a smaller neighborhood N_k of x such that

$$k^{-1}\|v\|_u \leq \|v\|_x \leq k\|v\|_u$$

for all u in N_k and all v in B. By duality (since for each fiber T_x^* of the cotangent bundle $T^*(X)$ of X, $T_x^*(X) = (T_x(X))^*$, the star on the right indicating the adjoint space), there exists a corresponding Finsler structure on the cotangent bundle $T^*(X)$, i.e. a norm $\|\cdot\|_x$ on each $T_x^*(X)$ satisfying the corresponding continuity conditions in x.

In terms of this Finsler structure, we may define a a metric on each component of X, where the distance $d(x, x_1)$ between two points x and x_1 is the infimum of the length of C^1 curves joining x to x_1 in X, the length being taken with respect to the Finsler structure and being defined for the C^1 curve $\{x(t): 0 \leq t \leq a\}$ as

$$\int_0^a \|x'(t)\|_{x(t)}\, dt.$$

We assume that this metric induces the same topology on each component of X as that given originally (as has been shown by PALAIS [14] if X is regular and a fortiori if X is metrizable).

Definition (4.1). *Let X be a connected C^1 manifold with a given Finsler structure on $T(X)$, h a C^1 function on X, K the set of critical points of h in X. Let q be the non-negative real-valued function on X given by*

$$q(x) = \inf_{z \in K} \{d(x, z) + |h(x) - h(z)|\}, \ (q(x) \equiv 1 \text{ if } K = \emptyset).$$

Let P be a continuous vector field on X. Then P is said to be a quasi-gradient vector field for h with respect to the given Finsler structure if the following three conditions all hold:

(a) *There exists a constant $k_0 \geqq 1$ such that for all x in X,*

$$\|P_x\|_x \leqq k_0\, q(x)\, \|d h_x\|_x,$$

the norm on the left being taken in $T_x(X)$, and that on the right in $T_x^*(X)$.

(b) *For each x in X,*

$$d h_x(P_x) \leqq -q(x)\, \|d h_x\|_x^2.$$

(c) *For each x_0 in $X - K$, we are given $\delta(x_0) > 0$, and on the interval $[0, \delta(x_0))$, a trajectory of the vector field P*

$$x(t) = \zeta_t(x_0)$$

starting at x_0 such that there exists a neighborhood $N(x_0)$ in $X - K$ with the mapping $[x, t] \to \zeta_t(x)$ well-defined and continuous from $N(x_0) \times [0, \frac{1}{2}\delta(x_0))$ into X, while for $0 \leqq t < t + s < \delta(x_0)$,

$$\zeta_{t+s}(x_0) = \zeta_s(\zeta_t(x_0)).$$

We shall construct our desired deformations by studying the properties of trajectories of quasi-gradient vector fields. We begin with the following result:

Proposition (4.1) *Let X be a connected C^1 manifold with a given Finsler structure such that X is complete in the metric induced by the Finsler structure. Let h be a C^1 real-valued function on X, K the set of critical points of h in X, and suppose that h is bounded from below on X and that for each $R > 0$, the set $K_R = K \cap \{x \mid h(x) \leqq R\}$ is compact. Suppose that P is a quasi-gradient vector field for h with respect to the given Finsler structure. Then:*

(a) *For each x in $X - K$, there exists $\beta(x) > 0$ (and possibly infinite) such that $[0, \beta(x))$ is the maximal interval on which there exists a trajectory $\{x(t) : 0 \leqq t < \beta(x)\}$ in $X - K$ of the vector field P such that $x(0) = x$,* *while for each t in* $[0, \beta(x))$ *and some* $\delta(t) > 0$,

$$x(t+s) = \zeta_s(x(t)), \ (0 \leqq s < \delta(t)).$$

This trajectory is uniquely determined by the above conditions. If $\beta(x)$ *is finite, then there exists a point* x' *of* K *such that* $x(t)$ *converges to* x' *as* t *approaches* $\beta(x)$.

(b) *For each* x *in* $X - K$, *there exists a constant* M *depending only on* $h(x)$ *such that for* $0 < t, t_1 < \beta(x)$,

$$d(x(t), x(t_1)) \leqq M|t - t_1|^{\frac{1}{2}}.$$

Proof of Proposition (4.1). By condition (c) of Definition (4.1), there exists a non-trivial interval $[0, \delta(x))$ on which the trajectory $x(t) = \zeta_t(x)$ satisfies the conditions of assertion (a) of the Proposition. The uniqueness of $x(t)$ in some small interval $[0, \delta)$ follows from the condition that $x(t) = \zeta_s(x(0))$ for s small.

Suppose now that $[0, \delta)$ is a given interval with $\delta > 0$ on which we have two trajectories $\{x(t): 0 \leqq t < \delta\}$ and $\{y(t): 0 \leqq t < \delta\}$ with

$$x(0) = x = y(0),$$
$$x(t+s) = \zeta_s(x(t)), \ (0 \leqq s < \delta(t)),$$
$$y(t+s) = \zeta_s(y(t)), \ (0 \leqq s < \delta(t)),$$

for all t in $[0, \delta)$. Let R_0 be the subset of all t in $[0, \delta)$ for which $x(t) = y(t)$. By the preceding paragraph, R_0 contains a small interval about zero. Since both curves are continuous, R_0 is a closed subset of $[0, \delta)$. Suppose that t_0 lies in R_0. Then for $0 < s < \delta(t_0)$,

$$x(t_0 + s) = \zeta_s(x(t_0)) = \zeta_s(y(t_0)) = y(t_0 + s),$$

i.e. the interval $[t_0, t_0 + \delta(t_0))$ is also contained in R_0. Let $[0, \delta_1)$ be the maximal right-open interval contained in R_0. Then δ_1 lies in R_0, and by the maximality of the interval, it follows that $\delta_1 = \delta$. Hence $R_0 = [0, \delta)$ and the two curves are identical on any common subinterval of their domain of definition. Thus the uniqueness assertion of part (a) of Proposition (4.1) follows.

From the uniqueness of trajectories satisfying the conditions of part (a), it follows that there exists an unique maximal trajectory satisfying those conditions and starting at x. To complete the proof of part (a), it suffices to show that if $\beta(x) < +\infty$, then $x(t)$ converges to some point x' of K as $t \to \beta(x)-$. On the given trajectory,

$$x'(t) = P_{x(t)},$$

and hence

$$\frac{d}{dt} h(x(t)) = dh_{x(t)}(P_{x(t)}) \leqq -q(x(t)) \|dh_{x(t)}\|^2_{x(t)} \leqq 0.$$

In particular, $h(x(t))$ does not increase with increasing t, and $h(x(t)) \leqq h(x)$ for all t in $[0, \beta(x))$. In particular,

$$\begin{aligned} q(x(t)) &= \inf_{z \in K} \{d(x(t), z) + |h(x(t)) - h(z)|\} \\ &\geqq \min(1, \operatorname{dist}(x(t), K_R)) \end{aligned}$$

where $R = h(x) + 1$. By hypothesis, K_R is compact. Since dh is continuous there exist constants d and M greater than 0 such that for all x in $N_d(K_R) = \{x | \operatorname{dist}(x, K_R) < d\}$, $\|dh_x\|_x \leqq M$.

We integrate the inequality

$$\frac{d}{ds} h(x(s)) \leqq -q(x(s)) \|dh_{x(s)}\|^2_{x(s)}, \quad (0 \leqq s < \beta(x)),$$

from 0 to t, noting that both sides are continuous, and obtain:

$$h(x(t)) - h(x) \leqq -\int_0^t q(x(s)) \|dh_{x(s)}\|^2_{x(s)} \, ds.$$

Let M_0 be the infimum of the values of h on X. By hypothesis, M_0 is finite. Hence

$$\int_0^t q(x(s)) \|dh_{x(s)}\|^2_{x(s)} \, ds \leqq h(x) - h(x(t)) \leqq h(x) - M_0.$$

By condition (a) in Definition (4.1) for quasi-gradient vector fields,

$$\|P_{x(s)}\|_{x(s)} \leqq k_0 \, q(x(s)) \|dh_{x(s)}\|_{x(s)}$$

so that

$$k_0^{-2} \int_0^t [q(x(s))]^{-1} \|P_{x(s)}\|^2_{x(s)} \, ds \leqq \int_0^t q((x)s) \|dh_{x(s)}\|^2_{x(s)} \, ds \leqq h(x) - M_0.$$

Let $d_0 = \min(1, d)$, $q_0 = \sup\{q(x) : x \in N_d(K_R)\} < \infty$. For any s for which $x(s)$ lies in $X - N_d(K_R)$, $q(x(s)) \geqq d_0$. For any s for which $x(s)$ lies in $N_d(K_R)$, we have

$$\|P_{x(s)}\|_{x(s)} \leqq k_0 \, q_0 \, M.$$

Hence

$$\begin{aligned} \int_0^t \|P_{x(s)}\|^2_{x(s)} \, ds &\leqq \int_{x(s) \in N_d(K_R)} + \int_{x(s) \in X - N_d(K_R)} \cdots \\ &\leqq k_0 \, q_0 \, M \beta(x) + k_0^2 \, d_0 [h(x) - M_0]. \end{aligned}$$

This bound is independent of t for $0<t<\beta(x)$. Hence

$$\int_0^{\beta(x)} \|x'(s)\|_{x(s)}^2\, ds = \int_0^{\beta(x)} \|P_{x(s)}\|_{x(s)}^2\, ds \leqq M_1 < \infty,$$

where the constant M_1 depends only on $h(x)$, (since R depends only on $h(x)$).

On the other hand, for $0<t<t_1<\beta(x)$,

$$d(x(t), x(t_1)) \leqq \int_t^{t_1} |x'(s)|_{x(s)}\, ds \leqq (t_1-t)^{\frac{1}{2}} \Big\{\int_{t_1}^{t} |x'(s)|_{x(s)}^2\, ds\Big\}^{\frac{1}{2}} \leqq M_1^{\frac{1}{2}}(t_1-t)^{\frac{1}{2}}.$$

In particular, $\{x(t): t\to\beta(x)-\}$ is a Cauchy sequence in the complete metric space X, so that $x(t)$ converges to some point x' of X as $t\to\beta(x)-$. If x' lies in $X-K$, we can extend the trajectory $x(t)$ to the interval $[\beta(x), \beta(x)+\delta(x'))$ by setting

$$x(t) = \zeta_{t-\beta(x)}(x'), \ (\beta(x) \leqq t < \beta(x)+\delta(x')).$$

The resulting extension is still a continuous trajectory of the vector field P satisfying our transition conditions, which contradicts the maximality of the interval $[0, \beta(x))$. Hence x' lies in K. Thus the proof of part (a) is complete.

For the proof of part (b), we note that for $\beta(x)<+\infty$, the inequality of the preceding paragraph establishes the desired result. Suppose $\beta(x)=+\infty$. Applying the argument used in the proof of part (a), let

$$R_1=\{t|\, x(t)\in N_d(K_R)\},\ R_2=\{t|\, x(t)\in X-N_d(K_R)\}.$$

There exists a constant k, depending only on $h(x)$, such that for t in R_1, $\|x'(t)\|_{x(t)} \leqq k$. By the argument of part (a),

$$\int_{R_2} \|x'(s)\|_{x(s)}^2 \leqq k_0^2\, d_0(h(x)-M_0).$$

Suppose first that $0<t<t_1$ with $(t_1-t)\leqq 1$. Then

$$d(x(t_1), x(t)) \leqq \int_t^{t_1} \|x'(s)\|_{x(s)}\, ds = \Big(\int_{[t,t_1]\cap R_1} + \int_{[t,t_1]\cap R_2}\Big) |x'(s)|_{x(s)}\, ds.$$

The first integral in the sum has its integrand $\leqq k$, and hence

$$\int_{[t,t_1]\cap R_1} |x'(s)|_{x(s)} ds \leqq k(t_1-t) \leqq k(t_1-t)^{\frac{1}{2}}.$$

For the second integral, we have

$$\int_{[t,t_1]\cap R_2} \|x'(s)\|_{x(s)}\, ds \leqq (t_1-t)^{\frac{1}{2}} \Big\{\int_{R_2} \|x'(s)\|_{x(s)}^2\, ds\Big\}^{\frac{1}{2}} \leqq k_2(t_1-t)^{\frac{1}{2}}.$$

Adding both inequalities, we obtain the desired result.

In the other case, $(t_1 - t) \geqq 1$. Let D_0 be the diameter of $N_d(K_R)$. If the interval $[t, t_1]$ contains no points of R_1, then the desired inequality follows as in the preceding paragraph. If R_1 does intersect $[t, t_1]$, let $t_2 = \inf\{t \mid t \in R_1 \cap [t, t_1]\}$, $t_3 = \sup\{t \mid t \in R_1 \cap [t, t_1]\}$. By the continuity of x,

$$d(x(t_2), x(t_3)) \leqq D_0.$$

If $t \neq t_2$, the closed interval $[t, t_2]$ lies in R_2, and we have

$$d(x(t), x(t_2)) \leqq k(t_2 - t)^{\frac{1}{2}}.$$

If $t_3 \neq t_1$, the closed interval $[t_3, t_1]$ lies in R_2, and we have

$$d(x(t_1), x(t_3)) \leqq k(t_1 - t_3)^{\frac{1}{2}}.$$

Obviously, if either of the equalities holds, i.e. $t = t_2$ or $t_3 = t_1$, the inequalities on the distances are still valid. Finally,

$$\begin{aligned} d(x(t_1), x(t)) &\leqq d(x(t_1), x(t_3)) + d(x(t_3), x(t_2)) + d(x(t_2), x(t)) \\ &\leqq k(t_1 - t_3)^{\frac{1}{2}} + D_0 + k(t_2 - t)^{\frac{1}{2}} \\ &\leqq 2k(t_1 - t)^{\frac{1}{2}} + D_0 \\ &\leqq (2k + D_0)(t_1 - t)^{\frac{1}{2}}. \end{aligned}$$

Hence the proof of part (b) is complete, and with it, the proof of Proposition (4.1). q.e.d.

Proposition (4.2). *Let X be a connected C^1 manifold with a given Finsler structure such that X is complete in the induced metric. Let h be a real-valued C^1 function on X, K the set of critical points of h in X. Suppose that h is bounded from below on X, and that for each $R > 0$, the set $K_R = \{K \cap \{x \mid h(x) \leqq R\}\}$ is compact. Suppose that P is a quasi-gradient vector field for h with respect the given Finsler structure. Then*:

(a) *We may define a family of mappings T_t of X into X as follows for $t \geqq 0$*:

(1) *$T_t(x) = x$ for all x in K, $t \geqq 0$.*

(2) *For x in $X - K$, $0 \leqq t < \beta(x)$, where $[0, \beta(x))$ is the maximal interval of existence of the trajectory $x(t)$ of part* (a) *of Proposition* (4.1), *we set $T_t(x) = x(t)$.*

(3) *For x in $X - K$, $t \geqq \beta(x)$, we set $T_t(x) = x'$, where x' is the point of K which is the limit of $x(t)$ as $t \to \beta(x)-$. (This case is possible only if $\beta(x) < \infty$.)*

(b) *For each $t \geqq 0$, T_t is a continuous self-mapping of X. The family $\{T_t : t \geqq 0\}$ forms a one-parameter semi-group, i.e.*

$$T_{t+s} = T_t T_s, \quad (0 \leqq s, t).$$

(c) *The mapping F of* $X \times R^+$ *into X given by*

$$F(x, t) = T_t(x)$$

is continuous.

(d) *For each x in X*, $t \geqq 0$,

$$(T_t\, x)' = P_{T_t x}.$$

Proof of Proposition (4.2). The assertion of (a) is merely the formulation of the definition of the mappings T_t, and the fact that they are well-defined follows from Proposition (4.1).

To prove the assertions of (b), (c), and (d), let us verify first that the family $\{T_t\}$ forms a one-parameter semigroup. If x is a point of K, x is a fixed point of all the mappings T_t so that the equality $T_{t+s}(x) = T_s(T_t(x))$ follows trivially. If x lies in $X - K$ and $0 \leqq t < \beta(x)$, then $T_t(x) = x(t)$, where $x(t)$ is the canonical trajectory of the vector field P described in Proposition (4.1). Suppose that $0 \leqq t < t+s < \beta(x)$. Then the two trajectories $T_s(x(t))$ and $T_{t+s}(x)$ both begin at the point $x(t)$ and both satisfy the transition condition

$$y(s+r) = \zeta_r(y(s)), \ (0 < r < \delta(s)).$$

Hence by the uniqueness result of Proposition (4.1), the two are the same, i.e. $T_s(x(t)) = T_{t+s}(x)$, or

$$T_{t+s}(x) = T_s(T_t(x)), \ 0 < t < t+s < \beta(x).$$

By continuity, this holds also for $t+s = \beta(x)$. Suppose finally that either $t \geqq \beta(x)$ or $t+s > \beta(x)$. If $t = \beta(x)$, $T_t(x) = x'$ which lies in K so that $T_s(T_t(x)) = x'$. On the other hand, $T_{t+s}(x) = x'$ by its definition, so that $T_{t+s}(x) = T_s(T_t(x))$ if $t \geqq \beta(x)$. If $t < \beta(x)$, $t+s > \beta(x)$, then $T_{t+s}(x) = x'$, while the canonical trajectory of P starting at $T_t(x)$ is simply $T_s(T_t(x)) = x(t+s)$. If $s \geqq \beta(x) - t$, as is the case. $T_s(T_t(x))$ is defined to be the limit of $x(t+s)$ as $s \to \beta(x) - t$, (i.e. x'). Hence, in this case, we also have $T_{t+s}(x) = x' = T_s(T_t(x))$. If we combine all these cases, we obtain the general conclusion

$$T_{t+s} = T_s\, T_t = T_t\, T_s, \ (0 \leqq s, t).$$

We turn now to proving the continuity of the individual mappings T_t as self-mappings of X. To show that T_t is continuous at a given point x of X, we break the proof down into several cases:

(1) x lies in $X - K$, $t < \beta(x)$.

(2) x lies in $X - K$, $t = \beta(x)$.

(3) x lies in K.

(4) x lies in $X - K$, $t > \beta(x)$.

To treat these four cases, we shall show first that the result in case (2) follows from that in case (1), while the result for case (4) follows from that in cases (2) and (3). To complete the argument, we then prove the continuity of T_t at x in cases (1) and (3).

Proof for case (2) using the result in case (1). Let $t=\beta(x)$ and consider $s<\beta(x)$. By the result in case (1), T_s is continuous at x. We choose a neighborhood U of x in X on which $h(u)$ is uniformly bounded, with U contained in $X-K$. For each u in U, it follows that for $t_1, t_2<\beta(u)$,

$$d(T_{t_1}(u), T_{t_2}(u)) \leqq M|t_2-t_1|^{\frac{1}{2}}$$

with a constant M depending only on $h(u)$ and therefore uniform over the neighborhood U. As $t_2 \to \beta(u)-$, we obtain the inequality

$$d(T_{t_2}(u), u') \leqq M|\beta(u)-t_1|^{\frac{1}{2}}, \quad (u' \in K).$$

If t_2 is greater that $\beta(u)$, $T_{t_2}(u)=u'$, and we obtain

$$d(T_{t_1}(u), T_{t_2}(u)) \leqq M|\beta(u)-t_1|^{\frac{1}{2}} \leqq M|t_2-t_1|^{\frac{1}{2}}.$$

Hence the inequality

$$d(T_{t_1}(u), T_{t_2}(u)) \leqq M|t_2-t_1|^{\frac{1}{2}}$$

holds for all u in U and all $t_1, t_2 \geqq 0$ with the same constant M.

For u in U,

$$\begin{aligned} d(T_t(u), T_t(x)) &\leqq d(T_t(u), T_s(u)) + d(T_s(u), T_s(x)) + d(T_s(x), T_t(x)) \\ &\leqq 2M(t-s)^{\frac{1}{2}} + d(T_s(u), T_s(x)). \end{aligned}$$

Given $\xi>0$, we first choose s so close to $t=\beta(x)$ that $2M(t-s)^{\frac{1}{2}}<\xi/2$. Then, s being fixed, we choose a neighborhood U_1 of x such that for u in U_1, $d(T_s(u), T_s(x))<\xi/2$. Then for u in $U\,U_1$,

$$d(T_t(u), T_t(x)) < \xi/2+\xi/2=\xi.$$

Since $\xi>0$ is arbitrary, it follows that T_t is continuous at x and the proof for case (2) is complete.

Proof for case (4) using cases (2) and (3). If $t>\beta(x)$, then $t=t_0+t_1$, where $t_0=\beta(x)$, $t_1=t-\beta(x)>0$. By the semigroup property of the family $\{T_t\}$, $T_t=T_{t_1}T_{t_0}$. Since $t_0=\beta(x)$, T_{t_0} is continuous at x by case (2). Since $T_{t_0}(x)=x'$, which lies in K, T_{t_1} is continuous at $T_{t_0}(x)$. Hence $T_t=T_{t_1}T_{t_0}$ is continuous at x, and the proof for case (4) is complete.

Proof for case (1). Let R be the subset of the interval $[0, \beta(x))$ of those values of t for which T_t is continuous at x. We note first that there exists $\delta(x)>0$ by Definition (4.1) and a neighborhood U of x such that for $0 \leqq s < \frac{1}{2}\delta(x)$, $T_s(u) = \zeta_s(u)$ for u in U, and $\zeta_s(u)$ is continuous in u on U. Hence R contains an interval of the form $[0, \delta)$ for some $\delta > 0$. If we shrink U to a neighborhood of x on which h is uniformly bounded, then it follows as in the proof for case (2) that there exists a constant M such that for all u in U and all $s, t \geqq 0$,

$$d(T_s(u), T_t(u)) \leqq M|t-s|^{\frac{1}{2}}.$$

It follows from the latter inequality that R is closed in $[0, \beta(x))$. Indeed, suppose we are siven a sequence $\{s_j\}$ in R with $s_j \to s$. Then for all u in U,

$$d(T_s(u), T_s(x)) \leqq 2M|s_j - s|^{\frac{1}{2}} + d(T_{s_j}(u), T_{s_j}(x)).$$

Given $\xi > 0$, we first choose s_j so close to s that $2m|s_j - s|^{\frac{1}{2}} < \xi/2$, and then for fixed j, choose u so close to x that $d(T_{s_j}(u), T_{s_j}(x)) < \xi/2$. For such a choice of u, $d(T_s(u), T_s(x)) < \xi$, and T_s is continuous at x.

Finally, if s_0 lies in R, let $v_0 = T_{s_0}(x)$. Since v_0 lies in $X - K$, there exists $\delta_1 > 0$ and a neighborhood V of v_0 such that for v in V and $0 \leqq s_1 < \delta_1$, $T_{s_1}(v) = \zeta_{s_1}(v)$. By the semigroup property for T_t, it follows since T_{s_0} is continuous at x that there exists a neighborhood U of x mapped by T_{s_0} into V and that on this neighborhood,

$$T_{s_0+s_1}(u) = \zeta_{s_1}(T_{s_0}(u)), \quad (u \in U).$$

Since T_{s_0} is continuous at x and ζ_{s_1} is continuous on V, it follows that $T_{s_0+s_1}$ is continuous at x for $0 \leqq s_1 < \delta_1$, i.e. $[s_0, s_0 + \delta_1)$ is contained in R. Finally, let $[0, \delta)$ be the maximal right open interval contained in R. Since R is closed, δ lies in R. If δ is different from $\beta(x)$, there exists a larger interval $[0, \delta + \delta_1)$ contained in R, contradicting the maximality of $[0, \delta)$. Thus, $R = [0, \beta(x))$, and T_t is continuous at x for all t in $[0, \beta(x))$. Thus the proof for case (1) is complete.

Proof for case (3). To prove that T_t is continuous at each point x of K, it suffices, since T_t can be written as the composition of $T_{t/n}$ n-times, to prove that there exists $\delta > 0$ such that T_t is continuous at x for $0 < t < \delta$.

Since dh_u is continuous in u, there exists a neighborhood U of x and a constant m_0 such that $\|dh_u\| \leqq m_0$ for all u in U. (In fact, since $dh_x = 0$, m_0 can be chosen as small as one pleases by suitable choice of U.) It follows that

$$|h(u) - h(x)| \leqq m_1 d(u, x)$$

with a suitable constant m_1 for all u in U. Since

$$q(u) \leqq d(u, x) + |h(u) - h(x)| \leqq (m_1 + 1)\, d(u, x),$$

we have

$$\|P_u\|_u \leqq k_0 (1 + m_1)\, d(u, x)\, m_0$$

for all u in U.

Let $\xi > 0$ be given. We shall show that there exists $\delta > 0$ such that δ is independent of ξ and for all t with $0 \leqq t < \delta$, T_t maps the ball of radius $\xi/2$ about x into the ball of radius ξ about x. Let u be a point with $d(u, x) < \xi/2$, and suppose that $T_t(u)$ does not always lie in the ball of radius ξ about x for all t with $0 \leqq t \leqq \delta$. Then by the continuity of $T_t(u)$ in t, it follows that there exists t in $[0, \delta]$ with $d(T_t(u), x) = \xi$ while $d(T_s(u), x) < \xi$ for all s in $[0, t)$. Since

$$\begin{aligned}\xi = d(T_t(u), x) &\leqq \int_0^t \|(T_s(u))'\|_{T_s(u)}\, ds = \int_0^t \|P_{T_s(u)}\|_{T_s(u)}\, ds \\ &\leqq \int_0^t k_0 (1 + m_1)\, m_0\, \xi\, ds \\ &\leqq k_0 (1 + m_1)\, m_0\, \xi\, \delta\end{aligned}$$

i.e.

$$\xi \leqq C\, \delta \xi, \quad (C \text{ a constant})$$

or

$$\delta \geqq C^{-1}.$$

For $\delta < C^{-1}$, therefore, we are lead to a contradiction by the assumption that $T_t(u)$ emerges from the ξ-ball about x for $0 \leqq t \leqq \delta$ for some u in the $(\xi/2)$ ball about. The continuity for T_t at x for every t follows and the proof of case (3) is thereby complete.

We have now completed the proof of parts (a) and (b) of Proposition (5.2). The proof of part (c) follows directly from the result of part (b). Indeed, let $[x, t]$ be a point of $X \times R^+$. By the argument given above, for each u in X, there exists a constant M depending only on $h(u)$ such that

$$d(T_t(u), T_s(u)) \leqq M |t - s|^{\frac{1}{2}}, \quad (t, s \geqq 0).$$

Hence

$$d(T_s(u), T_t(x)) \leqq d(T_s(u), T_t(u)) + d(T_t(u), T_t(x)).$$

If u is taken from a neighborhood U of x on which h is bounded, we have

$$d(T_s(u), T_t(u)) + d(T_t(u), T_t(x)) \leqq M |t - s|^{\frac{1}{2}} + d(T_t(u), T_t(x)).$$

Given $\xi > 0$, we choose s so close to t that

$$M |t - s|^{\frac{1}{2}} < \xi/2$$

and u so close to x that $d(T_t(u), T_t(x)) < \xi/2$, (which is possible by the continuity of T_t). For such choice of $[u, s]$ close to $[x, t]$, it follows that

$d(T_s(u), T_t(x)) < \xi$. Since $\xi > 0$ was arbitrary, this establishes the continuity of the mapping of $X \times R^+$ into X, completing the proof of part (c).

The assertion of part (d) is simply the observation that we have already established that for each fixed x, $T_t(x) = x(t)$ is indeed a trajectory of the continuous vector field P.

Thereby the proof of Proposition (4.2) is complete.

5. Lusternik-Schnirelmann theory on Finsler manifolds

In order to apply the deformations constructed in the results of Section 4 to obtain theorems of the Lusternik-Schnirelmann type for real-valued functions h on Finsler manifold, we must impose restrictions on the functions h involved of the following sort:

Definition (5.1). *Let X be a C^1 manifold with a given Finsler structure on its tangent bundle $T(X)$, h a real-valued C^1 function on X. Then h is said to satisfy condition* (C) *with respect to the Finsler structure if the following holds:*

(C) *For any sequence $\{x_j\}$ in X with $|h(x_j)|$ uniformly bounded and with $\|dh_{x_j}\|_{x_j} \to 0$, there exists an infinite subsequence $\{x_{j(k)}\}$ converging in the metric of X to some point x in X.*

In terms of this Definition, we may apply Propositions (5.1) and (5.2) to obtain the following theorem:

Theorem 5. *Let X be a C^1 manifold with a given Finsler structure and with X complete in the metric induced by the Finsler structure. Let h be a real-valued function on X of class C^1 which is bounded from below on X and which satisfies condition* (C) *with respect to the Finsler structure in the sense of Definition* (5.1). *Suppose that there exists a continuous vector field P on X which is a quasi-gradient vector field for h with respect to the Finsler structure (in the sense of Definition* (4.1).

Then:

(a) *The number of critical points of h is at least as large as the Lusternik-Schnirelmann category of X,* cat(X).

(b) *For each integer k, $1 \leq k \leq \mathrm{cat}(X)$, if*

$$m_k(h) = \inf_{A \in S_k} \sup_{x \in A} h(x),$$

then $m_1(h) < +\infty$, and for any integer k for which $m_k(h)$ is finite, there exists at least one critical point x_k of h such that

$$h(x_k) = m_k(h).$$

(c) *The sequence* $\{m_k(h)\}$ *does not decrease at any point with increasing* k. *If for positive integers* k *and* n *and some real number* r,

$$r = m_k(h) = m_{k+1}(h) = \cdots = m_{k+n}(h),$$

and if K *is the set of critical points of* H, *then*

$$\operatorname{cat}\left(K \cap h^{-1}(r); X\right) \geqq (n+1).$$

In particular,

$$\dim\left(K \cap h^{-1}(r)\right) \geqq n; \ \operatorname{cat}\left(K \cap h^{-1}(r)\right) \geqq n+1.$$

(d) *If* h *is bounded on the set of critical points* K, *then* $m_k(h)$ *is finite for every* $k \leqq \operatorname{cat}(X)$.

We derive the proof of Theorem 5 below from the following more technical result concerning the implications of the assumption of condition (C) on h for the nature of the deformation F which was constructed in Section 4:

Proposition (5.1). *Let* X *be a connected* C^1 *manifold with a Finsler structure such that* X *is complete in the induced metric. Let* h *be a real-valued function of class* C^1 *on* X *which satisfies condition* (C) *with respect to the Finsler structure and with* h *bounded from below on* X. *Suppose that there exists a quasi-gradient vector field* P *for* h *on* X *with respect to the given Finsler structure. Then*:

(a) *If* K *is the set of critical points of* h *on* X, *then for each* $R > 0$, *the set* $K_R = K \cap \{x \mid h(x) \leqq R\}$ *is compact.*

(b) *If* $F: X \times R^+ \to X$ *is the deformation defined in Propositions* (4.1) *and* (4.2) *with respect to the quasi-gradient vector field* P, *then* F *satisfies the properties of condition* $(c'_{h,K})$.

Before turning to the details of the proof of Proposition (5.1), we derive the proof of Theorem 5 from Proposition (5.1) and the results of Sections 1 and 2.

Proof of Theorem 5 using Proposition (5.1). X being a metrizable space is paracompact and HAUSDORFF. By hypothesis, h is continuous on X and bounded from below on X, while its set of critical points K is a closed subset of X. By part (a) of Proposition (5.1), $K_R = K \cap \{x \mid h(x) \leqq R\}$ is compact for each $R > 0$, and it follows that $h(K)$ is a closed subset of the reals. By part (b) of Proposition (5.1), the deformation F constructed in Section 4 with respect to the given vector field P satisfies the properties of condition $(c'_{h,K})$. Hence, by Theorem 2 of Section 2, the number of critical points in K is at least as large as $\operatorname{cat}(X)$. Hence part (a) of Theorem 5 follows.

Since the condition $(c'_{h,K})$ implies $(c_{h,K})$ and $h(K)$ is closed in the reals, it follows from Theorem 1 of Section 1 that for each integer k for which $m_k(h)$ is finite, there exists a critical point x_k in K such that $h(x_k) = m_k(h)$. Thus part (b) of Theorem 5 follows.

Since the classes S_k, consisting of the subsets A of X for which $\operatorname{cat}(A; X) \geqq k$, decrease with increasing k, the minimax values $m_k(h)$ do not decrease with increasing k. The first of these, $m_1(h)$, coincides with the infimum of h on X and is finite since h is bounded from below on X. X being a metrizable manifold is an absolute neighborhood retract with respect to the class of metrizable spaces. Since K_R is compact and F satisfies condition $(c'_{h,K})$ by Proposition (5.1), it follows from Theorem 3 that if for a real number r and two positive integers k and n

$$r = m_k(h) = \cdots = m_{k+n}(h),$$

then $\operatorname{cat}(K \cap h^{-1}(r); X) \geqq (n+1)$. Since

$$\operatorname{cat}(K \cap h^{-1}(r); X) \leqq \operatorname{cat}(K \cap h^{-1}(r)) \leqq \dim(K \cap h^{-1}(r)) + 1,$$

it follows that

$$\operatorname{cat}(K \cap h^{-1}(r)) \geqq (n+1), \ \dim(K \cap h^{-1}(r)) \geqq n.$$

Thereby, the assertions of part (c) of Theorem 5 are proved.

Similarly, the conclusion of part (d) of Theorem 5 follows from the results of Theorem 2. q.e.d.

We now turn to the proof of Proposition (5.1).

Proof of part (a) of Proposition (5.1). Let $\{x_j\}$ be an infinite sequence in K_R. If M is a lower bound for h on X, then for all j, $|h(x_j)| \leqq R + |M|$. Since each x_j is a critical point of h, we have $dh_{x_j} = 0$ and hence $\|dh_{x_j}\|_{x_j} = 0$ for all j. If we apply condition (C), it follows that there exists a convergent infinite subsequence $\{x_{j(k)}\}$ of the sequence $\{x_j\}$, with $x_{j(k)}$ converging to a point x of X as $k \to \infty$. Since K_R is closed in X, x lies in K_R. Hence K_R is compact for each $R > 0$. q.e.d.

Proof of part (b) of Proposition (5.1). To prove that the deformation D defined in Propositions (4.1) and (4.2) satisfies condition $(c'_{h,K})$, we must verify the two parts of that condition.

Let $[a, b]$ be a closed finite interval of the reals such that K does not intersect $h^{-1}([a, b])$. Since $h(K)$ is closed in the reals by the compactness of the sets K_R, there exists $d_0 > 0$ such that for all x in K, $h(x)$ lies outside the interval $[a - d_0, b + d_0]$. Hence for any point x in $h^{-1}([a, b])$ and any point z in K, $|h(x) - h(z)| \geqq d_0$. It follows that for all x in $h^{-1}([a, b])$

$$q(x) = \inf_{z \in K} \{d(x, z) + |h(x) - h(z)|\} \geqq d_0.$$

Let $x(t) = T_t(x)$ for a point x in $h^{-1}((-\infty, b])$. Then

$$\frac{d}{dt} h(x(t)) = dh_{x(t)}(x'(t)) = dh_{x(t)}(P_{x(t)}) \leqq -q(x(t)) \| dh_{x(t)} \|^2_{x(t)} \leqq 0.$$

In particular, $h(x(t))$ is a non-increasing function of t, and T_t maps $h^{-1}((-\infty, a])$ into itself for each $t \geqq 0$, and similarly for $h^{-1}((-\infty, b])$. If $x(t)$ lies in $h^{-1}([a, b])$, then

$$\frac{d}{dt} h(x(t)) \leqq -d_0 \| dh_{x(t)} \|^2_{x(t)}.$$

By condition (C), there exists a constant $d > 0$ such that for all x in $h^{-1}([a, b])$, $\|dh_x\|_x \geqq d$. Indeed, if this were not true, there would exist a sequence $\{x_j\}$ in $h^{-1}([a, b])$ with $\|dh_{x_j}\|_{x_j} \to 0$. Passing to the infinite subsequence which condition (C) guarantees, we may assume that x_j converges to x in X. By the continuity of $\|dh_u\|_u$ in u, we will have $dh_x = 0$, i.e. x lies in K. On the other hand, the continuity of h implies that x lies in $h^{-1}([a, b])$, contradicting the fact that there are no points of K in $h^{-1}([a, b])$.

Thus for all values of t for which a given $x(t)$ lies in the set $\{x | h(x) \geqq a\}$, we have $\frac{d}{dt} h(t) \leqq -d_0 d^2$. Let $T(x)$ be the length of the longest interval $[0, t]$ for which $x(s)$ lies in the set $\{x | h(x) \geqq a\}$, $(T(x) = +\infty$ if no longest interval exists). For any finite $t < T(x)$, we have

$$h(x(t)) - h(x) \leqq -d_0 d^2 t,$$

i.e.

$$t(d_0 d^2) \leqq h(x) - h(x(t)) \leqq (b - a);\ t \leqq d_0^{-1} d^2 (b - a).$$

Hence

$$T(x) \leqq d_0^{-1} d^{-2} (b - a)$$

for all x in $h^{-1}([a, b])$. For any $t_0 > d_0^{-1} d^{-2}(b - a)$ and any $t \geqq t_0$,

$$T_t(x) \text{ lies in } h^{-1}((-\infty, a]),$$

and the proof of part (1) of the condition $(c'_{h,K})$ is complete.

We now consider the proof of part (2) of the condition $(c'_{h,K})$. Let r be a real number, and consider a given neighborhood V in X of $K_0 = K \cap h^{-1}(r)$. We assert first that there exist real numbers a_0 and b_0 with $a_0 < r < b_0$ such that $K \cap h^{-1}([a_0, b_0])$ is contained in V. Indeed, if this were not true, we could find a sequence of real numbers r_j converging to r and a sequence of points $\{x_j\}$ in K such that for each j, $h(x_j) = r_j$ and x_j lies in $K - V$. Choosing $R > \sup h(x_j)$, the points x_j all lie in the compact set K_R and we may assume that x_j converges to some point x of K as $j \to \infty$. Since h is continuous, $h(x) = \lim_j h(x_j) = r$, so that x lies in $K \cap h^{-1}(r)$. Since each x_j lies in the closed set $K - V$, x lies in $K - V$, which contradicts the hypothesis that V is a neighborhood of $K \cap h^{-1}(r)$.

We may assume that $a_0 = r - d$, $b_0 = r + d$, for given $d > 0$. We consider real numbers a and b of the form $a = r - \delta$, $b = r + \delta$ with $0 < \delta < d/2$. We propose to show that for sufficiently small δ and an appropriately chosen $t > 0$,

$$T_t\left(h^{-1}((-\infty, b]\right) \subset h^{-1}((-\infty, a]) \cup V.$$

We note first that since $K \cap h^{-1}([a_0, b_0])$ is a compact subset of V, it has a positive distance d_1 from the closed set $X - V$, $d_1 > 0$. For any point x in $h^{-1}([a_0, b_0])$ which lies outside the $(d_1/2)$ neighborhood of $(K \cap h^{-1}([a_0, b_0])$, there exists a uniform constant $d_2 > 0$ by condition (C) such that $\|dh_x\|_x \geqq d_2$. On the other hand, for each point x in

$$h^{-1}\left(\left[r - \frac{d}{2}, r + \frac{d}{2}\right]\right)$$

with x lying outside of V', the $\frac{d_1}{2}$-neighborhood of $K \cap h^{-1}([a_0, b_0])$, we can estimate $q(x)$ from below by noting that for any z in K, either $h(z)$ lies in $[a_0, b_0]$ so that $d(x, z) \geqq \frac{d_1}{2}$ or else $h(z)$ lies outside of $[a_0, b_0]$ so that $|h(z) - h(z)| \geqq \frac{d}{2}$. In both cases,

$$d(x, z) + |h(x) - h(z)| \geqq \min\left(\frac{d}{2}, \frac{d_1}{2}\right) = d_3 > 0.$$

Hence $q(x) \geqq d_3$.

We may assume without loss of generality (by shrinking the neighborhood V to begin with) that $q(x)$ is bounded by a fixed constant q_0 on V and that there exists $k_1 > 0$ such that for all x in V, $\|dh_x\|_x \leqq k_1$. Then for all u in V, we have

$$\|P_u\|_u \leqq k_0\, q(u) \|dh_u\|_u \leqq k_0\, q_0\, k_1.$$

Consider a trajectory $\{y(t) : 0 \leqq t \leqq c\}$ of the vector field P with $y(0)$ lying in V', $y(s)$ lying in V for $s < c$, and with $y(c)$ lying in $X - V$. Then $d(y(c), y(0)) \geqq \frac{d_1}{2}$, and it follows that

$$d(y(0), y(c)) \leqq \int_0^c \|P_{y(s)}\|_{y(s)}\, ds \leqq k_0\, q_0\, k_1\, c.$$

We obtain the inequality

$$k_0\, q_0\, k_1\, c \geqq \frac{d_1}{2}$$

i.e.

$$c \geqq d_1 (2 k_0\, k_1\, q_0)^{-1}.$$

This is also valid if $y(0)$ lies not in V' itself, but on the boundary of V'.

Suppose now that the trajectory $\{y(t): 0 \leqq t \leqq c\}$ lies in $V - V'$, except for its final point $y(c)$. Then for all $y(s)$ with $s < c$, we have

$$\frac{d}{ds} h(y(s)) \leqq -q(y(s)) \, \|d h_{y(s)}\|^2_{y(s)} \, .$$

If we know in addition that $h(y(s))$ lies in $\left[r - \frac{d}{2}, r + \frac{d}{2}\right]$,

$\|d h_{y(s)}\|_{y(s)} \geqq d_2$ and $q(y(s)) \geqq d_3$ by the estimations already given. Hence

$$h(y(c)) - h(y(0)) \leqq -d_3 \, d_2^2 \, c \leqq -d_3 \, d_2^2 \, d_1 (2k_0 \, k_1 \, q_0)^{-1} = -d_4$$

with $d_4 > 0$.

We now choose $\delta < \min(\frac{1}{2} d, \frac{1}{2} d_4)$. Let x be any point in $h^{-1}((-\infty, r + \delta])$. We assert that if

$$t > 2\delta \, d_3^{-1} \, d_2^{-2},$$

then $T_t(x)$ lies in $h^{-1}((-\infty, r - \delta]) \cup V$. To establish this fact, we consider the trajectory $x(t) = T_t(x)$ for a given x. Let us assume that for $0 \leqq t \leqq t_0$, $h(x(t)) \geqq r - \delta > r - \frac{d}{2}$. For all values of t, $h(x(t)) \leqq r + \frac{d}{2}$ since h is non-increasing on the trajectory. We propose to show that $t_0 \leqq 2\delta \, d_3^{-1} \, d_2^{-2}$. If the trajectory lies completely outside of V', then for all t in $[0, t_0]$, $q(y(t)) \geqq d_3$ and $\|d h_{y(t)}\|_{y(t)} \geqq d_2$ for all t in $[0, t_0]$. Hence

$$h(y(t_0)) - h(y(0)) \leqq -\int_0^{t_0} q(y(s)) \, \|d h_{y(s)}\|^2 \, ds \leqq -t_0 \, d_3 \, d_2^2 .$$

Since $h(y(t_0)) \geqq r - \delta$, and $h(y(0)) = h(x) \leqq r + \delta$, we obtain

$$t_0 \, d_3 \, d_2^2 \leqq 2\delta ,$$

and hence $t_0 \leqq 2\delta \, d_3^{-1} \, d_2^{-2}$. If on the other hand, the trajectory contains a point of the closure of V', we can assume that its initial point $y(0)$ is a point of the closure of V'. The remainder of the trajectory then lies in $V \cup h^{-1}((-\infty, r - \delta])$. Indeed, otherwise we could restrict ourselves to the portion of the trajectory lying inside of V and assume that for $0 \leqq t < t_1$, $x(t)$ lies in $V - V'$ and that $x(t_1)$ lies in $X - V$. Applying our preceding calculations, it follows that

$$h(x(t_1)) \leqq h(x(0)) - d_4 < (r + \delta) - 2\delta = r - \delta .$$

Hence, the rest of the trajectory lies in $h^{-1}((-\infty, r - \delta])$.

With this, the proof of part (b) of Proposition (5.1) and thereby of Proposition (5.1) as a whole is complete. q.e.d.

Definition (5.2). *A C^1 manifold X modelled on a Banach space is said to be of class C^{2-} if the first derivatives of the coordinate transformations are locally Lipschitzian.*

Proposition (5.2). *Let X be a metrizable manifold of class C^{2-} with a given Finsler structure on its tangent bundle $T(X)$, and let h be a C^1 real-valued function on X, K the set of critical points of h on X. Then there exists a continuous quasi-gradient vector field P for h with respect to the Finsler structure.*

Proof of Proposition (5.2). We propose to construct the quasi-gradient vector field P to be locally Lipschitzian over $X-K$, which will automatically guarantee that the condition (c) of the definition of the quasi-gradient field is satisfied. The notion of P being locally Lipschitzian is meaningful since transformations from one trivialization of $T(X)$ over a neighborhood N as $N\times B$ to another can be assumed to be locally Lipschitzian by the hypothesis that X is of class C^{2-}.

We remark that it suffices to construct the field P locally in the neighborhood of each point x_0 of $X-K$. Indeed, by its definition, $P_x=0$ on K, and if we are given a covering of $X-K$ by neighborhoods U_j on each of which we have a vector field P_j which is locally Lipschitzian and satisfies the inequalities

$$\|P_{j,x}\|_x\leqq 2q(x)\|dh_x\|_x,\quad dh_x(P_{j,x})\leqq -q(x)\|dh_x\|^2,$$

then using the fact that $X-K$ is metrizable and hence paracompact and that we can find partitions of unity in any Banach space which are locally Lipschitzian, we can choose a locally finite partition of unity with locally Lipschitzian functions on $X-K$ $\{\xi_\beta\}$ with the property that the support of each ξ_β is contained in one of the neighborhoods U_j. We choose such a j for each β, and denote it by $j(\beta)$. Then it is trivial to verify that

$$P_x=\sum_\beta \xi_\beta(x)\, P_{j(\beta),x}$$

is the desired quasi-gradient field over all of $X-K$ and is locally Lipschitzian over $X-K$.

Let x_0 be a point of $X-K$. By the definition of the duality of norms in the two Banach spaces $T_{x_0}(X)$ and $T^*_{x_0}(X)=(T_{x_0}(X))^*$, we may find an element P_0 of T_{x_0} such that

$$\|P_0\|_{x_0}\leqq \tfrac{3}{2}\, q(x_0)\|dh_{x_0}\|_{x_0},\quad dh_{x_0}(P_0)\leqq -\tfrac{4}{3}\, q(x_0)\|dh_{x_0}\|^2_{x_0}.$$

Over a neighborhood U of x_0, we may consider a trivialization of $T(X)$ as $U\times B$, and consider the constant section P_0 over U. By the continuity of the functions q and dh, and the continuity of the Finsler norms, there exists a smaller neighborhood U' of x_0 such that for x in U',

$$\|P_0\|_x\leqq 2q(x)\|dh_x\|_x;\quad dh_x(P_0)\leqq -q(x)\|dh_x\|^2_x.$$

This is the desired local section of the quasi-gradient mapping over a neighborhood of x_0. q.e.d.

Theorem 6. *Let X be a C^{2-} manifold modelled on a Banach space B with a given Finsler structure on its tangent space $T(X)$ with X complete in the induced metric. Let h be a real-valued function on X of class C^1 with h bounded from below on X. Let K be the set of critical points of h on X, and suppose that h satisfies condition* (C) *with respect to the given Finsler structure on X. Then:*

(a) *The number of critical points of h on X is at least* $\operatorname{cat}(X)$.

(b) *For each integer k with $1 \leqq k \leqq \operatorname{cat}(X)$ for which $m_k(h) < +\infty$, there exists at least one critical point x_k of h such that $h(x_k) = m_k(h)$.*

(c) *If for two given positive integers k and n,*

$$r = m_k(h) = m_{k+1}(h) = \cdots = m_{k+n}(h) < +\infty,$$

then $\operatorname{cat}(K \cap h^{-1}(r); X) \geqq (n+1)$, *and hence*

$$\operatorname{cat}(K \cap h^{-1}(r)) \geqq (n+1), \ \dim(K \cap h^{-1}(r)) \geqq n.$$

(d) *If h is bounded on K, then $m_k(h)$ is finite for all $k \leqq \operatorname{cat}(X)$.*

Proof of Theorem 6. By Proposition (5.2), there exists a quasi-gradient field for h on X with respect to the given Finsler structure. Hence the conclusions of Theorem 6 follow from those of Theorem 5. q.e.d.

Theorem 7. *The conclusion of Theorem 6 remain valid if one replaces the hypothesis that X is a C^{2-} manifold by the weaker hypothesis that X is C^1-diffeomorphic with a C^{2-}-manifold X_1.*

Proof of Theorem 7. Let f be a C^1 diffeomorphism of X on X_1. We introduce a Finsler structure on X_1 by letting $\|df_x(u)\|_{f(x)} = \|u\|_x$ for each u in $T_x(X)$. With this Finsler structure, it follows that f is an isometry for the metric induced by the Finsler structure. Hence if X is complete in its induced metric, so is X_1.

Let h_1 be the real-valued C^1 function on X_1 given by $h_1(y) = h(f^{-1}(y))$ for each y in X_1. Then f maps critical points of h on critical points of h_1 and conversely, the minimax values of h and h_1 coincide, and the results of Theorem 7 for h will follow from those of Theorem 6 applied to the C^1 function h_1 on the C^{2-} manifold X_1 provided that h_1 satisfies condition (C) with respect to the Finsler structure on X_1.

Let $\{y_j\}$ be a sequence in X_1 with $h_1(y_j)$ uniformly bounded and $\|dh_{1,y_j}\|_{y_j} \to 0$. Let $x_j = f^{-1}(y_j)$ for each j. Then $h(x_j) = h_1(y_j)$ is uniformly bounded, and by the definition of the Finsler structure, $\|dh_{x_j}\|_{x_j} = \|(dh_1)_{y_j}\|_{y_j} \to 0$ as $j \to \infty$. Since h satisfies (C), it follows that there exists an infinite subsequence $\{x_{j(k)}\}$ converging to an element x of X. Then $y_{j(k)} = f(x_{j(k)})$ converges to $f(x)$ in Y as $k \to \infty$ by the continuity of f. q.e.d.

6. Nonlinear eigenvalue problems in Banach spaces

The study of nonlinear eigenvalue problems is brough within the context of the general Lusternik-Schnirelmann theory by the method of Lagrange multipliers.

Let B a be Banach space, g and h two real-valued functions of class C^1 on B. For c a real constant, we set

$$M_c = \{x \mid x \in B, \, g(x) = c\}.$$

The subset M_c of X inherits a manifold structure from the Banach space B if for each point x in M_c, $g'(x) \neq 0$. We note that in the Banach space case, the derivative g' is a continuous mapping from B to its adjoint space B^*. At each point x of M_c, its tangent space $T_x(M_c)$ can be identified with the subspace of B given by

$$T_x(M_c) \{v \mid v \in B, \, (g'(x), v) = 0\},$$

where we use the notation (w, v) for the pairing between w in the space B^* and v in the space B.

The injection of $T_x(M_c)$ into B which we have just described yields a canonical Finsler structure on the manifold M_c where the norm of v in $T_x(M_c)$ is the corresponding norm in B of its canonical image in B. We shall use this Finsler structure on M_c without further reference.

The restriction of the given function h to M_c yields a function on the manifold which we denote by h_c. To relate the differential dh_c of this function to the derivative h' of h on the space B, we use the following result:

Proposition (6.1). *Suppose that for each point x in M_c, $g'(x) \neq 0$, and that we are given a function N from M_c to B with the property that $(g'(x), N(x)) \neq 0$ for each x in M_c. Then:*

(a) *At a given point x of M_c, $(dh_c)_x = 0$ if and only if*

$$h'(x) = \xi(x) \, g'(x)$$

with

$$\xi(x) = (h'(x), N(x)) \, (g'(x), N(x))^{-1}.$$

(b) *Let*

$$\zeta(x) = \|g' \, x)\|_{B^*} \|N(x)\|_B \, |(g'(x), N(x))^{-1}|.$$

Then:

$$\|h'(x) - \xi(x) \, g'(x)\|_{B^*} \leq (1 + \zeta(x)^{-1}) \|(dh_c)_x\|_x$$

where the right-hand norm is taken in $T_x^*(M_c)$.

Remark. It follows from part (a) of Proposition (6.1) that to obtain eigenfunctions of $h'(x) = \xi g'(x)$ on M_c, it suffices to obtain critical points of the function h_c on M_c.

Proof of Proposition (6.1). An element w of B lies in the canonical image of $T_x(M_c)$ in B if and only if $(g'(x), w) = 0$. Hence each element w of B can be written in one and only one way in the form

$$w = v_w + \beta(w) N(x)$$

with v_w lying in $T_x(M_c)$ for a given x. Indeed, we observe that if such a representation exists, then $(g'(x), v_w) = 0$, and hence

$$(g'(x), w) = \beta(w)\, (g'(x), N(x))$$

so that

$$\beta(w) = (g'(x), w)\, (g'(x), N(x))^{-1}.$$

On the other hand, if we choose $\beta(w)$ to be this value, then

$$(g'(x), w - \beta(w) N(x)) = 0$$

so that $v_w = w - \beta(w) N(x)$ does lie in $T_x(M_c)$. In particular, v lies in $T_x(M_c)$ if and only if $\beta(v) = 0$.

If $(dh_c)_x = 0$ for a given x in M_c, then for each v in $T_x(M_c)$, $(dh_c)_x(v) = (h'(x), v) = 0$. Hence for each w in B,

$$\begin{aligned}(h'(x), w) &= (h'(x), v_w) + \beta(w)\, (h'(x), N(x)) \\ &= \beta(w)\, (h'(x), N(x)),\end{aligned}$$

i.e.

$$(h'(x), w) = (h'(x), N(x))\, (g'(x), N(x))^{-1} (g'(x), w), \quad (w \in B).$$

It follows that

$$h'(x) = \xi(x)\, g'(x).$$

On the other hand, if $h'(x) = \xi(x)\, g'(x)$ for a given x in M_c, then setting $v = v - \beta(v) N(x)$ for v in $T_x(M_c)$, we have

$$\begin{aligned}(dh_c)_x(v) &= (h'(x), v - \beta(v) N(x)) \\ &= (h'(x) - \xi(x)\, g'(x), v) = 0.\end{aligned}$$

Hence $(dh_c)_x = 0$. Thus the proof of part (a) is complete.

To prove part (b), we note first that

$$\begin{aligned}|\beta(w)| &= |(g'(x), w)\, (g'(x), N(x))^{-1}| \\ &\leqq \|g'(x)\|_{B^*} \|w\|_B |(g'(x), N(x))^{-1}.\end{aligned}$$

Hence for each w in B,

$$\|v_w\|_{T_x(M_c)} = \|v\|_B \leqq \|w\|_B + |\beta(w)| \cdot \|N(x)\|_B \leqq \|w\|_B (1 + \zeta(x)^{-1}).$$

It follows that for each w in B,

$$\begin{aligned}|(h'(x)-\xi(x)\,g'(x),w)| &= |(h'(x),w-\beta(w)N(x))| = |(h'(x),v_w)|\\ &= |(dh_c)_x(v_w)| \leqq \|(dh_c)_x\|_x \|v_w\|_{T_x(M_c)}\\ &\leqq (1+\zeta(x)^{-1})\,\|(dh_c)_x\|_x \|w\|_B .\end{aligned}$$

Since this is true for each w in B,

$$\|h'(x)-\xi(x)\,g'(x)\|_{B^*} \leqq (1+\zeta(x)^{-1})\,\|(dh_c)_x\|_x ,$$

and the proof of part (b) is complete. q.e.d.

Proposition (6.2). *Suppose that the following conditions hold on h and g:*

(1) *h' is compact on any subset of M_c on which h is bounded.*

(2) *g' is proper on any closed subset of M_c on which h is bounded.*

(3) *On any subset S of M_c on which h is bounded, there exists $d_S>0$ such that $\zeta(x) \geqq d_S$ for all x in S.*

(4) *On any subset S of M_c on which h is bounded, there exists $d'_S>0$ such that for all x in S, $|\xi(x)| \geqq d'_S$.*

Then: h_c satisfies condition (C) *on M_c.*

Proof of Proposition (6.2). Let $\{x_j\}$ be a sequence in M_c with $h_c(x_j) = h(x_j)$ uniformly bounded and with $\|(dh_c)_{x_j}\|_{x_j} \to 0$ as $j \to \infty$. By part (b) of Proposition (6.1), we know that

$$\|h'(x_j)-g'(x_j)\,\xi(x_j)\|_{B^*} \leqq (1+\zeta(x_j)^{-1}\|(dh_c)_{x_j}\|_{x_j} .$$

By hypothesis (3),

$$(1+\zeta(x_j)^{-1}) \leqq (1+d_S^{-1}) = c_1 < \infty .$$

Hence

$$\delta_j = \|h'(x_j)-g'(x_j)\,\xi(x_j)\|_{B^*} \to 0, \ (j\to\infty) .$$

By hypothesis (4),

$$\xi(x_j)^{-1} \leqq (d'_S)^{-1} \leqq c_2 < \infty .$$

Hence

$$\|\xi(x_j)^{-1}h'(x_j)-g'(x_j)\|_{B^*} \leqq \xi(x_j)^{-1}c_1 \leqq c_2\,\delta_j \to 0$$

where $h'(x_j)$ lies in a compact subset of B^*. Since $\xi(x_j)^{-1}$ is uniformly bounded, if we set $w_j = \xi(x_j)^{-1}h'(x_j)$, the set $\{w_j\}$ is relatively compact in B^*. Passing to an infinite subsequence, we may assume that w_j converges to an element w of B^* as $j\to\infty$. Thus for given integers j and k

$$\|g'(x_j)-g'(x_k)\|_{B^*} \leqq \|g'(x_j - w_j\| + \|w_j - w_k\| + \|w_k - g'(x_k)\| \to 0, \qquad (j, k\to\infty) .$$

Since the sequence $\{g'(x_j)\}$ is relatively compact and the sequence $\{x_j\}$ lies in M_c with $h(x_j)$ uniformly bounded, it follows from hypothesis (2) that $\{x_j\}$ is relatively compact in M_c. Hence h_c satisfies condition (C) on M_c. q.e.d.

Definition (6.1). *A mapping R of the Banach space B into B^* is said to satisfy condition* (S) *if the following holds:*

(S): *For any sequence $\{x_j\}$ in B with x_j converging weakly to x in X for which*

$$(R(x_j) - R(x),\ x_j - x) \to 0,$$

it follows that x_j converges strongly to x in B.

Proposition (6.3). *Let B be a reflexive Banach space, R a mapping of B into B^* which satisfies condition* (S). *Then R is a proper mapping on any bounded closed subset of B.*

Proof of Proposition (6.3). It suffices to show that if $\{x_j\}$ is a bounded sequence in B with $R(x_j)$ converging strongly to an element w in B^*, then we can extract a strongly convergent subsequence from the sequence $\{x_j\}$. Since B is reflexive and the sequence $\{x_j\}$ is bounded, we may assume without loss of generality that x_j converges weakly to an element x of B. Then:

$$(R(x_j) - R(x),\ x_j - x) \to (w - R(x),\ 0) = 0.$$

Applying the condition (S), we see that x_j converges strongly to x in B. q.e.d.

Proposition (6.4). *Let g be a C^1 function on B where the norm is of class C^1 on $B - \{0\}$ with $(g'(x), x) \neq 0$ for all x in M_c for a given c. Suppose that each ray from the origin intersects M_c in exactly one point. Let γ be the mapping of M_c into the unit sphere $S_1(B)$ given by $\gamma(x) = \|x\|^{-1} x$. Then γ is a C^1 diffeomorphism of M_c on $S_1(B)$, and $\|d\gamma_x\|_x$ is uniformly bounded on any subset of M_c on which $\|x\|_B$ is bounded from below.*

Proof of Proposition (6.4). Since by hypothesis each ray emanating from the origin hits M_c in exactly one point, it follows that γ is a one-to-one mapping of M_c onto $S_1(M)$. To show that γ is a diffeomorphism of M_c on $S_1(B)$, it suffices to prove that for each point x of M_c, $d\gamma_x$ is an isomorphism of $T_x(M_c)$ on $T_{\gamma(x)}(S_1(B))$.

Let x be a point of M_c, $y = \gamma(x)$. Obviously $x \neq 0$. The tangent space $T_y(S_1(B))$ can be identified with the subspace of B defined in the following way:

$$T_y(S_1(B)) = \{u \mid u \in B,\ (J(y), u) = 0\}$$

where J is the mapping of B into B^* given by

$$(J(y), y) = \|y\|_B^2, \; y \in B,$$
$$\|J(y)\|_{B^*} = \|y\|_B.$$

If B has a C^1-norm on the complement of the origin, J is continuous and $J(v)$ is simply the derivative of the function $f(v) = \frac{1}{2}\|v\|_B^2$ at the point v of B.

Let us now consider $d\gamma$, the differential of γ considered as a mapping of $B - \{0\}$ onto $S_1(B)$. A simple computation shows that

$$d\gamma_x(w) = \|x\|^{-1} w - (J(x), w)\|x\|^{-3} x.$$

An examination of the form of $d\gamma_x$ shows that $\|d\gamma_x\|_x$ is uniformly bounded on the complement of any neighborhood of 0 in B. Hence, to show that $d\gamma_x$ is an isomorphism of $T_x(M_c)$ with $T_y(S_1(B))$, it suffices to prove that it is surjective from $T_x(M_c)$ to $T_y(S_1(B))$.

Finally, let w be any element of $T_y(S_1(B))$. Since $(g'(x), x) \neq 0$, w can be written in form $w = v + \xi x$, $v \in T_x(M_c)$. Since $d\gamma_x(x) = 0$, it follows that $d\gamma_x(v) = d\gamma_x(w) = w\|x\|^{-1}$. Thus $d\gamma_x(\|x\| v) = w$, and $d\gamma_x$ is surjective. q.e.d.

Definition (6.2). *Let M be a C^1 manifold, h a real-valued C^1 function on M, and suppose that we are given a Finsler structure on the tangent bundle of M. Let ϕ be a C^1 diffeomorphism of M such that ϕ is periodic of a given prime period p, ($\phi^p = I$, the identity diffeomorphism of M). Let π be the cyclic groups of order p of transformations $\{I, \phi, \phi^2, \ldots, \phi^{p-1}\}$ and suppose that all the elements ϕ^j of π distinct from the identity act without fixed points on M. Let h be invariant under the action of ϕ, i.e. $h(\phi(x)) = h(x)$ for all x in M. Then:*

(a) *M/π is the C^1 manifold which is the quotient space of M under the group of transformations π, and $q\colon M \to M/\pi$ is the C^1 quotient mapping.*

(b) *The canonical Finsler structure on M/π is defined as follows: If z is a point of M/π and if v lies in $T_z(M/\pi)$, then we set*

$$\|v\|_z = \sup_{x \in q^{-1}(z)} \|(dq_x)^{-1}(v)\|_x.$$

(c) *The invariant function h on M induces the C^1 function h_π on M/π by $h_\pi(q(x)) = h(x)$.*

Proposition (6.5). *Let M, h, and ϕ be as in Definition* (6.2) *with a given Finsler structure on M. Then:*

(a) *If M is complete in the metric induced by its Finsler structure (assuming M to be connected), then M/π is complete in the metric induced by the Finsler structure defined in the canonical fashion by part* (b) *of Definition* (6.2).

(b) *Suppose that* $\|d\phi_x\|_x$ *is uniformly bounded over any set of* x *for which* $h(x)$ *is uniformly bounded. Then if* h *satisfies condition* (C) *with respect to the given Finsler structure on* M, *then* h_π *satisfies condition* (C) *with respect to the canonically induced Finsler structure on* M/π.

Proof of Proposition (6.5): Proof of part (a). By the definition of the Finsler structure on M/π, for any x in M and any v in $T_x(M)$,

$$\|(dq)_x(v)\|_{q(x)} \geqq \|v\|_x.$$

It follows that for any C^1 curve $\{x(t): 0 \leqq t \leqq a\}$ in M, the image of the curve under q has length at least as large as the original curve in M.

Let z_0 and z_1 be any pair of points in the connected manifold M/π. Given $\varepsilon > 0$, there exists a C^1 curve of the form $\{z(t): 0 \leqq t \leqq 1\}$ with $z(0) = z_0$, $z(1) = z_1$, and the length of the curve $\leqq (1+\varepsilon) d(z_0, z_1)$. Let x_0 be a point in M such that $q(x_0) = z_0$. Since M is a covering space of M/π, there exists one and only one curve $\{x(t): 0 \leqq t \leqq 1\}$ in M with $x(0) = x_0$ and $q(x(t)) = z(t)$ for all t in $[0, 1]$. The lifted curve is also of class C^1, and for each value in $[0, 1]$, $\|x'(t)\|_{x(t)} \leqq \|z'(t)\|_{z(t)}$ since $dq_{x(t)}(x'(t)) = z'(t)$. If C_0 is the lifted curve and C the original curve in M/π, it follows that

$$d(x_0, x_1) \leqq \text{length}(C_0) \leqq \text{length}(C) \leqq d(z_0, z_1)(1+\varepsilon).$$

On the other hand, the possible endpoints x_1 of such lifted curves C_0 must all lie in $q^{-1}(z_1)$. Since $\varepsilon > 0$ is arbitrary, there must exist at least one x_1 in the finite set $q^{-1}(z_1)$ such that

$$d(x_0, x_1) \leqq d(z_0, z_1).$$

Assume now that we are given a Cauchy sequence $\{z_j\}$ in M/π with respect to the induced metric. To show that the given sequence converges to a point z of M/π, it suffices to obtain a convergent infinite subsequence. We construct an infinite subsequence $\{z_{j(k)}\}$ such that $j(k)$ increases with k, while for $r > j(k)$,

$$d(z_{j(k)}, z_r) \leqq 2^{-k}.$$

In particular,

$$d(z_{j(k)}, z_{j(k+1)}) \leqq 2^{-k}.$$

We choose by iteration, a corresponding sequence $\{x_{j(k)}\}$ in M such that for each k,

$$d(x_{j(k)}, x_{j(k+1)}) \leqq 2^{-k}.$$

For this new sequence, we have for $m < k$,

$$d(x_{j(m)}, x_{j(k)}) \leqq \sum_{r=m+1}^{k} d(x_{j(k)}, x_{j(k-1)}) \leqq 2^{-m} \to 0, \quad (m \to \infty).$$

Hence the sequence $\{x_{j(k)}\}$ is a Cauchy sequence in the complete metrix space M and hence converges to a point x of M. It follows since q is continuous that $z_{j(k)}=q(x_{j(k)})\to q(x)=z$ in M/π. Hence M/π is a complete metric space. q.e.d.

Proof of part (b). Let x be a point in M with $|h(x)|\leqq R$, w an element of $T_x(M)$, $z=q(x)$, and $u=dq_x(w)$. Since

$$\|u\|_z=\sup_{0\leqq j\leqq(p-1)}\|(d\phi^j)_x(w)\|_{\phi^j(x)},$$

while

$$\|(d\phi^j)_x(w)\|_{\phi^j(x)}\leqq\prod_{k=0}^{j-1}\|d\phi_{\phi^k(x)}\|_{\phi^k(x)}\|w\|_x,$$

where $h(\phi^k(x))=h(x)$ so that there exists a constant $c=c(R)>1$ such that $\|d\phi_{\phi^k(x)}\|_{\phi^k(x)}\leqq c(R)$, it follows that

$$\|u\|_z\leqq c^p\|w\|_x.$$

Suppose that we are given a sequence $\{z_j\}$ in M/π with $|h_\pi(z_j)|\leqq R$ and with $\|(dh_\pi)_{z_j}\|_{z_j}\to 0$. For each j, we choose a point x_j in M such that $q(x_j)=z_j$. Then $h(x_j)=h_\pi(z_j)$ so that $|h(x_j)|\leqq R$. Since $h=h_\pi q$, we have

$$dh_{x_j}=(dh_\pi)_{z_j}dq_{x_j},$$

so that

$$\|dh_{x_j}\|\leqq\|(dh_\pi)_{z_j}\|_{z_j}\|dq_{x_j}\|_{x_j}.$$

By our preceding calculation, however, since $|h(x_j)|\leqq R$,

$$\|dq_{x_j}\|_{x_j}\leqq c(R)^p.$$

Hence $\|dh_{x_j}\|\leqq c(R)^p\|(dh_\pi)_{z_j}\|_{z_j}\to 0$. Applying condition (C) for the function h on M with respect to the given Finsler structure, we may find an infinite subsequence $\{x_{j(k)}\}$ of the sequence $\{x_j\}$ with $x_{j(k)}\to x$ as $k\to\infty$, for some point x of M. Since q is continuous from M into M/π, it follows that $z_{j(k)}=q(x_{j(k)})$ converges to $q(x)$ as $k\to\infty$. Hence h_π satisfies condition (C) with respect to the induced Finsler structure on M/π. q.e.d.

Thus the proof of Proposition (6.5) is complete.

We now combine the preceding results to derive a general theorem on the existence of infinitely many eigenfunctions for suitable nonlinear eigenvalue problems on infinite-dimensional Banach spaces B.

Theorem 8. *Let B be an infinite dimensional Banach space, g and h two real-valued C^1 functions on B such that for a given real constant c, and all u in B with $g(u)=c$, $(g'(u),u)\neq 0$. Let $M_c=\{u\mid u\in B,\ g(u)=c\}$ have the property that each ray emanating from the origin in B hits M_c in* *exactly one point and* $h(u) \geqq -k$ *on* M_c. *Suppose that the norm of* B *is a* C^{2-} *function on* $B - \{0\}$, *that* h' *is compact on any subset of* M_c *on which* h *is bounded, and that* g' *is proper on any closed subset of* M_c *on which* h *is bounded. Suppose that for each subset* S *of* M_c *on which* h *is bounded, there exists a constant* d_S *such that for each* u *in* S, *we can find an element* $N(u)$ *of* B *for which*

$$d_S |(g'(u), N(u))| \geqq \|g'(u)\| \cdot \|N(u)\|$$

and

$$|(g'(u), N(u)) \leqq d_S |(h'(u), N(u))|.$$

Let ϕ *be a* C^1 *diffeomorphism of* M_c *with itself such that* ϕ^p *is the identity for some prime* $p \geqq 2$, *with* ϕ^j *fixed-point free on* M_c *for each* j *with* $1 \leqq j \leqq p-1$, *and such that*

$$h(\phi(x)) = h(x), \ (x \in M_c).$$

Suppose that there exists a C^{2-} *diffeomorphism* ϕ_1 *of* $S_1(B)$ *such that for each* x *in* M_c,

$$\|\phi(x)\|^{-1} \phi(x) = \phi_1(\|x\|^{-1} x).$$

Then:

(a) *There exists an infinite number of distinct elements* x_k *of* M_c *such that for some real* ξ_k,

$$h'(x_k) = \xi_k \, g'(x_k).$$

(b) *Let* M_c/π *be the quotient space of* M_c *by the finite cyclic group* π *of transformations generated by* ϕ, q *the quotient map of* $M_c \to M_c/\pi$. *Then for every integer* $k \geqq 1$, *if we consider the* k*-th minimax value of* h *given by*

$$m_k(h) = \inf_{\operatorname{cat}(q(A); M_c/\pi) \geqq k} \ \sup_{x \in A} h(x),$$

and if $m_k(h)$ *is finite, then there exists an eigenfunction* x_k *on* M_c *with* $h'(x_k) = \xi_k \, g'(x_k)$ *such that* $h(x_k) = m_k(h)$.

(c) $\operatorname{cat}(M_c/\pi) = +\infty$.

(d) *Let* $K = \{x \mid x \in M_c$. *There exists a real* ξ *such that* $h'(x) = \xi g'(x)\}$. *Suppose that for a finite* r *and positive integers* k *and* n, $m_k(h) = m_{k+n}(h)$. *Then* $\operatorname{cat}(q(K \cap h^{-1}(r))) \geqq n+1$, *and* $\dim(q(K \cap h^{-1}(r))) \geqq n$.

(e) *If* h *is bounded on* K, *the set of eigenfunctions on* M_c, *then* $m_k(h)$ *is finite for every positive integer* k.

Proof of Theorem 8. By proposition (6.1), each critical point of the restricted function h_c on M_c is an eigenfunction of the pair $[g', h']$ on M_c, i.e. an element of the set K, and conversely. Hence, we consider the problem of determining the critical points of h_c on M_c. By hypothesis, the function h_t is invariant under the diffeomorphism ϕ of M_c on itself and hence induces a real-valued C^1 function $(h_c)_\pi$ on M_c/π. Moreover, x is a critical point of h_c if and only if $q(x)$ is a critical point of $(h_c)_\pi$. If we let K_π denote the set of critical points of $(h_c)_\pi$ on M_c/π, we have $K=q^{-1}(K_\pi)$, and it suffices for our various conclusions to study the set K_π.

We propose to prove Theorem 8 by applying Theorem 7 to the C^1 function $(h_c)_\pi$ on the C^1 manifold M_c/π with the Finsler structure given in Definition (6.2). By Proposition (6.4), the mapping γ which carries each x of M_c into $\gamma(x)$ in $S_1(B)$ by setting $\gamma(x)=\|x\|^{-1}x$ is a C^1 diffeomorphism of M_c onto $S_1(B)$ since $(g'(u), u) \neq 0$ for each u in M_c. By our hypothesis on ϕ, $\phi_1\gamma=\gamma\phi$ where ϕ_1 is a C^{2-} diffeomorphism of the C^{2-} manifold $S_1(B)$. Hence M_c/π has a C^1 diffeomorphism induced by γ on the C^{2-} manifold $S_1(B)/\pi_1$, where π_1 is the cyclic group of transformations of $S_1(B)$ generated by ϕ_1. Moreover, it follows from Theorem 4 of Section 3 that $\operatorname{cat}(M_c/\pi)=\operatorname{cat}(S_1(B)/\pi_1)=+\infty$.

Since h_c is bounded from below on M_c, $(h_c)_\pi$ is bounded from below on M_c/π. Since M_c is a closed subset of the complete metric space B, it is complete in its metric, which is dominated by the Finsler metric. Hence, it is complete in the Finsler metric. Thus, by Proposition (6.5) (a), M_c/π is complete in the metric induced by its canonical Finsler structure.

Thus to apply Theorem 7 to obtain the results asserted in Theorem 8, it suffices to prove that the function $(h_c)_\pi$ satisfies the condition (C) on M_c/π with respect to the canonical Finsler structure. By Proposition (6.5) (b), since $\|d\phi_x\|_x$ is bounded by hypothesis on each subset of M_c on which $h(u)$ is bounded, it suffices to prove that h_c itself satisfies condition (C) on the manifold M_c with respect to the natural Finsler structure on M_c induced by its imbedding in B. On the other hand, our hypotheses include the hypotheses of Proposition (6.2), (i.e. h' compact on subsets of M_c on which h is bounded, g' proper on such subsets, and $\zeta(x)$ and $|\xi(x)|$ bounded from below by positive constants on such subsets). Hence h satisfies condition (C) on M_c, and the proof of Theorem 8 is complete. q.e.d.

Theorem 9. *The conclusions of Theorem* 8 *remain valid if one drops the following hypotheses of that Theorem*:

$(H)_0$: (1) $(g'(u), u) \neq 0$, *for all* u *in* M_c.

(2) M_t *is intersected exactly once by each ray from the origin.*

(3) *There exists a* C^{2-} *diffeomorphism* ϕ_1 *of* $S_1(B)$ *such that* $\gamma\phi=\phi_1\gamma$ *for* γ *the radial projection of* M_c *on* $S_1(B)$;

and replaces them by the following:

$(H)_1$: (1) *Each compact subset of* M_c *is contractible in* M_c.

(2) *There exists a C^{2-} manifold M_1, a C^1 diffeomorphism ψ of M_c onto M_1, and a C^{2-} diffeomorphism ϕ_1 of M_1 such that*

$$\phi_1 \psi = \psi \phi .$$

Proof of Theorem 9. The hypotheses of the set (H_0) were used in the proof of Theorem 8 for only two purposes:

(1) To show that M_c/π is C^1 diffeomorphic to a C^{2-} manifold.

(2) To show that $\operatorname{cat}(M_c/\pi) = +\infty$.

Hypothesis (2) of $(H)_1$ implies that M_t/π is C^1 diffeomorphic to M_1/π_1, where π_1 is the cyclic group of transformations generated by ϕ_1, and the latter manifold is of class C^{2-}.

If each compact subset of M_c is contractible over M_c, then each homotopy group $\pi_k(M_c) = 0$ for $k \geqq 2$, and M_c is simply connected. Since M_c is a covering space of M_c/π, it has the same homotopy groups $\pi_k(M_c/\pi) = 0$, $(k \geqq 2)$ and $\pi_1(M_c/\pi) = \pi$. Hence M_c/π is an Eilenberg-MacLane space $K(\pi, 1)$, and the proof that $\operatorname{cat}(K(\pi, 1)) = +\infty$ for a cyclic group π of prime period $p \geqq 2$ is described in the proof of Theorem 4. q.e.d.

7. Nonlinear elliptic eigenvalue problems

We now apply the results of Section 6 concerning nonlinear eigenvalue problems in Banach spaces to the proof of the existence of infinitely many eigenfunctions for a general class of nonlinear elliptic eigenvalue problems.

Let Ω be a bounded open subset of the Euclidean n-dimensional space R^n. We let x denote the general point of R^n and dx, the element of Lebesgue n-measure on R^n. For each n-tulpe α of nonnegative integers $\alpha = (\alpha_1, \ldots, \alpha_n)$, we set

$$D^\alpha = \prod_{j=1}^{n} (\partial/\partial x_j)^{\alpha_j}, \quad |\alpha| = \sum_{j=1}^{n} \alpha_j ,$$

where for $\alpha = (0, \ldots, 0)$, D^α is the identity operator. For a given integer m, we let ξ_m denote the m-jet of a real-valued function u at a point of R^n, i.e.

$$\xi_m = \{\xi_\alpha \mid |\alpha| \leqq m\},$$

where ξ_m lies in a Euclidean space R^{s_m} with dimension depending on m. Similarly, we denote by ζ_m, a corresponding vector whose components are pure m-th derivatives

$$\zeta_m = \{\zeta_\alpha \mid |\alpha| = m\},$$

and let $\psi_{m-1} = \{\psi_\beta \mid |\beta| \leqq m-1\}$ represent the remaining components of ξ_m.

To obtain our nonlinear eigenvalue problems, we begin with two multiple integral variational problems of order m and $(m-1)$ respectively, of the form

$$g(u)=\int_{\Omega} G(x, u, Du, \ldots, D^m u)\, dx,$$

and

$$k(u)=\int_{\Omega} K(x, u, Du, \ldots, D^{m-1} u)\, dx,$$

for u a real-valued function on Ω, with G a C^1 function from $\Omega\times R^{s_m}$ to the reals, K a C^1 function from $\Omega\times R^{s_{m-1}}$ to the reals. We write

$$G(x, \xi_m)=G(x, \zeta_m, \psi_{m-1}),\ K(x, \psi_{m-1})$$

to indicate the variables upon which G and K depend. If we put $\xi_m(u)=\{D^\alpha u \mid |\alpha|\leq m\}$, and $\psi_{m-1}(u)=\{D^\beta u \mid |\beta|\leq m-1\}$, the functionals g and k can be written in a less suggestive but more precise way as

$$g(u)=\int G(x, \xi_m(u)(x))\, dx,$$
$$k(u)=\int K(x, \psi_{m-1}(x))\, dx.$$

Definition (7.1). *For each α with $|\alpha|\leq m$, let*

$$G_\alpha(x, \xi_m)=\frac{\partial G(x, \xi_m)}{\partial \xi_\alpha}$$

and for β with $|\beta|\leq m-1$, let

$$K_\beta(x, \psi_{m-1})=\frac{\partial K}{\partial \psi_\beta}(x, \psi_{m-1}).$$

Let A and B be the two partial differential operators of the form

$$A(u)=\sum_{|\alpha|\leq m}(-1)^{|\alpha|} D^\alpha G_\alpha(x, \xi_m(u)),$$
$$B(u)=\sum_{|\beta|\leq m-1}(-1)^{|\beta|} D^\beta K_\beta(x, \psi_{m-1}(u)).$$

Then by an eigenfunction of the pair $[A, B]$ is meant formally a function u such that for some real constant λ, we have

$$A(u)=\lambda B(u).$$

We shall interpet the above partial differential equation, as well as the corresponding boundary conditions, in a variational way by introducing the concept of a weak solution of the corresponding variational problem in a suitable subspace V of a Sobolev space $W^{m,p}(\Omega)$.

Let p be a real number with $1<p<+\infty$. We denote by $W^{m,p}(\Omega)$ the Banach space of functions on Ω given by

$$W^{m,p}(\Omega)=\{u \mid u\in L^p(\Omega),\ D^\alpha u\in L^p(\Omega) \text{ for } |\alpha|\leq m\},$$

where $L^p(\Omega)$ is the L^p space with respect to Lebesgue n-measure on Ω and the derivatives D^α are taken in the sense of the theory of distributions. If we introduce a norm on $W^{m,p}(\Omega)$ by setting

$$\|u\|_{m,p} = \Big(\sum_{|\alpha|\leq m} \|D^\alpha u\|^p_{L^p(\Omega)}\Big)^{1/p},$$

then with respect to this norm, $W^{m,p}(\Omega)$ is a separable Banach space which is uniformly convex and hence reflexive.

In order to define our given eigenvalue problem in terms of a functional equation on a closed subspace V of the Sobolev space $W^{m,p}(\Omega)$, we must introduce our basic analytic assumptions upon the functions G and K from which our differential operators A and B are derived. These assumptions are the following:

Assumptions on G and K:

(1) *For each fixed ξ_m, $G(x, \xi_m)$ is measurable in x on Ω. For each fixed ψ_{m-1}, $K(x, \psi_{m-1})$ is measurable in x on Ω. For each fixed x outside a nullset in Ω, $G(x, \xi_m)$ is C^1 in ξ_m on R^{s_m} and $K(x, \psi_{m-1})$ is C^1 in ψ_{m-1} on $R^{s_{m-1}}$.*

(2) *The functions G and K satisfy the following inequalities:*

$$|G(x, \xi_m) \leq c(\xi_b)(x) + c_1(\xi_b) \sum_{m-\frac{n}{p}\leq|\alpha|\leq m} |\xi_\alpha|^{s_\alpha},$$

$$|K(x, \psi_{m-1})| \leq c(\psi_b)(x) + c_1(\psi_b) \sum_{m-\frac{n}{p}\leq|\beta|\leq m-1} |\psi_\beta|^{t_\beta},$$

wnere

$$b = \left[m - \frac{n}{p}\right], \ \xi_b = \{\xi_\alpha \,|\, |\alpha| \leq b\}, \ \psi_b = \{\psi_\beta \,|\, |\beta| \leq b\},$$

c_1 is a continuous function from R^{s_b} to R^1, c is a continuous map from R^{s_b} to $L^p(\Omega)$, and $s_\alpha^{-1} \geq p^{-1} - n^{-1}(m - |\alpha|)$, $s_\beta < \infty$, $t_\beta < s_\beta$.

(3) *The functions G_α satisfy the following inequality:*

$$|G_\alpha(x, \xi_m)| \leq c_\alpha(\xi_b)(x) + c_1(\xi_b) \sum_{m-\frac{n}{p}\leq|\beta|\leq m} |\xi_\beta|^{p_{\alpha\beta}},$$

where: c_1 is a continuous function from R^{s_b} to R^1, c is a continuous function from R^{s_b} to $L^{p_\alpha}(\Omega)$, and the following inequalities hold for the exponents p_α and $p_{\alpha\beta}$: (r' = the conjugate exponent to r).

$$p_\alpha = p', \quad \textit{for} \quad |\alpha| = m,$$

$$p_\alpha > m'_\alpha, \quad \textit{for} \quad m - \frac{n}{p} \leq |\alpha| < m, \ m_\alpha^{-1} = p^{-1} - n^{-1}(m - |\alpha|),$$

$$p_\alpha = 1, \quad \textit{for} \ |\alpha| < m - \frac{n}{p}.$$

and

$$p_{\alpha\beta} \leqq p-1, \quad \text{for} \quad |\alpha|=|\beta|=m,$$

$$p_{\alpha\beta} \leqq m_\beta (m'_a)^{-1}, \quad \text{for} \quad m-\frac{n}{p} \leqq |\alpha| \leqq m, |\beta| \leqq m, |\alpha|+|\beta|<2m,$$

$$p_{\alpha\beta} \leqq m_\beta, \quad \text{for} \quad |\alpha|<m-\frac{n}{p}.$$

(4) *The functions K_β satisfy the following inequality:*

$$|K_\beta(x, \psi_{m-1})| \leqq c_\beta(\psi_b)(x)+c_1(\psi_b) \sum_{m-\frac{n}{p} \leqq |\phi| \leqq m-1} |\psi_\phi|^{p_{\beta\phi}},$$

where c_1 is a continuous function from R^{s_b} to R^1, c is a continuous function from R^{s_b} to $L^{p_\beta}(\Omega)$, and the exponents are defined as in (3).

(5) *For each x in Ω, each ψ_{m-1} in $R^{s_{m-1}}$, and two pure m-jets ζ_m and ζ'_m with $\zeta_m = \zeta'_m$, we have*

$$\sum_{|\alpha|=m} [G_\alpha(x, \zeta_m, \psi_{m-1}) - G_\alpha(x, \zeta'_m, \psi_{m-1})](\zeta_\alpha - \zeta'_\alpha) > 0.$$

(6) *There exist two continuous functions c_0 and c_1 from R^{s_b} to R^1 with $c_0(\psi_b) > 0$ for each ψ_b such that for all x, ζ_m, and ψ_{m-1}, we have*

$$\sum_{|\alpha|=m} G_\alpha(x, \zeta_m, \psi_{m-1})\zeta_\alpha \geqq c_0(\psi_b)|\zeta_m|^p - c(\psi_b) \sum_{m-\frac{n}{p} \leqq |\beta| \leqq m-1} |\psi_\beta|^{p_{m\beta}}$$

where $p_{m\beta}$ is an exponent for $p_{\alpha\beta}$ with $|\alpha|=m$.

Proposition (7.1). *Suppose that the functions G and K satisfy the above Assumptions, and for u and v in a given closed subspace V of $W^{m,p}(\Omega)$ for which the Sobolev Imbedding Theorem holds, let*

$$g(u) = \int G(x, \xi_m(u)(x))\, dx,$$
$$k(u) = \int K(x, \psi_{m-1}(u)(x))\, dx,$$

and letting $(w, w_1) = \int w(x) w_1(x)\, dx$, set

$$a(u, v) = \sum_{|\alpha| \leqq m} (G_\alpha(x, \xi_m(u)), D^\alpha v),$$
$$b(u, v) = \sum_{|\beta| \leqq m-1} (K_\beta(x, \psi_{m-1}(u)), D^\beta v).$$

Then:

(a) *The functionals g and k are well-defined, bounded on bounded subsets, and of class C^1 on V. For each u and v, in V, $a(u, v)$ and $b(u, v)$ are well-defined, and we have*

$$(g'(u), v) = a(u, v); \quad (k'(u), v) = b(u, v)$$

for all u and v in V.

(b) g' *satisfies the condition* (S) *of Section* 6, *Definition* (6.1) *and is bounded on bounded sets.*

(c) k' *is compact on each bounded closed subset of* V.

Proof of Proposition (6.1). For the proof, we refer to the Appendix to Section 1 of BROWDER [4].

Definition (6.1). *Let* G *and* K *be functions satisfying the Assumptions given above, and let* A *and* B *be the corresponding Euler-Lagrange operators*

$$A(u) = \sum_{|\alpha| \leq m} (-1)^{|\alpha|} D^\alpha G_\alpha(x, \xi_m(u)),$$
$$B(u) = \sum_{|\beta| \leq m-1} (-1)^{|\beta|} D^\beta K_\beta(x, \psi_{m-1}(u)).$$

If V *is a closed subspace of* $W^{m,p}(\Omega)$ *with the Sobolev Imbedding Theorem valid on* Ω, *then* u *from* V *is said to be an eigenfunction for* $A(u) = \lambda B(u)$, λ *real, with the natural boundary conditions corresponding to* V, *(or briefly,* u *is an eigenfunction for* (A, B) *with respect to* V *and with eigenvalue* λ*), if the following holds:*

$$a(u, v) = \lambda b(u, v), \quad \text{for all } v \text{ in } V.$$

Theorem 10. *Let* Ω *be a bounded open subset of* R^n *for which the Sobolev Imbedding Theorem is valid. Let* g *and* k *be two multiple integral functionals on* Ω *of the form*

$$g(u) = \int_\Omega G(x, \xi_m(u))\, dx, \quad k(u) = \int_\Omega K(x, \psi_{m-1}(u))\, dx,$$

where the functions G *and* K *satisfy the Assumptions stated above for a given real* p *with* $2 \leq p < \infty$. *Let* A *and* B *be the Euler-Lagrange operators for the functionals* g *and* k, *respectively, i.e.*

$$A(u) = \sum_{|\alpha| \leq m} (-1)^{|\alpha|} D^\alpha G_\alpha(x, \xi_m(u)),$$
$$B(u) = \sum_{|\beta| \leq m-1} (-1)^{|\beta|} D^\beta K_\beta(x, \psi_{m-1}(u)),$$

and let $a(u, v)$, $b(u, v)$ *be the generalized Dirichlet forms which correspond to the representation of* A *and* B *in generalized divergence form, i.e.*

$$a(u, v) = \sum_{|\alpha| \leq m} (G_\alpha(x, \xi_m(u)), D^\alpha v),$$
$$b(u, v) = \sum_{|\beta| \leq m-1} (K_\beta(x, \psi_{m-1}(u)), D^\beta v).$$

for u *and* v *in* V, *a closed subspace of* $W^{m,p}(\Omega)$. *Let* K *denote the set of eigenfunctions* u *in* V *of the pair* (A, B) *with respect to* V *and with real eigenvalues, and for a given real* c, *let* $M_c = \{u \mid u \in V, g(u) = c\}$.

Suppose the following additional conditions hold:

(1) *For u in M_c, $k(u) > 0$. Given $d > 0$, there exists a constant $d_1 > 0$ such that for u in M_c with $k(u) \geqq d$, $b(u, u) \geqq d_1$.*

(2) *There exist constants $d_0 > 0$ and $r_0 > 0$ such that for $\|u\| \geqq r_0$ or $g(u) \geqq c$, we have $a(u, u) \geqq d_0 > 0$.*

(3) *There exists a bounded linear mapping ϕ of $W^{m,p}(\Omega)$ into itself which is continuously invertible such that $\phi^p = I$ for a prime $q \geqq 2$ with ϕ^j having no fixed points except for 0 for $1 \leqq j \leqq q-1$, such that ϕ maps M_c onto itself, and $k(\phi(u)) = k(u)$ for all u in M_c.*

Then there exists an infinite number of distinct elements of K in M_c, i.e. an infinite sequence $\{u_k\}$ of elements of V with $g(u_k) = c$ such that each u_k is an eigenfunction in the sense of Definition (6.1) with the natural boundary conditions of V of

$$A(u_k) = \lambda_k B(u_k).$$

Proof of Theorem 10. We shall apply Theorem 8 with the Banach space $B = V$, with the functional g as given and with

$$h(u) = 1/k(u).$$

V, being a closed subspace of a reflxive Banach space, is reflexive.

By Proposition (7.1), g and k are C^1 real-valued functions on V. Since $k(u) > 0$ for u in M_c, h is of class C^1 on a neighborhood of M_c in V, and no other use of the differentiability hypotheses on h is made use of in the proof of Theorem 8. By proposition (6.1), the eigenfunctions of K correspond to the solutions of the equation $g'(u) = \lambda k'(u)$, while

$$h'(u) = [k(u)]^{-2}\, k'(u).$$

Hence, eigenfunctions of K in M_c correspond to the eigenfunctions u in M_c of $h'(u) = \xi g'(u)$ with $\xi \neq 0$.

It follows from hypothesis (2) of Theorem 10 that M_c is bounded and that each ray from the origin intersects M_c in exactly one point. Since M_c is bounded, it follows from Proposition (6.1) that g' is proper on M_c since g' satisfies condition (S) on a reflexive Banach space and hence is proper on bounded closed subsets of M_c by Proposition (6.3).

By the hypothesis of Theorem 10, given $d > 0$, there exists $d_1 > 0$ such that for any u in M_c with $k(u) \geqq d$, we have $b(u, u) \geqq d_1$. By Proposition (6.1), $b(u, v) = (k'(u), v)$, and by definition $h(u) = k(u)^{-1}$. Hence if $h(u) \leqq d^{-1}$, we have

$$(h'(u), u) = k(u)^{-2}(k'(u), u) \geqq d_1 M^{-2},$$

where M is an upper bound for the bounded functional k on M_c. (Indeed, k is bounded on bounded subsets of V.)

Since ϕ is itself obviously a diffeomorphism of class C^{2-} on $S_1(V)$ being a bounded linear mapping on the Banach space V, whose norm is of class C^{2-} because of the assumption that $p \geqq 2$, and since

$$\phi(\|x\|^{-1} x) = \|x\|^{-1} \phi(x)$$

and $\|\phi(x)\|^{-1} \phi(x)$ are both positive multiples of $\phi(x)$ lying on $S_1(V)$, we have $\phi(\|x\|^{-1} x) = \|\phi(x)\|^{-1} \phi(x)$ for all x in M_c.

Finally, since k' is compact on bounded subsets of V and hence on M_c, while $h' = k^{-2} k'$, h' is compact on each closed subset of M_c on which h is bounded.

Thus all the hypotheses of Theorem 8 are satisfied for this case, and the conclusion of Theorem 10 follows if we set $N(u) = u$ for each u in M_c, and note that

$$(g'(u), u) \geqq d_0, \; \|g'(u)\| \leqq k_1, \; \|u\| \leqq k_2, \; (u \in M_c),$$

and for any u with $h(u) \leqq d^{-1}$

$$(h'(u), u) \geqq d_3 > 0. \quad \text{q.e.d.}$$

8. Galerkin approximations for nonlinear eigenvalue problems

If we wish to extend the results of Theorem 10 to the case of p with $1 < p < 2$, we can no longer assume that the problem can be transplanted to a suitable C^{2-} manifold since the Banach space V has a norm whose second derivatives do not exist. Thus it is of interest to obtain another method of establishing results of the Lusternik-Schnirelmann type for C^1 functions on C^1 manifolds without any sort of regularization in terms of differentiability class. The technique of argument which we shall apply in the present Section accomplishes such objectives in the treatment of nonlinear eigenvalue problems and is based upon the use of GALERKIN or (more properly speaking) Rayleigh-Ritz approximation of the given eigenvalue problem by finite dimensional eigenvalue problems.

Let B be an infinite dimensional Banach space, g and h two continuous real-valued functions defined on B. Let c be a real number, and as before,

$$M_c = \{u \mid u \in B, \; g(u) = c\}.$$

We assume that 0 does not lie in M_c, that both g and h are of class C^1 on a neighborhood of M_c, and that for all u in M_c, $g'(u) \neq 0$. Then M_c is a closed submanifold of B of class C^1. We are concerned with determining the critical points on M_c of the function $h_c = h|_{M_c}$.

Proposition (8.1). *Let B be a separable Banach space, ϕ a continuous, invertible linear mapping of B on B such that for a given prime q, $\phi^q = I$, while for $1 \leqq j \leqq q-1$, ϕ^j has only 0 as a fixed point.*

Then there exists an increasing sequence $\{B_n\}$ *of finite dimensional subspace of* B *such that each* B_n *is invariant under* ϕ, *the union of the* B_n *is dense in* B, *and if*

$$M_{c,n} = M_c \cap B_n,$$

the union of the sets $M_{c,n}$ *is dense in* M_c.

Proof of Proposition (8.1). Let $\{x_j\}$ be a dense sequence in B containing a dense sequence in M_c. For each n, let B_n be the finite dimensional subspace of B spanned by $\{\phi^k(x_j): 1 \leqq j \leqq n,\ 0 \leqq k \leqq q-1\}$. Then the sequence of subspaces B_n satisfies the conditions of Proposition (8.1). q.e.d.

Proposition (8.2). *Let* B *be a reflexive Banach space,* g *a real-valued* C^1 *function on* B *such that* g' *satisfies the condition* (S) *and maps bounded subsets of* B *into bounded subsets of* B^*. *Suppose that* B *is separable, that* M_c *is bounded, and that* $\{B_n\}$ *is the sequence of finite dimensional subspaces of* B *constructed with respect to some mapping* ϕ *as in Proposition* (8.1). *Let* g_n *be the restriction of* g *to* B_n, *and* $g'_n: B_n \to B^*_n$ *its derivative. Then there exists* $n_0 \geqq 1$ *such that for all* $n \geqq n_0$, $g'_n(u) \neq 0$ *for all* u *in* $M_{c,n}$ *and* $M_{c,n}$ *is a closed* C^1 *submanifold of* B_n.

Proof of Proposition (8.2). It suffices to prove that for $n \geqq n_0$ for some n_0, $g'_n(u) \neq 0$ for all u in $M_{c,n}$. Suppose this to be false. Then for an infinite sequence of integers, which we can identify with our original sequence, there exist elements x_n of $M_{c,n}$ with $g'_n(x_n) = 0$.

For each n, let j_n be the injection map of B_n into B, and j^*_n the dual projection map of B^* onto B^*_n. For each u and v in B_n,

$$(g'_n(u), v) = (g'(u), v) = (g'\, j_n(u), j_n(v)) = (j^*_n\, g'\, j_n(u), v).$$

Hence $g'_n = j^*_n\, g'\, j_n$, the n-th Galerkin approximant of g' with respect to the sequence $\{B_n\}$ of finite dimensional subspaces of B.

Let v be an element of the dense subset $\bigcup_n B_n$ of B. Then v lies in B_m for some m, and for $n \geqq m$,

$$0 = (g'_n(x_n), v) = (j^*_n\, g'(x_n), v) = (g'(x_n), v).$$

Since all the x_n belong to M_c, which is bounded, $g'(x_n)$ is uniformly bounded for all n. Hence, since $g'(x_n)$ converges weakly to 0 against a dense subset of B, $g'(x_n)$ converges weakly to 0 as $n \to \infty$.

Consider

$$(g'(x_n) - g'(x), x_n - x) = (g'(x_n), x_n) - (g'(x_n), x) - (g'(x), x_n - x).$$

Since x_n converges weakly to x, $(g'(x), x_n - x) \to 0$. Since $g'(x_n)$ converges weakly to 0, $(g'(x_n), x) \to 0$. Finally, $(g'(x_n), x_n) = (g'_n(x_n), x_n) = 0$.

Combining these results, we see that

$$(g'(x_n) - g'(x), x_n - x) \to 0.$$

Since g' satisfies condition (S), it follows that x_n converges strongly to x in X, x lies in M_c, and $g'(x) = \lim_n g'(x_n)$. However, $g'(x_n)$ converges weakly to 0, so that $g'(x)$ must equal 0. This contradicts the assumption that $g'(x) = 0$ for x in M_c. This contradiction establishes the validity of the Proposition. q.e.d.

Definition (8.1). *If X is a topological space, the compact category of X is defined by*

$$\textit{comp}\ \mathrm{cat}(X) = \sup\{\mathrm{cat}(K) : K \textit{ is a compact subset of } X\}.$$

Proposition (8.3). *Let X be the quotient of $S_1(B)$ by a cyclic group π of transformations of prime order, B an infinite dimensional Banach space. Then comp* $\mathrm{cat}(X) = +\infty$.

More generally let $X = Z/\pi$, where Z is a metrizable absolute neighborhood retract such that every compact subset of Z is contractible in Z. Then comp $\mathrm{cat}(X) = +\infty$.

Proof of Proposition (8.3). In both cases, X is an Eilenberg-MacLane space $K(\pi, 1)$, and the result follows by a slight modification of the argument of Section 3.

Proposition (8.4). *Let B be a separable Banach space with norm of class C^1 on $B - \{0\}$, ϕ a bounded linear operator in B such that $\phi^q = I$ for a given prime $q \geqq 2$, and ϕ^j has no non-null fixed points for $1 \leqq j \leqq (q-1)$. Let $\{B_n\}$ be an increasing sequence of finite dimensional subspaces of B such that each B_n is invariant under ϕ and the union of the B_n is dense in B. Let g be a C^1 function on B, and for a real c, let $M_c = \{u \mid g(u) = c\}$ and $M_{c,n} = M_c \cap B_n$. Assume that M_c is bounded and invariant under φ, while the union of the $M_{c,n}$ is dense in M_c. We assume also that for each u in M_c, $(g'(u), u) \neq 0$, and that each ray from the origin intersects M_c in exactly one point.*

Let $X = M_c/\pi$, $X_n = M_{c,n}/\pi$, where π is the cyclic group of transformations generated by ϕ. Then:

(a) *If K is a compact subset of X, and $\varepsilon > 0$ is given, there exists an integer n and a compact subset K_n of X_n such that K can be deformed into K_n through X and the distance in X of K from K_n is less than ε.*

(b) *For each n,* $\mathrm{cat}(X_n, X)$ *is finite, and*

$$\mathrm{cat}(X_n; X) \leqq \mathrm{cat}(X_{n+1}; X) \leqq \mathrm{comp\ cat}(X),$$
$$\mathrm{cat}(X_n; X) \to \mathrm{comp\ cat}(X).$$

(c) *If h_c is a continuous real-valued function on M_c invariant under ϕ, and if*

$$m_{k,\mathrm{comp}}(h) = \inf_{A\in S'_k} \sup_{x\in A} h(x),$$

where

$$S'_k = \{A \mid A \text{ compact in } M_c,\ q(A) \text{ of category} \geqq k \text{ in } X\}$$

where q is the quotient map of M_c on $X = M_c/\pi$, and if

$$m_{k,n}(h) = \inf_{A\in S'_k,\ A\subset B_n} \sup_{x\in A} h(x),$$

then:

$$m_{k,n}(h) \geqq m_{k,n+1}(h) \geqq m_{k,\mathrm{comp}}(h),$$

and

$$m_{k,n}(h) \to m_{k,\mathrm{comp}}(h),\ (n\to\infty).$$

Proof of Proposition (8.5): Proof of part (a). It suffices to show that for given ε and for $K' = q^{-1}(K)$, there exists an ε-deformation of K' into B_n which is invariant under the mapping ϕ for n sufficiently large, i.e. a mapping $F\colon K'\times[0,1]\to M_c$ such that F is continuous, $F(K'\times\{1\})\subset B_n$,

$$F(\phi(x), t) = \phi(F(x,t)),\ ([x,t]\in K'\times[0,1])$$

and

$$\|F(x,t) - x\| \leqq \varepsilon,$$

for all x in K' and each t in $[0, 1]$. Indeed, if such a deformation F exists, we take $K_n = q(f_1(K'))$, where $f_1(x) = F(x, 1)$, and let the deformation G of $K\times[0,1]$ through X be given by

$$G(v,t) = q(F(q^{-1}(v), t)).$$

If G is a deformation, the result of part (a) will then follow. If F satisfies the commutation condition with ϕ stated above, G is well defined, and it suffices to show that G is continuous. Since q is a covering mapping, however, for each point v_0 in K, we can choose a neighborhood N on which there exists a continuous single-valued branch of q^{-1}. Using the continuity of this branch, the continuity of G on $N\times[0,1]$ follows by the definition of G. Hence, it suffices to prove the existence of the deformation F as described above.

By hypothesis, the norm of B is of class C^1 on $B-\{0\}$, each ray from the origin intersects M_c in exactly one point, and $(g'(u), u) \neq 0$ for each u in M_c. Hence by Proposition (6.4), the mapping γ of M_c onto $S_1(B)$ given by radial projection is a C^1 diffeomorphism of M_c onto $S_1(B)$, $(\gamma(x) = \|x\|^{-1}x)$. Let ψ_0 be the inverse mapping of $S_1(B)$ into N_c. If we extend γ to the whole of $B-\{0\}$ by the same definition, then $\psi = \psi_0\gamma$ is a continuous retraction of $B-\{0\}$ into M_c, where $\psi(x)$ for each x is the unique point of M_c on the ray from the origin passing through the point x. Since ϕ is linear and M_c is invariant under ϕ, it follows that $\phi\psi=\psi\phi$ where both mappings are assumed to act on $B-\{0\}$.

Let $\varepsilon>0$ be given. Since ψ is continuous and K' is compact, there exists ε_1 with $0<\varepsilon_1<\frac{\varepsilon}{2}$ such that if x lies in the ε_1-neighborhood of K', then $\|\psi(x)-x\|\leqq\varepsilon/2$. By hypothesis, the union of the spaces B_n is dense in B. Since K' is compact, there exists an integer n and a finite subset $S=\{x_1,\ldots,x_r\}$ in B_n invariant under ϕ such that for each x in K' and some index k with $1\leqq k\leqq r$, we have $\|x-x_k\|<\varepsilon_2$, where $\varepsilon_2=\|\phi\|^{-q}\varepsilon_1$. (We note that since $\phi q=I$, $\|\phi\|\geqq 1$.) Let β be a continuous function from R^+ to R^+ such that $\beta(r)=0$ for $r\geqq\varepsilon_2$, $\beta(r)>0$ for $0\leqq r<\varepsilon_2$. For each k, let

$$\beta_k(x)=\sum_{j=0}^{q-1}\beta(\|\varphi^j(x)-\varphi^j(x_k)\|)\,.$$

β_k is a continuous non-negative function from B to the reals, and since the summand for $j=0$ is simply $\beta(\|x-x_k\|)$, $\beta_k(x)>0$ on the open ε_2-ball about x_k. The support of β_k consists of a subset of the union on j of $\{x|\ \phi^j(x)$ lies in the ε_2 ball about $\phi^j(x_k)\}$, i.e.

$$\operatorname{supp}(\beta_k)\subset\bigcup_{j=0}^{q-1}\phi^{-j}(B_{\varepsilon_2}(\phi^j(x_k))=\bigcup_{j=0}^{q-1}\phi^{q-j}(B_{\varepsilon_2}(\phi^j(x_k))\,.$$

Since $\varepsilon_2\|\phi\|^q\leqq\varepsilon_1$, it follows that

$$\phi^{q-j}(B_{\varepsilon_2}(\phi^j(x_k))\subset B_{|\phi|^{q-j}\varepsilon_2}(\phi^q(x_k))\subset B_{\varepsilon_1}(x_k)$$

for each j in the range $[0,q-1]$. Hence the support of each function β_k is contained in the closed ε_1-ball about the point x_k in B.

For two given integers k and m, let $x_r=\phi(x_m)$. Then

$$\beta_k(x)=\sum_{j=0}^{q-1}\beta(\|\phi^j(x)-\phi^j(x_k)\|=\sum_{j=0}^{q-1}\beta(\|\phi^{j-1}\phi x-\phi^{j-1}x_m\|=\beta_m(\phi(x))\,.$$

Hence the system of functions $\{\beta_k\}$ is invariant under the mapping ϕ.

For each k with $1\leqq k\leqq r$, we set

$$\alpha_k(x)=\beta_k(x)\Big(\sum_{k_1=1}^{r}\beta_{k_1}(x)\Big)^{-1}\,.$$

For every point x of K', at least one of the functions β_k is different from zero at x, and hence $\sum_{k_1=1}^{r}\beta_{k_1}(x)>0$. It follows that the functions α_k are all well-defined and continuous on K' with $\operatorname{supp}(\alpha_k)\subset\operatorname{supp}(\beta_k)$ for each

k, $0 \leqq \alpha_k(x) \leqq 1$ for all x in K', and $\sum_{k=1}^{r} \alpha_k(x) = 1$ for all x in K'. Moreover, it follows as before that if $x_k = \phi(x_m)$, then $\alpha_k(x) = \alpha_m(\phi(x))$.

We now define the deformation $F: K' \times [0, 1] \to M_c$ by setting

$$F(x, t) = \psi\left((1-t)\,x + t\sum_{k=1}^{r} \alpha_k(x)\, x_k\right).$$

Let f be the mapping of K' into B defined by

$$f(x) = \sum_{k=1}^{r} \alpha_k(x)\, x_k.$$

Then

$$f(\phi(x)) = \sum_{k=1}^{r} \alpha_k(\phi(x))\, x_k = \sum_{k=1}^{r} \phi\,[\alpha_m(x)\, x_m]_{x_m} = \phi^{-1}(x_r)$$
$$= \phi(f(x)),\ x \in K'.$$

It follows that for each x in K' and t in $[0, 1]$,

$$F(\phi(x), t) = \psi((1-t)\,\phi(x) + t f(\phi(x))) = \psi(\phi((1-t)\,x + t f(x)))$$
$$= \phi(\psi((1-t)\,x + t f(x))) = \phi F(x, t).$$

For a given x in K', those x_k in S for which $\alpha_k(x) > 0$ are all contained in the ball of radius ε_1 about x. Since $f(x)$ is a convex linear combination of such points x_k, it follows that $f(x) \in B_{\varepsilon_1}(x)$ and for each t in $[0, 1]$, the same is true for the convex linear combination $(1-t)\,x + t f(x)$. By our choice of ε_1, since $(1-t)\,x + t f(x)$ lies in the ε_1 neighborhood of K',

$$\|\psi((1-t)\,x + t f(x)) - ((1-t)\,x + t f(x))\| \leqq \frac{\varepsilon}{2}.$$

Hence

$$\|F(x, t) - x\| \leqq \|F(x, t) - ((1-t)\,x + t f(x))\| + \|(1-t)\,x + t f(x)) - x\|$$
$$\leqq \varepsilon_2 + \varepsilon_1 < \varepsilon.$$

Hence F is the desired deformation, and the proof of part (a) is complete. q.e.d.

Proof of part (b) of Proposition (8.4). Since each $M_{c,n}$ is a closed bounded set in a finite dimensional space, it is compact. Since the $M_{c,n}$ increase with n, it follows immediately that $\operatorname{cat}(X_n; X)$ do not decrease with increasing n, and that for every n

$$\operatorname{cat}(X_n; X) \leqq \operatorname{comp\,cat}(X).$$

It suffices therefore to show that $\operatorname{cat}(X_n; X) \to \operatorname{comp\,cat}(X)$. This follows directly from the result of part (a), however, since given a compact subset K of X with $\operatorname{cat}(K; X) \geqq k$ for a given integer k, we may deform it over X into a compact subset K_n of X_n. Hence $\operatorname{cat}(K_n; X) \geqq k$, and the limit relation follows. q.e.d.

Proof of part (c) of Proposition (8.4). It follows immediately from the definitions that since X_n is increasing, the minimax numbers $m_{k,n}(h)$ decrease with n and are all bounded from below by $m_{k,\mathrm{comp}}(h)$. Hence, we need only prove that $m_{k,n}(h) \to m_{k,\mathrm{comp}}(h)$ as $n \to \infty$.

By the definition of $m_{k,\mathrm{comp}}(h)$, if we are given $\varepsilon > 0$, we may find a compact subset K of X such that $\operatorname{cat}(K; X) \geqq k$, and

$$\sup_{x \in K} h(x) \leqq m_{k,\mathrm{comp}}(h) + \varepsilon .$$

Since h is continuous and K is compact, we may find $\varepsilon_0 > 0$ such that if x lies in K and x_1 lies in the ε_0 ball about x, then $|h(x) - h(x_1)| < \varepsilon$. If we apply the result of part (a), we may deform K over X into K_n, a subset of X_n, such that for each x_1 in K_n there exists x in K such that $\|x - x_1\| \leqq \varepsilon_0$. It follows that $\operatorname{cat}(K_n; X) \geqq k$, and

$$\sup_{x_1 \in K} h(x_1) \leqq m_{k,\mathrm{comp}}(h) + 2\varepsilon .$$

Hence

$$m_{k,n}(h) \leqq m_{k,\mathrm{comp}}(h) + 2\varepsilon ,$$

and since $\varepsilon > 0$ is arbitrary, it follows that $m_{k,n}(h) \to m_{k,\mathrm{comp}}(h)$ as $n \to \infty$. Hence the proof of part (c) is complete, and with it the proof of Proposition (8.4). q.e.d.

Theorem 11. *Let B be an infinite dimensional reflexive Banach space whose norm is of class C^1 on $B - \{0\}$, ϕ a bounded linear mapping of B on itself such that for a given prime $q \geqq 2$, $\phi^q = I$, the identity, while for $1 \leqq j \leqq q-1$, ϕ^j has no non-null fixed points. Let g be a real-valued function of class C^1 on B with g' satisfying the condition* (S) *of Definition* (6.1). *For a given real number c, let $M_c = \{u \mid u \in B, g(u) = c\}$ and suppose that each ray from the origin hits M_c in exactly one point while for u in M_c, $(g'(u), u) \neq 0$. Suppose that h is a C^1 function from a neighborhood of M_c to the reals, and that ϕ maps M_c on itself with $h(\phi(x)) = h(x)$ for all x in M_c. Suppose that M_c is bounded and that h' is compact on subsets of M_c on which h is bounded. Suppose that g' is bounded on bounded sets, and that on any subset S_0 of M_c on which h is bounded, we have a positive constant $d_{S_0} > 0$ such that for all u in S_0,*

$$d_{S_0}(g'(u), u) \leqq |(h'(u), u)| .$$

Then:

(a) *Let B_n be a sequence of finite dimensional subspaces of B whose union is dense in B with each B_n invariant under ϕ as constructed on*

Proposition (8.1) *with* $B_n \subset B_{n+1}$ *for each* n. *For each* n *and each integer* $k \leqq \operatorname{cat}(X_n; X)$, *where* $X = M_c/\pi$ *and* $X_n = M_{c,n}/\pi$ *as in Proposition* (8.4), *there exists an element* $u_{k,n}$ *of* $M_{c,n}$ *such that*

$$g'_n(u_{k,n}) = \lambda_{k,n} h'_n(u_{k,n}), \; h(u_{k,n}) = m_{k,n}(h)$$

where $m_{k,n}$ *is the restricted minimax defined in Proposition* (8.4) *and* g_n *and* h_n *are the restriction of the functions* g *and* h *to* B_n.

(b) *For each* k *for which* $m_{k,\text{comp}}(h) < +\infty$, *there exists an element* u_k *in* M_c *with*

$$g'(u_k) = \lambda_k h'(u_k), \; h(u_k) = m_{k,\text{comp}}(h).$$

and for any weakly convergent sequence $\{u_{k,n_j}\}$ *with* $n_j \to \infty$, *the weak limit is an element* u_k *satisfying the conditions of part* (b), *and the sequence* u_{k,n_j} *converges strongly to* u_k.

(c) *If* $m_{k,\text{comp}}(h) \to \infty$, *then there exist an infinite number of distinct* u_k *in* M_c *such that* $g'(u_k) = \lambda_k h'(u_k)$ *for real* λ_k.

Proof of Theorem 11: Proof of part (a). The result of the finite dimensional case follows from a special case of Theorem 8, with the modification that we consider the deformation invariant class of subsets of X_n whose category with respect to X is $\geqq n$.

Proof of part (b). We know that $m_{k,n}(h) \to m_{k,\text{comp}}(h)$ by Proposition (8.4). We consider the sequence $\{u_{k,n}\}$ defined by the finite dimensional problems in part (a), which lies in the bounded subset M_t of B. Hence, using the reflexivity of the Banach space B, we can assume that u_{k,n_j} converges weakly to an element u_k as $j \to \infty$ for an infinite subsequence u_{k,n_j}. It suffices to prove that u_k satisfies the desired conditions and that u_{k,n_j} converges strongly to u_k. To simplify our notation, we may identify the sequence $\{n_j\}$ with our original sequence, and assume that $u_{k,n}$ converges to the element u_k weakly.

For each n,

$$g'_n(u_{k,n}) = \lambda_{k,n} h'_n(u_{k,n}), \; h(u_{k,n}) = m_{k,n}.$$

By a previously given argument for g, $g'_n = j^*_n g' j_n$, $h'_n = j^*_n h' j_n$, where j_n is the injection map of B_n into B, and j^*_n is the dual projection mapping of B^* onto B^*_n. Then for each v in B_n, we have

$$(g'_n(u_{k,n}), v) = (g'(u_{k,n}), v),$$
$$(h'_n(u_{k,n}), v) = (h'(u_{k,n}), v).$$

In particular,

$$(g'(u_{k,n}), u_{k,n}) = \lambda_{k,n} (h'(u_{k,n}), u_{k,n}),$$

and therefore,

$$\lambda_{k,n} = (g'(u_{k,n}), u_{k,n}) \, (h'(u_{k,n}), u_{k,n})^{-1}.$$

Since by our construction $h(u_{k,n}) = m_{k,n} \to m_{k,\text{comp}}(h) < +\infty$, the $u_{k,n}$ lie on a fixed S_0 on which h is bounded, and by our hypotheses, there exists a constant c_0 such that $|\lambda_{k,n}| \leqq c_0$, $n \geqq 1$.

To show that our given sequence $\{u_{k,n}\}$ converges strongly, it suffices to show that for any such sequence, one can extract a strongly convergent subsequence. (Indeed, if the original sequence did not converge, one could extract an infinite subsequence whose distance from the weak limit u_k remained above a fixed bound. If one extracts a strongly convergent infinite subsequence of this subsequence, its limit must be u_k since its weak limit is u_k. However, no subsequence of the subsequence first chosen can converge to u_k.)

Since $|\lambda_{k,n}|$ is uniformly bounded and $h(u_{k,n})$ is uniformly bounded for all n, we may extract an infinite subsequence which we identify with our original sequence and assume that $\lambda_{k,n} \to \lambda_k$ for some λ_k as $n \to \infty$ and that $h'(u_{k,n})$ converges strongly to an element w of B^* since h' is compact on subsets on which h is bounded.

Let v be any element of the dense subset $u_n B_n$ of B. Then v lies in some B_m, and for $n \geqq m$, we have

$$(g'(u_{k,n}), v) = (g'_n(u_{k,n}), v) = \lambda_{k,n}(h'_n(u_{k,n}), v) = \lambda_{k,n}(h'(u_{k,n}), v).$$

Hence

$$(g'(u_{k,n}), v) \to \lambda_k(w, v), \ (n \to \infty).$$

Since $g'(u_{k,n})$ is uniformly bounded for all n, it follows that $g'(u_{k,n})$ converges weakly to $\lambda_k w$ as $n \to \infty$. As a consequence

$$(g'(u_{k,n}) - g'(u_k), u_{k,n} - u_k) \to (\lambda_k w - g'(u_k), 0) = 0$$

as $n \to \infty$ since the sequence in one term of the pairing converges strongly and the other weakly. Since g' satisfies condition (S), it follows that $u_{k,n}$ converges strongly to u_k. Hence $h'(u_{k,n}) \to h'(u_k)$ and $h'(u_k) = \lambda_k w$. Finally $g'(u_{k,n})$ converges strongly to $g'(u_k)$, and $g'(u_k) = \lambda_k w = \lambda_k h'(u_k)$ by the uniqueness of weak limits. Finally, $h(u_k) = \lim h(u_{k,n}) = \lim_n m_{k,n}(h) = m_{k,\text{comp}}(h)$. q.e.d.

The proof of part (c) follows obviously from the result of part (b), and thereby the proof of Theorem 11 is complete. q.e.d.

References

1. Berstein, L., and T. Ganea: Homotopical nilpotency. Illinois J. Math. **5**, 99—130 (1961).
2. Browder, F. E.: Infinite dimensional manifolds and nonlinear elliptic eigenvalue problems. Ann. of Math. **82**, 459—477 (1965).
3. — Nonlinear eigenvalue problems and Galerkin approximations. Bull. Am. Math. Soc. **74**, 651—656 (1968).

4. — Existence theorems for nonlinear partial differential equations. Proceedings of Am. Math. Soc. Summer Institute on Global Analysis 1968 (to appear).
5. — Nonlinear operators and nonlinear equations of evolution in Banach spaces. Proceedings of Am. Math. Soc. Symposium on Nonlinear Functional Analysis (to appear).
6. EELLS, J.: A setting for global analysis. Bull. Am. Math. Soc. **72**, 751—807 (1966).
7. EILENBERG, S., and T. GANEA: On the Lusternik-Schnirelmann category of abstract groups. Ann. of Math. **65**, 517—518 (1957).
8. FOX, R. H.: On the Lusternik-Schnirelmann category. Ann. of Math. **42**, 333—370 (1941).
9. GANEA, T.: Sur quelques invariants numeriques du type d'homotopie. Cahiers Topologie et Geometrie Differentielle **1**, 181—241 (1967).
10. KRASNOSELSKI, M. A.: Topological methods in the theory of nonlinear integral equations, 1956, Gostekhteoretizdat, Moscow (Engl. trans. 1964, Pergamon Press).
11. LUSTERNIK, L., and T. SCHNIRELMANN: Methodes topologiques dans les problems variationnels. Paris: Hermann 1934.
12. — The topology of the calculus of variations in the large. Trudi Mat. Inst. im. V. A. Steklova, No. 19, Izdat. Akad. Nauk SSSR, Moscow, 1947 (Engl. trans. Am. Math. Soc. Transl. of Math. Monographs Vol. 16, 1966).
13. PALAIS, R.: Lusternik-Schnirelmann category of Banach manifolds. Topology **5**, 115—132 (1966).
14. — Critical point theory and the minimax principle. Proceedings of Am. Math. Soc. Summer Institute on Global Analysis, 1968 (to appear).
15. — Foundations of global non-linear analysis. New York: W. A. Benjamin 1968.
16. —, and S. SMALE: A generalized Morse theory. Bull. Am. Math. Soc. **70**, 165—171 (1964).
17. SCHWARTZ, J. T.: Generalizing the Lusternik-Schnirelmann theory of critical points. Comm. Pure Appl. Math. **17**, 307—315 (1964).
18. STEENROD, N. E., and D. B. A. EPSTEIN: Cohomology operations. Ann. of Math. Study No. 50, Princeton, 1962.
19. VAINBERG, M. M.: Variational methods for the study of nonlinear operators. Moscow 1956 (Engl. trans. Holden-Day, San Francisco, 1964).
20. WEISS, S.: Nonlinear eigenvalue problems. Ph. D. Dissertation, University of Chicago, August, 1969.

Minimal Submanifolds of a Sphere with Second Fundamental Form of Constant Length

By S. S. Chern*, M. do Carmo**, and S. Kobayashi***

University of California, Berkeley, and IMPA, Rio de Janeiro

1. Introduction

Let M be an n-dimensional manifold which is minimally immersed in a unit sphere S^{n+p} of dimension $n+p$. Let h be the second fundamental form of this immersion; it is a certain symmetric bilinear mapping $T_x \times T_x \to T_x^\perp$ for $x \in M$, where T_x is the tangent space of M at x and $T_x^\perp$ is the normal space to M at x. We denote by S the square of the length of h. It is known that if M is moreover compact, then

$$\int_M \left(\left(2-\frac{1}{p}\right) S - n\right) S \cdot *1 \geqq 0,$$

where $*1$ denotes the volume element of M, (Simons [3]; a slightly more general formula will be proved in §§ 2 and 3). It follows that if $S \leqq n \Big/ \left(2-\frac{1}{p}\right)$ everywhere on M, then either

1) $S=0$ (i.e., M is totally geodesic)

or

2) $S = n \Big/ \left(2-\frac{1}{p}\right)$.

The purpose of the present paper is to determine all minimal submanifolds M of S^{n+p} satisfying $S = n \Big/ \left(2-\frac{1}{p}\right)$. The proof depends upon the results presented by the first named author in his lectures on minimal submanifolds in Berkeley in the Winter of 1968, in which an exposition of the work of Simons [3] was made by the use of moving frames. To describe our result, we begin with examples of minimal submanifolds.

In general, let $S^q(r)$ denote a q-dimensional sphere in $\boldsymbol{R}^{q+1}$ with radius r. Let m and n be positive integers such that $m < n$ and let

* Work done under partial support by NSF Grant GP-6974;

** Work done under partial support by NSF Grant GP-6974 and Guggenheim Foundation;

*** Work done under partial support by NSF Grant GP-8008.

$M_{m,\,n-m}=S^m\left(\sqrt{\frac{m}{n}}\right)\times S^{n-m}\left(\sqrt{\frac{n-m}{n}}\right)$. We imbed $M_{m,\,n-m}$ into $S^{n+1}=S^{n+1}(1)$ as follows. Let (u, v) be a point of $M_{m,\,n-m}$ where u (resp. v) is a vector in $\boldsymbol{R}^{m+1}$ (resp. $\boldsymbol{R}^{n-m+1}$) of length $\sqrt{\frac{m}{n}}\left(\text{resp.}\sqrt{\frac{n-m}{n}}\right)$. We can consider (u, v) as a unit vector in $\boldsymbol{R}^{n+2}=\boldsymbol{R}^{m+1}\times\boldsymbol{R}^{n-m+1}$. It will be shown that $M_{m,\,n-m}$ is a minimal submanifold of S^{n+1} satisfying $S=n$.

We shall now define the Veronese surface. Let (x, y, z) be the natural coordinate system in $\boldsymbol{R}^3$ and $(u^1, u^2, u^3, u^4, u^5)$ the natural coordinate system in $\boldsymbol{R}^5$. We consider the mapping defined by

$$u^1=\frac{1}{\sqrt{3}}\,yz,\quad u^2=\frac{1}{\sqrt{3}}\,zx,\quad u^3=\frac{1}{\sqrt{3}}\,xy,\quad u^4=\frac{1}{2\sqrt{3}}\,(x^2-y^2),$$

$$u=\frac{1}{6}\,(x^2+y^2-2z^2).$$

This defines an isometric immersion of $S^2(\sqrt{3})$ into $S^4=S^4(1)$. Two points (x, y, z) and $(-x, -y, -z)$ of $S^2(\sqrt{3})$ are mapped into the same point of S^4, and this mapping defines an imbedding of the real projective plane into S^4. This real projective plane imbedded in S^4 will be called the *Veronese surface*. It will be shown that the Veronese surface is a minimal submanifold of S^4 satisfying $S=4/3$.

Main theorem. *The Veronese surface in S^4 and the submanifolds $M_{m,\,n-m}$ in S^{n+1} are the only compact minimal submanifolds of dimension n in S^{n+p} satisfying $S=n\Big/\left(2-\frac{1}{p}\right)$.*

The proof is by local argument and the corresponding local result also holds: An *n-dimensional* minimal submanifold of S^{n+p} satisfying $S=n\Big/\left(2-\frac{1}{p}\right)$ is locally $M_{m,\,n-m}$ or the Veronese surface.

For $p=1$, the theorem was proved independently by B. LAWSON [2].

It should be pointed out that, by the equation of GAUSS, $S=n(n-1)-R$ for an n-dimensional minimal submanifold M of S^{n+p}, where R is the scalar curvature of M. In particular, S is an intrinsic invariant when M is minimal.

2. Local formulas for a minimal submanifold

In this section we shall compute the Laplacian of the second fundamental form of a minimal submanifold of a symmetric space.

Let M be an n-dimensional manifold immersed in an $(n+p)$-dimensional riemannian manifold N. We choose a local field of orthonormal frames $e_1, \ldots, e_{n+p}$ in N such that, restricted to M, the vectors $e_1, \ldots, e_n$ are tangent to M (and, consequently, the remaining vectors $e_{n+1}, \ldots, e_{n+p}$

are normal to M). We shall make use of the following convention on the ranges of indices:

$$1 \leqq A, B, C, \ldots, \leqq n+p; \quad 1 \leqq i, j, k, \ldots, \leqq n;$$
$$n+1 \leqq \alpha, \beta, \gamma, \ldots, \leqq n+p,$$

and we shall agree that repeated indices are summed over the respective ranges. With respect to the frame field of N chosen above, let $\omega^1, \ldots, \omega^{n+p}$ be the field of dual frames. Then the structure equations of N are given by

$$d\omega^A = -\sum \omega^A_B \wedge \omega^B, \quad \omega^A_B + \omega^B_A = 0, \tag{2.1}$$

$$d\omega^A_B = -\sum \omega^A_C \wedge \omega^C_B + \Phi^A_B, \quad \Phi^A_B = \tfrac{1}{2}\sum K^A{}_{BCD}\,\omega^C \wedge \omega^D,$$
$$K^A{}_{BCD} + K^A{}_{BDC} = 0. \tag{2.2}$$

We restrict these forms to M. Then

$$\omega^\alpha = 0. \tag{2.3}$$

Since $0 = d\omega^\alpha = -\sum \omega^\alpha_i \wedge \omega^i$, by CARTAN's lemma we may write

$$\omega^\alpha_i = \sum h^\alpha_{ij}\,\omega^j, \quad h^\alpha_{ij} = h^\alpha_{ji}. \tag{2.4}$$

From these formulas, we obtain

$$d\omega^i = -\sum \omega^i_j \wedge \omega^j, \quad \omega^i_j + \omega^j_i = 0, \tag{2.5}$$

$$d\omega^i_j = -\sum \omega^i_k \wedge \omega^k_j + \Omega^i_j, \quad \Omega^i_j = \tfrac{1}{2}\sum R^i{}_{jkl}\,\omega^k \wedge \omega^l, \tag{2.6}$$

$$R^i{}_{jkl} = K^i{}_{jkl} + \sum_\alpha (h^\alpha_{ik} h^\alpha_{jl} - h^\alpha_{il} h^\alpha_{jk}), \tag{2.7}$$

$$d\omega^\alpha_\beta = -\sum \omega^\alpha_\gamma \wedge \omega^\gamma_\beta + \Omega^\alpha_\beta, \quad \Omega^\alpha_\beta = \tfrac{1}{2}\sum R^\alpha{}_{\beta kl}\,\omega^k \wedge \omega^l, \tag{2.8}$$

$$R^\alpha{}_{\beta kl} = K^\alpha{}_{\beta kl} + \sum_i (h^\alpha_{ik} h^\beta_{il} - h^\alpha_{il} h^\beta_{jk}). \tag{2.9}$$

The riemannian connection of M is defined by (ω^i_j). The form (ω^α_β) defines a connection in the normal bundle of M. We call $\sum h^\alpha_{ij}\,\omega^i\,\omega^j\,e_\alpha$ the *second fundamental form* of the immersed manifold M. Sometimes we shall denote the second fundamental form by its components h^α_{ij}. We call $\sum_\alpha \frac{1}{n}\left(\sum_i h^\alpha_{ii}\right) e_\alpha$ the *mean curvature normal* or the *mean curvature vector*. An immersion is said to be *minimal* if its mean curvature normal vanishes identically, i.e., if $\sum_i h^\alpha_{ii} = 0$ for all α.

We take exterior differentiation of (2.4) and define h^α_{ijk} by

$$\sum h^\alpha_{ijk}\,\omega^k = dh^\alpha_{ij} - \sum h^\alpha_{il}\,\omega^l_j - \sum h^\alpha_{lj}\,\omega^l_i + \sum h^\beta_{ij}\,\omega^\alpha_\beta. \tag{2.10}$$

Then

$$\sum (h^\alpha_{ijk} + \tfrac{1}{2} K^\alpha_{ijk})\, \omega^j \wedge \omega^k = 0, \tag{2.11}$$

$$h^\alpha_{ijk} - h^\alpha_{ikj} = K^\alpha{}_{ikj} = -K^\alpha{}_{ijk}. \tag{2.12}$$

We take exterior differentiation of (2.10) and define h^α_{ijkl} by

$$\sum h^\alpha_{ijkl}\, \omega^l = d h^\alpha_{ijk} - \sum h^\alpha_{ljk}\, \omega^l_i - \sum h^\alpha_{ilk}\, \omega^l_j - \sum h^\alpha_{ijl}\, \omega^l_k + \sum h^\beta_{ijk}\, \omega^\alpha_\beta. \tag{2.13}$$

Then

$$\sum (h^\alpha_{ijkl} - \tfrac{1}{2} \sum h^\alpha_{im} R^m{}_{jkl} - \tfrac{1}{2} \sum h^\alpha_{mj} R^m{}_{ikl} + \tfrac{1}{2} \sum h^\beta_{ij} R^\alpha{}_{\beta kl})\, \omega^k \wedge \omega^l = 0, \tag{2.14}$$

$$h^\alpha_{ijkl} - h^\alpha_{ijlk} = \sum h^\alpha_{im} R^m{}_{jkl} + \sum h^\alpha_{mj} R^m{}_{ikl} - \sum h^\beta_{ij} R^\alpha{}_{\beta kl}. \tag{2.15}$$

We stated earlier that (ω^i_j) defines a connection in the tangent bundle $T = T(M)$ [and, hence, a connection in the cotangent bundle $T^* = T^*(M)$ also] and that (ω^α_β) defines a connection in the normal bundle $T^\perp = T^\perp(M)$. Consequently, we have covariant differentiation which maps a section of $T^\perp \otimes T^* \otimes \dots \otimes T^*$, $(T^*{:}k$ times), into a section of $T^\perp \otimes T^* \otimes \dots \otimes T^* \otimes T^*$, $(T^*{:}k+1$ times). The second fundamental form h^α_{ij} is a section of the vector bundle $T^\perp \otimes T^* \otimes T^*$, and h^α_{ijk} is the covariant derivative of h^α_{ij}. Similarly, h^α_{ijkl} is the covariant derivative of h^α_{ijk}.

Similarly, we may consider $K^\alpha{}_{ijk}$ as a section of the bundle $T^\perp \otimes T^* \otimes T^* \otimes T^*$. Its covariant derivative $K^\alpha{}_{ijkl}$ is defined by

$$\begin{aligned} \sum K^\alpha{}_{ijkl}\, \omega^l = d K^\alpha{}_{ijk} - \sum K^\alpha{}_{mjk}\, \omega^m_i - \sum K^\alpha{}_{imk}\, \omega^m_j \\ - \sum K^\alpha{}_{ijm}\, \omega^m_k + \sum K^\beta{}_{ijk}\, \omega^\alpha_\beta. \end{aligned} \tag{2.16}$$

This covariant derivative of $K^\alpha{}_{ijkl}$ must be distinguished from the covariant derivative of $K^A{}_{BCD}$ as a curvature tensor of N, which will be denoted by $K^A{}_{BCD;E}$. Restricted to M, $K^\alpha{}_{ijk;l}$ is given by

$$\begin{aligned} K^\alpha{}_{ijk;l} = K^\alpha{}_{ijkl} - \sum K^\alpha{}_{\beta jk}\, h^\beta_{il} - \sum K^\alpha{}_{i\beta k}\, h^\beta_{jl} - \sum K^\alpha{}_{ij\beta}\, h^\beta_{kl} \\ + \sum K^m{}_{ijk}\, h^\alpha_{ml}. \end{aligned} \tag{2.17}$$

In this section, *we shall assume that N is locally symmetric, i.e.,* $K^A{}_{BCD;E} = 0$.

The Laplacian Δh^α_{ij} of the second fundamental form h^α_{ij} is defined by

$$\Delta h^\alpha_{ij} = \sum_k h^\alpha_{ijkk}. \tag{2.18}$$

From (2.12) we obtain

$$\Delta h^\alpha_{ij} = \sum_k h^\alpha_{ikjk} - \sum_k K^\alpha{}_{ijkk} = \sum_k h^\alpha_{kijk} - \sum_k K^\alpha{}_{ijkk}. \tag{2.19}$$

From (2.15) we obtain

$$h^\alpha_{kijk} = h^\alpha_{kikj} + \sum h^\alpha_{km} R^m{}_{ijk} + \sum h^\alpha_{mi} R^m{}_{kjk} - \sum h^\beta_{ki} R^\alpha{}_{\beta jk}. \tag{2.20}$$

Replace h_{kikj} in (2.20) by $h_{kkij} - K^{\alpha}{}_{kikj}$ [see (2.12)] and then substitue the right hand side of (2.20) into h^{α}_{kijk} of (2.19). Then

$$\begin{aligned}\Delta h^{\alpha}_{ij} = &\sum_k (h^{\alpha}_{kkij} - K^{\alpha}{}_{kikj} - K^{\alpha}{}_{ijkk}) \\ &+ \sum_k \Big(\sum_m h^{\alpha}_{km} R^m{}_{ijk} + \sum_m h^{\alpha}_{mi} R^m{}_{kjk} - \sum_\beta h^{\beta}_{ki} R^{\alpha}{}_{\beta jk}\Big).\end{aligned} \tag{2.21}$$

From (2.7), (2.9), (2.17) and (2.21) we obtain

$$\begin{aligned}\Delta h^{\alpha}_{ij} = &\sum_k h^{\alpha}_{kkij} + \sum_{\beta, k} (-K^{\alpha}{}_{ij\beta} h^{\beta}_{kk} + 2K^{\alpha}{}_{\beta ki} h^{\beta}_{jk} - K^{\alpha}{}_{k\beta k} h^{\beta}_{ij} \\ &+ 2K^{\alpha}{}_{\beta kj} h^{\beta}_{ki}) + \sum_{m,k} (K^m{}_{kik} h^{\alpha}_{mj} + K^m{}_{kjk} h^{\alpha}_{mi} \\ &+ 2K^m{}_{ijk} h^{\alpha}_{mk}) + \sum_{\beta, m, k} (h^{\alpha}_{mi} h^{\beta}_{mj} h^{\beta}_{kk} + 2h^{\alpha}_{km} h^{\beta}_{ki} h^{\beta}_{mj} \\ &- h^{\alpha}_{km} h^{\beta}_{km} h^{\beta}_{ij} - h^{\alpha}_{mi} h^{\beta}_{mk} h^{\beta}_{kj} - h^{\alpha}_{mj} h^{\beta}_{ki} h^{\beta}_{mk}).\end{aligned} \tag{2.22}$$

Now, *we assume that M is minimal in N so that $\sum h^{\beta}_{kk} = 0$ for all β.* Then, from (2.22) we obtain

$$\begin{aligned}\sum h^{\alpha}_{ij} \cdot \Delta h^{\alpha}_{ij} = &\sum_{\alpha, \beta, i, j, k} (4K^{\alpha}{}_{\beta ki} h^{\beta}_{jk} h^{\alpha}_{ij} - K^{\alpha}{}_{k\beta k} h^{\alpha}_{ij} h^{\beta}_{ij}) \\ &+ \sum_{\alpha, m, i, j, k} (2K^m{}_{kik} h^{\alpha}_{mj} h^{\alpha}_{ij} + 2K^m{}_{ijk} h^{\alpha}_{mk} h^{\alpha}_{ij}) \\ &- \sum_{\alpha, \beta, i, j, k, l} (h^{\alpha}_{ik} h^{\beta}_{jk} - h^{\alpha}_{jk} h^{\beta}_{ik})(h^{\alpha}_{il} h^{\beta}_{jl} - h^{\alpha}_{jl} h^{\beta}_{il}) \\ &- \sum_{\alpha, \beta, i, j, k, l} h^{\alpha}_{ij} h^{\alpha}_{kl} h^{\beta}_{ij} h^{\beta}_{kl}.\end{aligned} \tag{2.23}$$

3. Minimal submanifolds of a space of constant curvature

Throughout this section we shall *assume that the ambient space N is a space of constant curvature c.* Then

$$K^A{}_{BCD} = c(\delta_{AC}\delta_{BD} - \delta_{AD}\delta_{BC}).$$

Then (2.23) reduces to

$$\begin{aligned}\sum h^{\alpha}_{ij} \cdot \Delta h^{\alpha}_{ij} = &-\sum (h^{\alpha}_{ik} h^{\beta}_{kj} - h^{\beta}_{ik} h^{\alpha}_{kj})(h^{\alpha}_{il} h^{\beta}_{lj} - h^{\beta}_{il} h^{\alpha}_{lj}) \\ &- \sum h^{\alpha}_{ij} h^{\alpha}_{lk} h^{\beta}_{ij} h^{\beta}_{kl} + nc \sum (h^{\alpha}_{ij})^2.\end{aligned} \tag{3.1}$$

For each α, let H_α denote the symmetric matrix (h^{α}_{ij}), and set

$$S_{\alpha\beta} = \sum_{i,j} h^{\alpha}_{ij} h^{\beta}_{ij}. \tag{3.2}$$

Then the $(p \times p)$-matrix $(S_{\alpha\beta})$ is symmetric and can be assumed to be diagonal for a suitable choice of $e_{n+1}, \ldots, e_{n+p}$. We set

$$S_\alpha = S_{\alpha\alpha}. \tag{3.3}$$

We denote the square of the length of the second fundamental form by S, i.e.,

$$S=\sum h_{ij}^{\alpha}\, h_{ij}^{\alpha}=\sum_{\alpha} S_{\alpha}. \tag{3.4}$$

In general, for a matrix $A=(a_{ij})$ we denote by $N(A)$ the square of the norm of A, i.e.,

$$N(A)=\operatorname{trace} A\cdot {}^{t}A=\sum (a_{ij})^2.$$

Clearly, $N(A)=N(T^{-1}AT)$ for any orthogonal matrix T. Now, (3.1) may be rewritten as follows:

$$\sum h_{ij}^{\alpha}\cdot \Delta h_{ij}^{\alpha}=-\sum_{\alpha,\beta} N(H_{\alpha}H_{\beta}-H_{\beta}H_{\alpha})-\sum_{\alpha} S+ncS. \tag{3.5}$$

We need the following algebraic lemma.

Lemma 1. *Let A and B be symmetric $(n\times n)$-matrices. Then*

$$N(AB-BA)\leqq 2N(A)\cdot N(B),$$

and the equality holds for nonzero matrices A and B if and only if A and B can be transformed simultaneously by an orthogonal matrix into scalar multiples of $\tilde{A}$ and $\tilde{B}$ respectively, where

$$\tilde{A}=\left(\begin{array}{cc|c} 0 & 1 & \\ 1 & 0 & 0 \\ \hline \multicolumn{2}{c|}{0} & 0 \end{array}\right); \qquad \tilde{B}=\left(\begin{array}{cc|c} 1 & 0 & \\ 0 & -1 & 0 \\ \hline \multicolumn{2}{c|}{0} & 0 \end{array}\right).$$

Moreover, if A_1, A_2 and A_3 are $(n\times n)$-symmetric matrices and if

$$N(A_{\alpha}A_{\beta}-A_{\beta}A_{\alpha})=2N(A_{\alpha})\cdot N(A_{\beta}) \qquad 1\leqq \alpha,\beta\leqq 3,$$

then at least one of the matrices A_{α} must be zero.

Proof. We may assume that B is diagonal and we denote by $b_1,\ldots,b_n$ the diagonal entries in B. By a simple calculation we obtain

$$N(AB-BA)=\sum_{i\neq k} a_{ik}^2\cdot (b_i-b_k)^2,$$

where $A=(a_{ij})$. Since $(b_i-b_k)^2\leqq 2(b_i^2+b_k^2)$, we obtain

$$\begin{aligned} N(AB-BA)&=\sum_{i\neq k} a_{ik}^2 (b_i-b_k)^2\leqq 2\sum_{i\neq k} a_{ik}^2 (b_i^2+b_k^2)\\ &\leqq 2\Big(\sum_{i,k} a_{ik}^2\Big)\Big(\sum_{i} b_i^2\Big)=2N(A)\cdot N(B). \end{aligned} \tag{3.6}$$

Now, assume that A and B are nonzero matrices and that the equality holds. Then the equality must holds everywhere in (3.6). From the second equality in (3.6), it follows that

$$a_{11} = \dots = a_{nn} = 0,$$

and that

$$b_i + b_k = 0 \quad \text{if} \quad a_{ik} \neq 0.$$

Without loss of generality, we may assume that $a_{12} \neq 0$. Then $b_1 = -b_2$. From the third equality, we now obtain

$$b_3 = \dots = b_n = 0.$$

Since $B \neq 0$, we must have $b_1 = -b_2 \neq 0$ and we conclude that $a_{ik} = 0$ for $(i, k) \neq (1, 2)$. To prove the last statement, let A_1, A_2, A_3 be all nonzero symmetric matrices. From the second statement we have just proved, we see that one of these matrices can be transformed to a scalar multiple of $\tilde{A}$ as well as to a scalar multiple of $\tilde{B}$ by orthogonal matrices. But this is impossible since $\tilde{A}$ and $\tilde{B}$ are not orthogonally equivalent. q.e.d.

Applying Lemma 1 to (3.5), we obtain

$$\begin{aligned} -\sum h^{\alpha}_{ji} \cdot \Delta h^{\alpha}_{ij} &\leqq 2 \sum_{\alpha \neq \beta} N(H_\alpha) \cdot N(H_\beta) + \sum_{\alpha} S^2_\alpha - ncS \\ &= 2 \sum_{\alpha \neq \beta} S_\alpha S_\beta + \sum_{\alpha} S^2_\alpha - ncS \\ &= \Big(\sum_{\alpha} S_\alpha\Big)^2 + 2 \sum_{\alpha < \beta} S_\alpha S_\beta - ncS \\ &= p\sigma_1^2 + p(p-1)\,\sigma_2 - ncS, \end{aligned} \tag{3.7}$$

where

$$p\,\sigma_1 = \sum_{\alpha} S_\alpha = S, \qquad \frac{p(p-1)}{2}\,\sigma_2 = \sum_{\alpha<\beta} S_\alpha S_\beta. \tag{3.8}$$

It can be easily seen that

$$p^2(p-1)\,(\sigma_1^2 - \sigma_2) = \sum_{\alpha<\beta} (S_\alpha - S_\beta)^2 \geqq 0, \tag{3.9}$$

and therefore

$$\begin{aligned} -\sum h^{\alpha}_{ij} \cdot \Delta h^{\alpha}_{ij} &\leqq p^2 \sigma_1^2 + p(p-1)\,\sigma_2 - ncS \\ &= (2p^2 - p)\,\sigma_1^2 - p(p-1)\,(\sigma_1^2 - \sigma_2) - ncS \\ &\leqq p(2p-1)\,\sigma_1^2 - ncS \\ &= \Big(2 - \frac{1}{p}\Big) S^2 - ncS. \end{aligned} \tag{3.10}$$

Theorem 1. *Let M be an n-dimensional compact oriented manifold which is minimally immersed in an $(n+p)$-dimensional space of constant curvature c. Then*

$$\int_M \Big[\Big(2 - \frac{1}{p}\Big) S - nc\Big] S * 1 \geqq 0, \tag{3.11}$$

where $*1$ *denotes the volume element of* M.

Proof. This follows from (3.10) and the following lemma.

Lemma 2. *If M is an n-dimensional oriented compact manifold immersed in an $(n+p)$-dimensional riemannian manifold N, then*

$$\int_M \left(\sum h_{ij}^\alpha \cdot \Delta h_{ij}^\alpha\right) * 1 = -\int_M \sum (h_{ijk}^\alpha)^2 * 1 \leqq 0.$$

Proof of Lemma 2. We have

$$\tfrac{1}{2}\Delta\left(\sum (h_{ij}^\alpha)^2\right) = \sum (h_{ijk}^\alpha)^2 + \sum h_{ij}^\alpha \cdot \Delta h_{ij}^\alpha. \tag{3.12}$$

Integrating (3.12) over M and applying Green's theorem to the left hand side, we see that the integral of the left hand side and hence that of the right hand side also vanish. q.e.d.

Corollary. *Let M be a compact manifold minimally immersed in a space N of constant curvature c. If M is not totally geodesic and if $S \leqq nc\Big/\left(2-\frac{1}{p}\right)$ everywhere on M, then $S = nc\Big/\left(2-\frac{1}{p}\right)$.*

Assume that $S = \sum (h_{ij}^\alpha)^2$ is a constant. Whether M is compact or not, (3.12) implies

$$0 = \sum (h_{ijk}^\alpha)^2 + \sum h_{ij}^\alpha \cdot \Delta h_{ij}^\alpha.$$

This combined with (3.10) yields

$$\left[\left(2-\frac{1}{p}\right)S - nc\right] S \geqq \sum (h_{ijk}^\alpha)^2.$$

We may therefore conclude that if $S = nc\Big/\left(2-\frac{1}{p}\right)$, then $h_{ijk}^\alpha = 0$, i.e., the second fundamental form h_{ij}^α is parallel.

4. Minimal submanifolds of a unit sphere with $S = n\Big/\left(2-\frac{1}{p}\right)$.

Throughout this section, we shall assume that N is a space of constant curvature 1, that M is not totally geodesic and that

$$S = \sum (h_{ij}^\alpha)^2 = n\Big/\left(2-\frac{1}{p}\right).$$

At the end of § 3 we proved that $h_{ijk}^\alpha = 0$. Then $\Delta h_{ij}^\alpha = 0$, and the terms at the both ends of (3.10) vanish. It follows that all inequalities in (3.7), (3.9) and (3.10) are actually equalities. In deriving (3.7) from (3.5), we made use of the inequality $N(H_\alpha H_\beta - H_\beta H_\alpha) \leqq 2N(H_\alpha)\cdot N(H_\beta)$. Hence,

$$N(H_\alpha H_\beta - H_\beta H_\alpha) = 2N(H_\alpha)\cdot N(H_\beta) \qquad \alpha \neq \beta. \tag{4.1}$$

From (3.9) we obtain

$$p(p-1)(\sigma_1^2 - \sigma_2) = 0. \tag{4.2}$$

From (4.1) and Lemma 1, we conclude that at most two of the matrices H_α are nonzero, in which case they can be assumed to be scalar multiples of $\tilde{A}$ and $\tilde{B}$ in Lemma 1. We now consider the cases $p=1$ and $p \geqq 2$ separately.

Case $p=1$. *We* set

$$h_{ij} = h_{ij}^{n+1}.$$

We choose our frame field in such a way that

$$h_{ij} = 0 \quad \text{for} \quad i \neq j, \tag{4.3}$$

and we set

$$h_i = h_{ii}.$$

Lemma 3. *After a suitable renumbering of the basis elements* $e_1, \ldots, e_n$, *we have*

(a) $h_1 = \ldots = h_m = \lambda =$ *constant,*

$h_{m+1} = \ldots = h_n = \mu =$ *constant,* $\quad (1 < m < n)$,

$\lambda\mu = -1$,

(b) $\omega_j^i = 0$ *for* $1 \leqq i \leqq m$ *and* $m+1 \leqq j \leqq n$.

Proof. Since $h_{ijk}=0$, setting $i=j$ in (2.10) and noting (4.3) we obtain

$$0 = dh_i - 2\sum h_{il}\,\omega_i^l = dh_i, \tag{4.4}$$

which shows that h_i is a constant. Since $h_{ijk}=0$ and $dh_{ij}=0$, (2.10) implies

$$0 = \sum h_{il}\,\omega_j^l + \sum h_{lj}\,\omega_i^l = (h_i - h_j)\,\omega_j^i,$$

which shows that $\omega_j^i = 0$ whenever $h_i \neq h_j$. Thus, if $h_i \neq h_j$, then

$$0 = d\omega_j^i = -\sum \omega_k^i \wedge \omega_j^k - \omega_{n+1}^i \wedge \omega_j^{n+1} + \omega^i \wedge \omega^j.$$

The first sum of the equation above is zero, because $\omega_k^i \neq 0$ and $\omega_j^k \neq 0$ would imply $h_i = h_k = h_j$, contradicting the hypothesis. Hence,

$$\begin{aligned} 0 &= -\omega_{n+1}^i \wedge \omega_j^{n+1} + \omega^i \wedge \omega^j \\ &= \sum h_{ik}\, h_{jl}\,\omega^k \wedge \omega^l + \omega^i \wedge \omega^j \\ &= (h_i\, h_j + 1)\,\omega^i \wedge \omega^j. \end{aligned}$$

This shows that if $h_i \neq h_j$, then $h_i\, h_j = -1$. Set $\lambda = h_1$. By renumbering the indices of $e_1, \ldots, e_n$, let $\lambda = h_1 = \ldots = h_m$ and $\lambda \neq h_j$ for $j \geqq m+1$. Since $\sum h_i = 0$ and M is not totally geodesic, not all $h_1, \ldots, h_n$ are equal to λ. Since $h_1\, h_j = -1$ for $j \geqq m+1$, we obtain $h_{m+1} = \ldots = h_n = -\frac{1}{\lambda}$. We set $\mu = -\frac{1}{\lambda}$. q.e.d.

From (b) of Lemma 3, it follows that the two distributions defined by $\omega^1 = \ldots = \omega^m = 0$ and $\omega^{m+1} = \ldots = \omega^n = 0$ are both integrable and give a local decomposition of M. Then every point of M has a neighborhood U which is a riemannian product $V_1 \times V_2$ with $\dim V_1 = m$ and $\dim V_2 = n - m$. The curvatures of V_1 and V_2 are given by [see (2.7)]

$$R^i_{jkl} = (1 + \lambda^2)(\delta_{ik}\delta_{jl} - \delta_{il}\delta_{jk}) \quad \text{for} \quad 1 \leqq i, j, k, l \leqq m; \tag{4.5}$$

$$R^i_{jkl} = (1 + \mu^2)(\delta_{ik}\delta_{jl} - \delta_{il}\delta_{jk}) \quad \text{for} \quad m+1 \leqq i, j, k, l \leqq n. \tag{4.6}$$

If $m \geqq 2$ (resp. $n - m \geqq 2$), then V_1 (resp. V_2) is a space of constant curvature $1 + \lambda^2$ (resp. $1 + \mu^2$). If $m = 1$ (resp. $n - m = 1$), then V_1 (resp. V_2) is a curve and hence is also a space of constant curvature.

The minimality of the immersion implies

$$0 = \sum h_i = m\lambda + (n - m)\mu.$$

On the other hand, the assumption $S = n\Big/\left(2 - \frac{1}{p}\right) = n$ implies

$$n = S = \sum h_i^2 = n\lambda^2 + (n - m)\mu^2.$$

These two relations together with $\lambda\mu = -1$ imply

$$\lambda = \sqrt{(n-m)/m}, \qquad \mu = -\sqrt{m/(n-m)}$$

or

$$\lambda = -\sqrt{(n-m)/m}, \qquad \mu = \sqrt{m/(n-m)}.$$

Replacing e_{n+1} by $-e_{n+1}$ if necessary, we may assume that $\lambda = \sqrt{(n-m)/m}$ and $\mu = -\sqrt{m/(n-m)}$. In summary, we have

Theorem 2. *Let M be a minimal hypersurface immersed in an $(n+1)$-dimensional space N of constant curvature 1 satisfying $S = n$. Then M is locally a riemannian direct product $M \supset U = V_1 \times V_2$ of spaces V_1 and V_2 of constant curvature, $\dim V_1 = m \geqq 1$ and $\dim V_2 = n - m \geqq 1$. With respect to an adapted frame field, the connection form (ω^A_B) of N, restricted to M, is given by*

$$\left(\begin{array}{ccc|ccc|c}
\omega^1_1 & \cdots & \omega^1_m & & & & \lambda\omega^1 \\
\vdots & \vdots & \vdots & & 0 & & \vdots \\
\omega^m_1 & \cdots & \omega^m_m & & & & \lambda\omega^n \\
\hline
 & & & \omega^{m+1}_{m+1} & \cdots & \omega^{m+1}_n & \mu\omega^{m+1} \\
 & 0 & & \vdots & \vdots & \vdots & \vdots \\
 & & & \omega^n_{m+1} & \cdots & \omega^n_n & \mu\omega^n \\
\hline
-\lambda\omega^1 & \cdots & -\lambda\omega^m & -\mu\omega^{m+1} & \cdots & -\mu\omega^n & 0
\end{array}\right) \tag{4.7}$$

where $\lambda = \sqrt{(n-m)/m}$ and $\mu = -\sqrt{n/(n-m)}$.

We consider now the submanifold $M_{m,\,n-m}$ of S^{n+1} defined in § 1 and shall prove that the connection form of S^{n+1}, restricted to $M_{m,\,n-m}$, is given by (4.7). Let $f_0, f_1, \ldots, f_m$ be an orthonormal frame field for $\boldsymbol{R}^{m+1}$ such that f_0 is normal to $S^m\left(\sqrt{\frac{m}{n}}\right)$ and let $\varphi^0, \varphi^1, \ldots, \varphi^{m+1}$ be the dual frame field. Similarly, for $S^{n-m}\left(\sqrt{\frac{n-m}{n}}\right)$ in $\boldsymbol{R}^{n-m+1}$, we choose an orthonormal frame field $f_{m+1}, \ldots, f_{n+1}$ such that f_{n+1} is normal to $S^{n-m}\left(\sqrt{\frac{n-m}{n}}\right)$ and let $\varphi^{m+1}, \ldots, \varphi^{n+1}$ be the dual frame field. Let $(\varphi^A_B)_{A,\,B=0,\,1,\cdots,\,n+1}$ be the connection form for $\boldsymbol{R}^{n+2}$ with respect to the dual frame field $(\varphi^A)_{A=0,\,1,\cdots,\,n+1}$. These forms, restricted to $M_{m,\,n-m}$, satisfy

$$\varphi^0 = \varphi^{n+1} = 0,$$

$$\varphi^0_i = -\varphi^i_0 = -\sqrt{\frac{n}{m}}\,\varphi^i \qquad i = 1, \ldots, m,$$

$$\varphi^j_{n+1} = -\varphi^{n+1}_j = \sqrt{\frac{n}{n-m}}\,\varphi^j, \qquad j = m+1, \ldots, n,$$

$$\varphi^A_B = -\varphi^B_A = 0 \quad \text{for} \quad A = 0, 1, \ldots, m \quad \text{and} \quad B = m+1, \ldots, n+1.$$

The image of the imbedding $M_{m,\,n-m} \to \boldsymbol{R}^{n+2}$ lies in the unit sphere S^{n+1}. We take a new frame field $e_0, \ldots, e_{n+1}$ for $\boldsymbol{R}^{n+2}$ as follows:

$$e_0 = \sqrt{\frac{m}{n}}\,f_0 + \sqrt{\frac{n-m}{n}}\,f_{n+1},$$

$$e_i = f_i, \qquad i = 1, \ldots, n,$$

$$e_{n+1} = \sqrt{\frac{n-m}{n}}\,f_0 - \sqrt{\frac{m}{n}}\,f_{n+1}.$$

Then e_0 is normal to S^{n+1} and e_{n+1} is normal to $M_{m,\,n-m}$. Let $\omega^0, \ldots, \omega^{n+1}$ be the dual frame field. Then

$$\omega^0 = \sqrt{\frac{m}{n}}\,\varphi^0 + \sqrt{\frac{n-m}{n}}\,\varphi^{n+1},$$

$$\omega^i = \varphi^i, \qquad i = 1, \ldots, n,$$

$$\omega^{n+1} = \sqrt{\frac{n-m}{n}}\,\varphi^0 - \sqrt{\frac{m}{n}}\,\varphi^{n+1}.$$

The connection form $(\omega^A_B)_{A,\,B=0,1,\cdots,\,n}$ for $\boldsymbol{R}^{n+2}$ with respect to the dual frame field (ω_A) is then given by

$$\omega^0_j = -\omega^j_0 = \sqrt{\frac{m}{n}}\,\varphi^0_j + \sqrt{\frac{n-m}{n}}\,\varphi^{n+1}_j \quad \text{for} \quad j = 1, \ldots, n,$$

$$\omega^0_{n+1} = -\omega^{n+1}_0 = -\varphi^0_{n+1},$$

$$\omega^i_j = \varphi^i_j \quad \text{for} \quad i, j = 1, \ldots, n,$$

$$\omega^i_{n+1} = -\omega^{n+1}_i = \sqrt{\frac{n-m}{n}}\,\varphi^i_0 - \sqrt{\frac{m}{n}}\,\varphi^i_{n+1}.$$

We restrict these forms to $M_{m,\,n-m}$. Then by a straightforward calculation we can verify easily that the connection form $(\omega_B^A)_{A,\,B=1,\cdots,\,n+1}$ of S^{n+1}, restricted to $M_{m,\,n-m}$ coincides with the form in (4.7). We may therefore conclude that a minimal hypersurface of S^{n+1} satisfying $S=n$ coincides locally with $M_{m,\,n-m}$. If it is compact, then it coincides with $M_{m,\,n-m}$.

Case $p\geqq 2$. In this case, (4.2) implies

$$\sigma_1^2=\sigma_2.$$

We know that at most two of H_α, $\alpha=n+1,\ldots,n+p$, are different from zero. Assume that only one of them, say H_α, is different from zero. Then we have $\sigma_1=\frac{1}{p}S_\alpha$ and $\sigma_2=0$, in contradiction to $\sigma_1^2=\sigma_2$. We may therefore assume that

$$H_{n+1}=\lambda\tilde{A},\quad H_{n+2}=\mu\tilde{B},\quad \lambda,\mu\neq 0,$$
$$H_\alpha=0\quad\text{for}\quad \alpha\geqq n+3,$$

where $\tilde{A}$ and $\tilde{B}$ are defined in Lemma 1. In other words,

$$\omega_1^{n+1}=\lambda\omega^2,\quad \omega_2^{n+1}=\lambda\omega^1,\quad \omega_i^{n+1}=0\quad\text{for}\quad i=3,\ldots,n,$$
$$\omega_1^{n+2}=\mu\omega^1,\quad \omega_2^{n+2}=-\mu\omega^2,\ \omega_i^{n+2}=0\quad\text{for}\quad i=3,\ldots,n,$$
$$\omega_i^\alpha=0\quad\text{for}\quad \alpha=n+3,\ldots,n+p\quad\text{and}\quad i=1,\ldots,n.$$

Since $h_{ijk}^\alpha=0$, we have [see (2.10)]

$$d h_{ij}^\alpha=\sum h_{ik}^\alpha\,\omega_j^k+\sum h_{kj}^\alpha\,\omega_i^k-\sum h_{ij}^\beta\,\omega_\beta^\alpha. \tag{4.8}$$

Setting $\alpha=n+1$, $i=1$ and $j=2$, we see that $d\lambda=dh_{12}^{n+1}=0$, i.e., λ is constant. Setting $\alpha=n+1$, $i=1$ and $j\geqq 3$, we see that

$$\omega_j^2=0\quad\text{for}\quad j\geqq 3. \tag{4.9}$$

Setting $\alpha=n+1$, $i=2$ and $j\geqq 3$, we see that

$$\omega_j^1=0\quad\text{for}\quad j\geqq 3. \tag{4.10}$$

Similarly, setting $\alpha=n+2$, and $i=j=1$, we see that μ is a constant. From (4.8), (4.9) and (4.10), it follows that if $j\geqq 3$, then

$$0=d\omega_j^1=-\sum\omega_k^1\wedge\omega_j^k+\omega^1\wedge\omega^j=\omega^1\wedge\omega^j.$$

Since $\omega^1,\ldots,\omega^n$ are orthonormal, $\omega^1\wedge\omega^j=0$ implies $\omega^j=0$ for $j\geqq 3$. This shows that $\dim M=2$. From

$$p\,\sigma_1=2(\lambda^2+\mu^2)\quad\text{and}\quad p(p-1)\,\sigma^2=8\lambda^2\mu^2,$$

we obtain

$$p^2(p-1)(\sigma_1^2-\sigma_2)=4[(p-1)\lambda^4-2\lambda^2\mu^2+(p-1)\mu^4].$$

Since the left hand side is zero by (4.2), the discriminant of the right hand side must be non-negative, i.e.,

$$1-(p-1)^2\geqq 0.$$

Since $p\geqq 2$, p must be 2. Hence, dim $N=4$. From $0=\lambda^4-2\lambda^2\mu^2+\mu^4$, it follows that $\lambda^2=\mu^2$. Since $\frac{4}{3}=S=4\lambda^2$, we have

$$\lambda^2=\mu^2=\tfrac{1}{3}. \tag{4.11}$$

Replacing e_3 by $-e_3$ and e_4 by $-e_4$ if necessary, we may assume that $-\lambda=\mu=\sqrt{1/3}$.

Setting $\alpha=3$ and $i=j=1$, we obtain

$$\omega_4^3=\frac{2\lambda}{\mu}\omega_1^2=-2\omega_1^2. \tag{4.12}$$

The curvature of M is given by

$$\Omega_2^1=\omega^1\wedge\omega^2+\omega_1^3\wedge\omega_2^3+\omega_1^4\wedge\omega_2^4=(1-\lambda^2-\mu^2)\,\omega^1\wedge\omega^2=\tfrac{1}{3}\omega^1\wedge\omega^2. \tag{4.13}$$

In summary, we have

Theorem 3. *Let M be an n-dimensional manifold immersed minimally in an $(n+p)$-dimensional space N of constant curvature 1 satisfying $S=n\Big/\left(2-\frac{1}{p}\right)$. If $p\geqq 2$, then $n=p=2$. With respect to a an adapted dual orthonormal frame field $\omega^1,\omega^2,\omega^3,\omega^4$, the connection form (ω_B^A) of N, restricted to M, is given by*

$$\begin{pmatrix} 0 & \omega_2^1 & \mu\omega^2 & -\mu\omega^1\\ \omega_1^2 & 0 & \mu\omega^1 & \mu\omega^2\\ \lambda\omega^2 & \lambda\omega^1 & 0 & 2\omega_2^1\\ -\lambda\omega^1 & \lambda\omega^2 & 2\omega_1^2 & 0\end{pmatrix},\qquad -\lambda=\mu=\sqrt{\frac{1}{3}} \tag{4.14}$$

We consider now the Veronese surface defined in § 1. We shall compute its structure equations by group theoretic means. Let

$$u^1=\frac{1}{\sqrt{3}}yz,\quad u^2=\frac{1}{\sqrt{3}}zx,\quad u^3=\frac{1}{\sqrt{3}}xy,\quad u^4=\frac{1}{2\sqrt{3}}(x^2-y^2),$$
$$u^5=\frac{1}{6}(x^2+y^2-2z^2)$$

be as in § 1. These equations define an immersion of $S^2(\sqrt{3})$ into S^4 and induce an action of $SO(3)$ on S^4 so that the immersion $S^2(\sqrt{3})\to S^4$

is equivariant. In other words, we obtain a representation of $SO(3)$ into $SO(5)$. This induces a representation of the Lie algebra $so(3)$ into the Lie algebra $so(5)$. By a straightforward, simple calculation, we see that this representation maps a matrix of the form

$$\begin{pmatrix} 0 & -\alpha & -\gamma \\ \alpha & 0 & -\beta \\ \gamma & \beta & 0 \end{pmatrix} \in so(3)$$

into a matrix of the form

$$\begin{pmatrix} 0 & 0 & \gamma & -\beta & \sqrt{3}\beta \\ -\alpha & 3 & \beta & \gamma & \sqrt{3}\gamma \\ -\gamma & -\beta & 0 & 2\alpha & 0 \\ \beta & -\gamma & -2\alpha & 0 & 0 \\ -\sqrt{3}\beta & -\sqrt{3}\gamma & 0 & 0 & 0 \end{pmatrix} \in so(5).$$

Let (ω_B^A) be the Maurer-Cartan form for $SO(5)$ and set

$$\omega^i = \omega_5^i \qquad i = 1, \ldots, 4.$$

Then the restriction of (ω_B^A) to the image of $SO(3)$ in $SO(5)$ is given by

$$\begin{pmatrix} 0 & \omega_2^1 & \mu\omega^2 & -\mu\omega^1 & \omega^1 \\ \omega_1^2 & 0 & \mu\omega^1 & \mu\omega^2 & \omega^2 \\ \lambda\omega^2 & \lambda\omega^1 & 0 & 2\omega_2^1 & 0 \\ -\lambda\omega^1 & \lambda\omega^2 & 2\omega_1^2 & 0 & 0 \\ -\omega^1 & -\omega^2 & 0 & 0 & 0 \end{pmatrix}, \qquad -\lambda = \mu = \sqrt{\frac{1}{3}}.$$

Comparing this with (4.14), we may conclude that a minimal surface in S^4 satisfying $S = \frac{4}{3}$ coincides locally with the Veronese surface. If it is compact, it coincides with the Veronese surface. This completes the proof of the main theorem.

5. Related examples

Example 1.

$$S^m \times S^q \to S^{m+q+mq}$$

Let

$$S^m = \{(x_0, x_1, \ldots, x_m) \in \boldsymbol{R}^{m+1};\ \textstyle\sum x_i^2 = 1\},$$
$$S^q = \{(y_0, y_1, \ldots, y_q) \in R^{q+1};\ \textstyle\sum y_i^2 = 1\},$$
$$S^{m+q+mq} = \{(u_{ij});\ i = 0, 1, \ldots, m;\ j = 0, 1, \ldots, q;\ \textstyle\sum u_{ij}{}^2 = 1\}.$$

Then the mapping $S^m \times S^q \to S^{m+q+mq}$ defined by $u_{ij} = x_i y_j$ is an isometric immersion. Two points (x, y) and $(-x, -y)$ of $S^m \times S^q$ are mapped into the same point. We have

$$R = \text{scalar curvature of } S^m \times S^q = m(m-1) + q(q-1),$$
$$S = (m+q)(m+q-1) - R = 2mq.$$

On the other hand, if we denote $m+q$ by n and the codimension mq by p, then

$$n\Big/\left(2 - \frac{1}{p}\right) = mq(m+q)/(2mq-1).$$

Consider the case $m = q = 1$. Then $S = 2 = n\Big/\left(2 - \frac{1}{p}\right)$, and this minimal immersion satisfies the assumption in our main theorem. We shall show that this immersion of $S^1 \times S^1 \to S^3$ is a double covering of the immersion of $M_{1,1} = S^1\left(\frac{1}{\sqrt{2}}\right) \times S^1\left(\frac{1}{\sqrt{2}}\right)$ into S^3 defined in § 1. The immersion

$$(x_0, x_1, y_0, y_1) \in S^1 \times S^1 \to (v_0, v_1 \cdot w_0, w_1) \in S^1\left(\frac{1}{\sqrt{2}}\right) \times S^1\left(\frac{1}{\sqrt{2}}\right) \in S^3$$

defined by

$$v_0 = \frac{1}{\sqrt{2}}(x_0 y_0 - x_1 y_1), \qquad v_1 = \frac{1}{\sqrt{2}}(x_1 y_0 + x_0 y_1),$$
$$w_0 = \frac{1}{\sqrt{2}}(x_0 y_0 + x_1 y_1), \qquad w_1 = \frac{1}{\sqrt{2}}(x_1 y_0 - x_0 y_1),$$

differs from the immersion

$$(x_0, x_1, y_0, y_1) \in S^1 \times S^1 \to (x_0 y_0, x_0 y_1, x_1 y_0, x_1 y_1) \in S^3$$

by a rigid motion of S^3.

Example 2. $S^n\left(\sqrt{\frac{2(n+1)}{n}}\right) \to S^{n+p}$ with $p = \frac{1}{2}(n-1)(n+2)$.

Let

$$S^n\left(\sqrt{\frac{2(n+1)}{n}}\right) = \left\{(x_0, x_1, \ldots, x_n) \in \mathbf{R}^{n+1}; \sum x_i^2 = \frac{2(n+1)}{n}\right\}.$$

Let E be the space of $(n+1) \times (n+1)$ symmetric matrices (u_{ij}), $(i, j = 1, \ldots, n)$, such that $\sum u_{ii} = 0$; it is a vector space of dimension $\frac{1}{2}n(n+3)$. We define a norm in E by $\|(u_{ij})\|^2 = \sum u_{ij}{}^2$. Let S^{n+p} with $p = \frac{1}{2}(n-1)(n+2)$ be the unit hypersphere in E. The mapping of $S^n\left(\sqrt{\frac{2(n+1)}{n}}\right)$ into S^{n+p} defined by

$$u_{ij} = \frac{1}{2}\sqrt{\frac{n}{n+1}}\left(x_i x_j - \frac{2}{n}\delta_{ij}\right)$$

is an isometric minimal immersion. (Actually, this gives an imbedding of the real projective space of n-dimension into S^{n+p}.) We have

$$R = \text{scalar curvature of } S^n\left(\sqrt{\frac{2(n+1)}{n}}\right) = n^2(n-1)/2(n+1),$$

$$S = n(n-1) - R = n(n-1)(n+2)/2(n+1).$$

On the other hand,

$$n\Big/\left(2-\frac{1}{p}\right) = n(n-1)(n+2)/2(n^2+n-3).$$

For $n=2$, we recover the Veronese surface.

Example 3. $S^2(\sqrt{6}) \to S^6$.

Making use of harmonic polynomials of degree 3, we consider the following minimal immersion of $S^2(\sqrt{6})$ into S^3 defined by

$$u_0 = \frac{\sqrt{6}}{72} z(-3x^2-3y^2+2z^2), \qquad u_1 = \frac{1}{24} x(-x^2-y^2+4z^2),$$

$$u_2 = \frac{\sqrt{10}}{24} z(x^2-y^2), \qquad u_3 = \frac{\sqrt{15}}{72} x(x^2-3y^2),$$

$$u_4 = \frac{1}{24} y(-x^2-y^2+4z^2), \qquad u_5 = \frac{\sqrt{10}}{12} xyz,$$

$$u_6 = \frac{\sqrt{15}}{72} y(3x^2-y^2).$$

We have

$$R = \text{scalar curvature of } S^2(\sqrt{6}) = \tfrac{1}{3},$$

$$S = 2 - R = \tfrac{5}{3},$$

whereas

$$n\Big/\left(2-\frac{1}{p}\right) = \frac{8}{7}.$$

For each positive integer k, the space of harmonic polynomials of degree k in variables x, y, z is a vector space of dimension $2k+1$. Introducing an inner product in this vector space in a well known manner, we get a minimal immersion of a 2-sphere into the hypersphere S^{2k} in a natural manner. The case $k=2$ gives the Veronese surface. The case $k=3$ was described above. Generally, for every positive integer k, we have an isometric minimal immersion

$$S^2\left(\sqrt{\frac{k(k+1)}{2}}\right) \to S^{2k},$$

for which $S = 2 - \dfrac{4}{k(k+1)}$. For a systematic study of minimal immersions of 2-spheres obtained in this manner, see BORUVKA [1] and a forthcoming paper of DO CARMO and WALLACH.

Example 4. $S^{m_1}\left(\sqrt{\frac{m_1}{n}}\right) \times \cdots \times S^{m_k}\left(\sqrt{\frac{m_k}{n}}\right) \to S^{n+k-1}$, $n = \sum_{i=1}^{k} m_i$.

We can generalize the construction of $M_{m,\,n-m}$ as follows. Let $m_1, \ldots, m_k$ be positive integers and $n = m_1 + \cdots + m_k$. Let x_i be a point of $S^{m_i}\left(\sqrt{\frac{m_i}{n}}\right)$, i.e., a vector of length $\sqrt{\frac{m_i}{n}}$ in $\boldsymbol{R}^{m_i+1}$. Then $(x_1, \ldots, x_k)$ is a unit vector in $\mathbb{R}^{n+k}$. This defines a minimal immersion of $M_{m_1,\ldots,m_k} = \prod S^{m_i}\left(\sqrt{\frac{m_i}{n}}\right)$ into S^{n+k-1}. We have

$$R = \text{scalar curvature of } M_{m_1,\ldots,m_k} = (n-k)\,n,$$
$$S = n(n-1) - R = (k-1)\,n,$$
$$n\Big/\left(2-\frac{1}{p}\right) = (k-1)\,n/(2k-3).$$

6. Some questions

The above discussions seem to show the interest of the study of compact minimal submanifolds on the sphere with $S =$ constant. With fixed n and p the question naturally arises as to the possible values for S. We proved in the above that S does not take values in the open interval $\left(0, n\Big/\left(2-\frac{1}{p}\right)\right)$. It is plausible that the set of values for S is discrete, at least for S not arbitrarily large. If this is the case, an estimate of the value for S next to $n\Big/\left(2-\frac{1}{p}\right)$ should be of interest. This problem can be restricted by imposing further conditions on M, such as M be topologically or metrically a sphere.

Another natural question is that of uniqueness. At least for compact minimal hypersurfaces (codim 1) it seems likely that the values of S should determine the hypersurface up to a rigid motion in the ambient sphere S^{n+1}.

References

1. BORUVKA, O.: Sur les surfaces representées par les fonctions sphériques de première espèce. J. Math. pure et appl. **12**, 337—383 (1933).
2. LAWSON, B.: Local rigidity theorems for minimal hypersurfaces. Ann. of Math. **89**, 187—197 (1969).
3. SIMONS, J.: Minimal varieties in riemannian manifolds. Ann. of Math. **88**, 62—105 (1968).

Eisenstein Series over Finite Fields

By Harish-Chandra
The Institute for Advanced Study, Princeton

1. Introduction

Let me begin by recalling the definition and some properties of the classical Eisenstein Series. Let $G = SL(2, \boldsymbol{R})$ and $\Gamma = SL(2, \boldsymbol{Z})$. Then $G = KAU$ where

$$K = SO(2), \quad A = \left\{ a_t = \begin{pmatrix} e^t & 0 \\ 0 & e^{-t} \end{pmatrix}, t \in \boldsymbol{R} \right\}, \quad U = \left\{ \begin{pmatrix} 1 & u \\ 0 & 1 \end{pmatrix}, u \in \boldsymbol{R} \right\}.$$

Let P be the normalizer of U in G. Then $P = MAU$ where $M = \{1, -1\}$. Every element x of G can be written uniquely as $x = kau$ ($k \in K$, $a \in A$, $u \in U$). Put $t(x) = t$ where $a = a_t$. For $\lambda \in \boldsymbol{C}$ with* $\Re \lambda < -1$, put

$$E(\lambda : x) = \sum_{\gamma \in \Gamma / \Gamma \cap P} e^{(\lambda - 1) t(x\gamma)} = \tfrac{1}{2} \sum_{\gamma \in \Gamma / \Gamma_\infty} e^{(\lambda - 1) t(x\gamma)}$$

where $\Gamma_\infty = \Gamma \cap U$. If

$$x = \begin{pmatrix} a & b \\ c & d \end{pmatrix}, \quad x_1 = \begin{pmatrix} a \\ c \end{pmatrix} \quad \text{and} \quad |x_1| = (a^2 + c^2)^{\frac{1}{2}},$$

then

$$E(\lambda : x) = \tfrac{1}{2} \sum_{\gamma \in \Gamma / \Gamma_\infty} |(x\gamma)_1|^{\lambda - 1}.$$

For any $\varepsilon > 0$, this series converges absolutely and uniformly if x remains in a compact subset of G and $\Re \lambda \leqq -1 - \varepsilon$. Hence the function E is holomorphic in λ for $\Re \lambda < -1$. Moreover $E(\lambda : kx) = E(\lambda : x)$ ($k \in K$) and

$$e^t \int_{U/\Gamma_\infty} E(\lambda : a_t u)\, du = e^{\lambda t} + c(\lambda) e^{-\lambda t} \quad (t \in \boldsymbol{R}, \Re \lambda < -1).$$

Here $c(\lambda)$ is a holomorphic function of λ for $\Re \lambda < -1$ and the Haar measure du on U is so normalized that the total measure of U/Γ_∞ is 1. The functions c and E have the following properties.

1) c extends to a meromorphic function on $\boldsymbol{C}$ satisfying the functional equation $c(\lambda) c(-\lambda) = 1$. Moreover $|c(\lambda)| = 1$ when $\Re \lambda = 0$.

---

* $\Re \lambda$ denotes the real part of a complex number λ.

2) For any $x \in G$, $E(\lambda : x)$ extends to a meromorphic function of $\lambda \in \boldsymbol{C}$ with the functional equation

$$E(\lambda : x) = c(\lambda) E(-\lambda : x).$$

In fact

$$c(\lambda) = \pi^{\lambda} \Gamma(-\lambda/2) \zeta(-\lambda) \{\Gamma(\lambda/2) \zeta(\lambda)\}^{-1}$$

where ζ is the Riemann Zeta function.

Similar results hold in other cases. During the past fifteen years, such series have been studied by SELBERG [5]. More recently LANGLANDS [4(a), (b)] has proved corresponding results for the Eisenstein Series on G/Γ where G is any semisimple algebraic group defined over $\boldsymbol{Q}$ and Γ an arithmetic subgroup of G.

In 1966—1967 I gave some lectures on LANGLANDS' work in Princeton (see [3]) and thereby learnt, what I now call, the philosophy of cusp forms. In the case of G/Γ this philosophy is certainly implicit in SELBERG [5]. Moreover it has been expounded, in a more general context, by GELFAND in his 1962 Stockholm address [2], although I could not understand it then. Actually I realized its full scope only when I tried to relate LANGLANDS' work to my own on harmonic analysis on G. This philosophy works in the following four cases.

1) G/Γ. In fact it originated there.

2) A real semisimple group G.

3) I believe it is also applicable to a reductive $\mathfrak{p}$-adic group G, although this case has not yet been sufficiently investigated.

4) It works for reductive algebraic groups defined over a finite field.

In this lecture we shall be concerned with 4). Our main object is to define the Eisenstein Series and prove their functional equations (see §7, Theorem 3).

2. Bruhat's Lemma and its consequences

We begin by recalling some known facts (see [1]). Let K be a field which will be kept fixed throughout. By a K-group we mean a (linear) algebraic group defined over K. Let $\boldsymbol{G}$ be a connected reductive K-group. By a parabolic subgroup $\boldsymbol{P}$ of $\boldsymbol{G}$ we mean an algebraic subgroup which contains a Borel subgroup of $\boldsymbol{G}$. We say that $\boldsymbol{P}$ is K-parabolic if it is parabolic and defined over K. Fix a K-parabolic subgroup $\boldsymbol{P}$ and let $\boldsymbol{U}$ denote the unipotent radical of $\boldsymbol{P}$. Then $\boldsymbol{U}$ is a K-subgroup. By a Levi K-subgroup $\boldsymbol{M}$ of $\boldsymbol{P}$, we mean a reductive K-subgroup such that the mapping $(m, u) \to mu$ $(m \in \boldsymbol{M}, u \in \boldsymbol{U})$ defines a K-isomorphism of the algebraic varieties $\boldsymbol{M} \times \boldsymbol{U}$ and $\boldsymbol{P}$. Such a subgroup $\boldsymbol{M}$ always exists and is connected. Fix $\boldsymbol{M}$ and let $\boldsymbol{A}$ be a maximal K-split torus lying in the center of $\boldsymbol{M}$. Then $\boldsymbol{A}$ is unique and $\boldsymbol{M}$ is the centralizer of $\boldsymbol{A}$ in $\boldsymbol{G}$. We call $\boldsymbol{A}$ a split component of $\boldsymbol{P}$. Let G denote the group of K-rational points of $\boldsymbol{G}$. For any split component $\boldsymbol{A}'$ of $\boldsymbol{P}$ there exists a unique element $u\in U=\boldsymbol{U}\cap G$ such that* $\boldsymbol{A}'=\boldsymbol{A}^u$. Hence dim $\boldsymbol{A}$ depends only on $\boldsymbol{P}$. We call it the parabolic rank of $\boldsymbol{P}$ and denote it by prk $\boldsymbol{P}$.

A cuspidal subgroup P of G is a group of the form $P=G\cap\boldsymbol{P}$, where $\boldsymbol{P}$ is a K-parabolic subgroup of G. P determines $\boldsymbol{P}$ completely. By a split component $\boldsymbol{A}$ of P, we mean a split component of $\boldsymbol{P}$. We write prk $P=\dim\boldsymbol{A}$ and call $(P,\boldsymbol{A})$ a cuspidal pair in G. Once $\boldsymbol{A}$ is fixed, we have the corresponding Levi K-decompositions $\boldsymbol{P}=\boldsymbol{M}\boldsymbol{U}$ and $P=MU$ where $M=\boldsymbol{M}\cap G$. We shall call U the unipotent radical of P.

Let $(P_i,\boldsymbol{A}_i)$ $(i=1,2)$ be two cuspidal pairs in G. We write $(P_1,\boldsymbol{A}_1)>(P_2,\boldsymbol{A}_2)$ if $P_1\supset P_2$ and $\boldsymbol{A}_1\subset\boldsymbol{A}_2$. A cuspidal pair is called mincuspidal if it is minimal with respect to this partial order. Let $\mathfrak{w}(\boldsymbol{A}_1,\boldsymbol{A}_2)$ denote the set of all bijections $s:\boldsymbol{A}_1\to\boldsymbol{A}_2$ with the following property. There should exist an element $y\in G$ such that $a^s=a^y$ for all $a\in\boldsymbol{A}_1$. P_1, P_2 are called associated if $\mathfrak{w}(\boldsymbol{A}_1,\boldsymbol{A}_2)\neq\emptyset$ i.e. if $\boldsymbol{A}_1$ and $\boldsymbol{A}_2$ are conjugate under G. (The two pairs $(P_i,\boldsymbol{A}_i)$ $i=1,2$, are then also called associated.) It is known that $\mathfrak{w}(\boldsymbol{A}_1,\boldsymbol{A}_2)$ is a finite set. Fix $s\in\mathfrak{w}(\boldsymbol{A}_1,\boldsymbol{A}_2)$. We say that $y\in G$ is a representative of s in G if $a^s=a^y$ for all $a\in\boldsymbol{A}_1$. In case $\boldsymbol{A}_1=\boldsymbol{A}_2=\boldsymbol{A}$, we write $\mathfrak{w}(\boldsymbol{A})=\mathfrak{w}(\boldsymbol{A},\boldsymbol{A})$. Then $\mathfrak{w}(\boldsymbol{A})$ is a finite group. The following important result has been proved in [1, p. 100].

Bruhat's Lemma. *Let $(P,\boldsymbol{A})$ be a mincuspidal pair in G and $P=MU$ the corresponding Levi K-decomposition. For each $s\in\mathfrak{w}(\boldsymbol{A})$, fix a representative y_s of s in G. Then*

$$G=\bigcup_{s\in\mathfrak{w}(\boldsymbol{A})}Uy_sP$$

where the union is disjoint.

Fix two cuspidal pairs $(P_i,\boldsymbol{A}_i)$ $(i=1,2)$ in G and let $P_i=M_iU_i$ be the corresponding decompositions. The following facts are simple consequences of BRUHAT's lemma.

Lemma 1. *$P_2\backslash G/P_1$ is a finite set. Fix $x\in G$. Then $P_2\cap U_1^x\subset U_2$ if and only if there exist elements $u_i\in U_i$ $(i=1,2)$ such that $\boldsymbol{A}_1^{xu_1}\subset\boldsymbol{A}_2^{u_2}$.*

Corollary. *The following two statements are equivalent.*

1) $P_2\cap U_1^x\subset U_2$, $U_2^{x^{-1}}\cap P_1\subset U_1$.

2) *There exist $u_i\in U_i$ $(i=1,2)$ such that $\boldsymbol{A}_1^y=\boldsymbol{A}_2$ for $y=u_2xu_1$.*

* We write $x^y=yxy^{-1}$ for $x,y\in\boldsymbol{G}$. If S is any subset of G then $S^y=ySy^{-1}$.

3. The cusp forms

Let S be a finite set. By the standard measure on S, we mean the measure which assigns to each point of S the mass 1. Let $C(S)$ denote the space of all complex-valued functions on S and $[S]$ the number of elements of S.

Now assume that K is a finite field. Then G is a finite group. For any $f \in C(G)$ and any cuspidal subgroup P of G, put

$$f_P(x) = \int_U f(xu)\, du \qquad (x \in G)$$

where du is a Haar measure on the unipotent radical U of P. We say that f is a cusp form if $f_P = 0$ for every cuspidal subgroup $P \neq G$ of G.

Let ${}^0C(G)$ be the space of all cusp forms. It is easy to see that ${}^0C(G)$ is stable under both left- and right-translations of G. Hence ${}^0C(G)$ is a two-sided ideal in the group algebra $C(G)$.

Let λ denote the left-regular representation of G on $C(G)$ and ${}^0\lambda$ the restriction of λ on ${}^0C(G)$. Let $\mathscr{E}(G)$ denote the set of all equivalence classes of irreducible (complex) representations of G and ${}^0\mathscr{E}(G)$ the subset of those classes which occur in the reduction of ${}^0\lambda$. It is easy to see that an element $\omega \in \mathscr{E}(G)$ lies in ${}^0\mathscr{E}(G)$ if and only if the character of ω is a cusp form. For any $\omega \in \mathscr{E}(G)$, let $C(G, \omega)$ denote the space of all elements in $C(G)$ which transform under λ according to ω.

Let $(P, \boldsymbol{A})$ be a cuspidal pair in G and $P = MU$ the corresponding decomposition. Fix $f \in C(G)$. Then we write $f_P \sim 0$ if*

$$\int_M \operatorname{conj} \phi(m) \cdot f_P(xm)\, dm = 0$$

for all $\phi \in {}^0C(M)$ and $x \in G$. (Here dm is the standard measure on M.) It is easy to verify that this definition is independent of the choice of $\boldsymbol{A}$.

Lemma 2. *Let f be an element in $C(G)$ such that $f_P \sim 0$ for all cuspidal subgroups P of G (including $P = G$). Then $f = 0$.*

This is entirely analogous to a result of LANGLANDS [4(a), p. 3.24].

4. The irreducibility of induced representations

Let $(P, \boldsymbol{A})$ be a cuspidal pair in G and $P = MU$ the corresponding decomposition. Fix $\omega \in {}^0\mathscr{E}(M)$ and let ω^* denote the class contragredient to ω. Let σ be a right representation of M on a finite-dimensional complex vector space L such that ω^* is the class of σ. (This means that $\operatorname{tr} \sigma(m) = \operatorname{conj} \theta(m)$ $(m \in M)$ where θ is the character of ω.) Let $D(P, \omega)$ denote the space of all functions $f: G \to L$ such that $f(xmu) = f(x)\, \sigma(m)$

* conj c denotes the complex conjugate of a number $c \in \boldsymbol{C}$.

$(x \in G, m \in M, u \in U)$. We get a representation λ of G on $D(P, \omega)$ by setting

$$(\lambda(x)f)(y) = f(x^{-1}y) \qquad (y \in G)$$

for $x \in G$ and $f \in D(P, \omega)$. Let $\Omega(P, \omega)$ denote the class of λ.

Let (P_i, A_i) $(i = 1, 2)$ be two cuspidal pairs and $P_i = M_i U_i$ the corresponding decompositions. Fix $\omega \in {}^0\mathscr{E}(M_1)$, $s \in \mathfrak{w}(A_1, A_2)$ and let y be a representative of s in G. Then $M_2 = M_1^y$. If σ is a representation of M_1 in ω, we define the representation σ^y of M_2 by $\sigma^y(m^y) = \sigma(m)$ $(m \in M_1)$. Let ω^y denote the class of σ^y. Then ω^y depends only on s and we also denote it by ω^s. It is easy to verify that $\omega^s \in {}^0\mathscr{E}(M_2)$.

The following theorem and its corollary play an important role in our discussion.

Theorem 1. *Fix* $\omega_i \in {}^0\mathscr{E}(M_i)$ $(i = 1, 2)$ *and let* $\mathfrak{w}_0$ *be the set of all* $s \in \mathfrak{w}(A_1, A_2)$ *such that* $\omega_1^s = \omega_2$. *Then*

$$I(\Omega(P_1, \omega_1), \Omega(P_2, \omega_2)) = [\mathfrak{w}_0].$$

Here I denotes the intertwining number.

It follows in particular that if P_1, P_2 are not associated, then this intertwining number is zero.

Let (P, A) be a cuspidal pair $(P = MU)$. Fix $\omega \in {}^0\mathscr{E}(M)$ and let $\mathfrak{w}(\omega)$ denote the subgroup of all $s \in \mathfrak{w}(A)$ such that $\omega^s = \omega$. Then $r = [\mathfrak{w}(\omega)]$ is called the ramification index of ω. We say that ω is unramified (in G) if $r = 1$.

Corollary. $\Omega(P_1, \omega_1) = \Omega(P_2, \omega_2)$ *if* $\mathfrak{w}_0 \neq \emptyset$. *Moreover* $\Omega(P_1, \omega_1)$ *is irreducible if* ω_1 *is unramified.*

5. Definition and some properties of Eisenstein Series

Now fix, once for all, a subgroup Γ of G. Let (P, A) be a cuspidal pair $(P = MU)$. We denote by ${}^0\mathscr{D}(P)$ the space of all functions $f \in C(G)$ such that:

1) $f(xy) = f(x)$ for all $x \in G$ and $y \in (\Gamma \cap P)U$.

2) For every $x \in G$, the function $m \to f(xm)$ $(m \in M)$ lies in ${}^0C(M)$.

It is easy to see that the space ${}^0\mathscr{D}(P)$ does not depend on the choice of A. For $f \in {}^0\mathscr{D}(P)$, define

$$E_f(x) = \sum_{\gamma \in \Gamma/\Gamma \cap P} f(x\gamma) \qquad (x \in G).$$

Then $E_f \in C(G/\Gamma)$. For any $g \in C(G/\Gamma)$, put

$$g_P(x) = \int_{U/\Gamma \cap U} g(xu)\, d^0u \qquad (x \in G)$$

where d^0u is the Haar measure on U so normalized that the total measure of $U/\Gamma \cap U$ is 1.

Theorem 2. *Let (P, A) and (P', A') be two cuspidal pairs ($P = MU$, $P' = M'U'$). Fix $f \in {}^0\mathscr{D}(P)$. Then $(E_f)_{P'} \sim 0$ unless P' is associated to P. Now suppose P and P' are associated. Put $\mathfrak{w} = \mathfrak{w}(A, A')$ and let d^0u' denote the Haar measure on U' so normalized that the total measure of $U'/\Gamma \cap U'$ is 1. Then*

$$\int_{U'/\Gamma \cap U'} E_f(xu')\, d^0u' = \sum_{s \in \mathfrak{w}} \big(c_{P'|P}(s) f\big)(x) \qquad (x \in G).$$

Here $c_{P'|P}(s)$ is a linear transformation of ${}^0\mathscr{D}(P)$ into ${}^0\mathscr{D}(P')$ defined as follows. Put

$$\Gamma(s) = \Gamma \cap (P' y P)$$

where y is a representative of s in G. Then

$$\big(c_{P'|P}(s) f\big)(x) = \int_{U'/U' \cap \Gamma} \sum_{\gamma \in \Gamma(s)/\Gamma \cap P} f(xu'\gamma)\, d^0u'$$

for $f \in {}^0\mathscr{D}(P)$ and $x \in G$.

Fix $\omega \in {}^0\mathscr{E}(M)$ and let $\mathscr{D}(P, \omega)$ be the space of all $f \in {}^0\mathscr{D}(P)$ with the following property. For every $x \in G$, the function $m \to f(xm)$ $(m \in M)$ lies in $C(M, \omega)$. Then $c_{P'|P}(s)$ maps $\mathscr{D}(P, \omega)$ into $\mathscr{D}(P', \omega.)$. In fact $c_{P'|P}(s)$ is the zeta-function of Gelfand [2, §8].

For $f, g \in {}^0\mathscr{D}(P)$, define

$$(f, g)_{G/\Gamma \cap P} = \int_{G/\Gamma \cap P} \operatorname{conj} f(x) \cdot g(x)\, dx$$

where dx is the standard measure on $G/\Gamma \cap P$. In this way ${}^0\mathscr{D}(P)$ becomes a Hilbert space.

Lemma 3. *Let (P_i, A_i) $(i = 1, 2)$ be two associated cuspidal pairs. Then*

$$\big(f_2, c_{P_2|P_1}(s) f_1\big)_{G/\Gamma \cap P_2} = \big(c_{P_1|P_2}(s^{-1}) f_2, f_1\big)_{G/\Gamma \cap P_1}$$

for $f_i \in {}^0\mathscr{D}(P_i)$ $(i = 1, 2)$ and $s \in \mathfrak{w}(A_1, A_2)$.

We regard $C(G/\Gamma)$ as a Hilbert space in the usual way so that

$$(\psi, \phi)_{G/\Gamma} = \int_{G/\Gamma} \operatorname{conj} \psi(x) \cdot \phi(x)\, dx \qquad (\psi, \phi \in C(G/\Gamma))$$

where dx is the standard measure on G/Γ.

Corollary. *Let (P_i, A_i) $(i = 1, 2)$ be two cuspidal pairs in G. Fix $f_i \in {}^0\mathscr{D}(P_i)$ and put*

$$E_{f_i}(x) = \sum_{\gamma \in \Gamma/\Gamma \cap P_i} f_i(x\gamma) \qquad (x \in G).$$

Then

$$(E_{f_2}, E_{f_1})_{G/\Gamma} = \sum_{s \in \mathfrak{w}(\mathbf{A}_1, \mathbf{A}_2)} (f_2, c_{P_2|P_1}(s) f_1)_{G/\Gamma \cap P_2}.$$

In particular,

$$(E_{f_2}, E_{f_1})_{G/\Gamma} = 0$$

if P_1, P_2 are not associated.

We return to the notation of Theorem 2.

Lemma 4. *Let $g \in C(G/\Gamma)$. Then*

$$(E_f, g)_{G/\Gamma} = (f, g_P)_{G/\Gamma \cap P}$$

for all $f \in {}^0\mathcal{D}(P)$.

Combining this with Lemma 2 we get the following corollary.

Corollary. *$C(G/\Gamma)$ is spanned by E_f for all $f \in {}^0\mathcal{D}(P)$ and all cuspidal subgroups P of G.*

6. The series $\mathbf{E}(P:\phi)$

Consider a cuspidal pair $(P, \mathbf{A})$ in G with $P = MU$. Fix $\omega \in {}^0\mathcal{E}(M)$ and let y_i $(1 \leqq i \leqq q)$ be a complete system of representatives for $\Gamma \backslash G/P$ in G. Define

$$(P_i, \mathbf{A}_i, M_i, U_i, \omega_i) = (P^{y_i}, \mathbf{A}^{y_i}, M^{y_i}, U^{y_i}, \omega^{y_i})$$

and put

$$\mathfrak{D}(P, \omega) = \prod_{1 \leqq i \leqq q} \mathcal{D}(P_i, \omega_i), \quad {}^0\mathfrak{D}(P) = \prod_{1 \leqq i \leqq q} {}^0\mathcal{D}(P_i),$$

where the products are direct. Then ${}^0\mathfrak{D}(P)$ is a Hilbert space with the scalar product

$$(\phi, \psi) = \sum_{1 \leqq i \leqq q} (\phi_i, \psi_i)_{G/\Gamma \cap P_i} \quad (\phi, \psi \in {}^0\mathfrak{D}(P)).$$

(Here ϕ_i is the component of ϕ in ${}^0\mathcal{D}(P_i)$, $1 \leqq i \leqq q$. Similarly for ψ.) Put

$$\mathbf{E}(P:\phi:x) = \sum_{1 \leqq i \leqq q} \sum_{\gamma \in \Gamma/\Gamma \cap P_i} \phi_i(x\gamma) \quad (x \in G)$$

for $\phi \in {}^0\mathfrak{D}(P)$.

Lemma 5. *Let $f \in C(G/\Gamma)$ and $\phi \in {}^0\mathfrak{D}(P)$. Then*

$$(\mathbf{E}(P:\phi), f)_{G/\Gamma} = \sum_{1 \leqq i \leqq q} (\phi_i, f_{P_i})_{G/\Gamma \cap P_i}.$$

7. Functional equations of the Eisenstein Series

Let $(P_i, \mathbf{A}_i)$ $(1 \leqq i \leqq r)$ be a set of associated cuspidal pairs in G with $P_i = M_i U_i$. For each i, fix a complete system of representatives y_{ij} $(1 \leqq j \leqq q_i)$ for $\Gamma \backslash G/P_i$. Then we can define the Hilbert spaces

${}^0\mathfrak{D}(P_i)$ and $\mathfrak{D}(P_i, \omega_i)$ $(\omega_i \in {}^0\mathcal{E}(M_i))$ as in §6. For any $s \in \mathfrak{w}(\boldsymbol{A}_i, \boldsymbol{A}_j)$, define a linear transformation $\boldsymbol{c}_{P_j|P_i}(s)$ of ${}^0\mathfrak{D}(P_i)$ into ${}^0\mathfrak{D}(P_j)$ as follows. If $\phi \in {}^0\mathfrak{D}(P_i)$, then

$$\big(\boldsymbol{c}_{P_j|P_i}(s)\,\phi\big)_l = \sum_{1 \leq i \leq q_i} c_{P_{jl}|P_{ik}}(y_{jl} \circ s \circ y_{ik}^{-1})\,\phi_k \qquad (1 \leq l \leq q_j).$$

Here $y_{jl} \circ s \circ y_{ik}^{-1}$ denotes the element $t \in \mathfrak{w}(\boldsymbol{A}_i^{y_{ik}}, \boldsymbol{A}_j^{y_{jl}})$ given by

$$(a^{y_{ik}})^t = (a^s)^{y_{jl}} \qquad (a \in \boldsymbol{A}_i).$$

(We note that $(P_{ik}, \boldsymbol{A}_{ik}) = (P_i^{y_{ik}}, \boldsymbol{A}_i^{y_{ik}})$ and similarly for $(P_{jl}, \boldsymbol{A}_{jl})$.)

Theorem 3. *Fix i, j, k $(1 \leq i, j, k \leq r)$ and $\omega \in {}^0\mathcal{E}(M_i)$. Then if ω is not ramified in G,*

$$\boldsymbol{E}(P_i : \phi) = \boldsymbol{E}\big(P_j : \boldsymbol{c}_{P_j|P_i}(s)\,\phi\big)$$

and

$$\boldsymbol{c}_{P_k|P_j}(t)\, \boldsymbol{c}_{P_j|P_i}(s)\,\phi = \boldsymbol{c}_{P_k|P_i}(ts)\,\phi$$

for $\phi \in \mathfrak{D}(P_i, \omega)$, $s \in \mathfrak{w}(\boldsymbol{A}_i, \boldsymbol{A}_j)$ and $t \in \mathfrak{w}(\boldsymbol{A}_j, \boldsymbol{A}_k)$. Moreover $\boldsymbol{c}_{P_j|P_i}(s)$ then defines a unitary transformation of $\mathfrak{D}(P_i, \omega)$ onto $\mathfrak{D}(P_j, \omega^s)$.

The situation when ω is ramified seems to be complicated. However, not all ramified ω behave badly. For example suppose prk $P_i = 1$ and ω is ramified. Then $\Omega(P_i, \omega) = \Omega_1 + \Omega_2$ where Ω_1, Ω_2 are two distinct elements of $\mathcal{E}(G)$. Let us further assume that the degrees of Ω_1 and Ω_2 are the same. Then one can show that the statements of Theorem 3 continue to hold in this case.

8. Proof of Theorem 3

Let me briefly sketch the proof of Theorem 3.

Lemma 6. *Let $\phi \in {}^0\mathfrak{D}(P_i)$. Then*

$$\big(\boldsymbol{E}(P_i : \phi)\big)_{P_{jl}} = \sum_{s \in \mathfrak{w}(\boldsymbol{A}_i, \boldsymbol{A}_j)} \big(\boldsymbol{c}_{P_j|P_i}(s)\,\phi\big)_l$$

for $1 \leq l \leq q_j$.

This follows from Theorem 2.

We may regard $C(G/\Gamma)$ as a subspace of $C(G)$. For any $\Omega \in \mathcal{E}(G)$, put $C(G/\Gamma, \Omega) = C(G/\Gamma) \cap C(G, \Omega)$. Now let us use the notation of §6 and suppose

$$\Omega(P, \omega) = \sum_{1 \leq i \leq N} n_i\, \Omega_i$$

where n_i are positive integers and Ω_i $(1 \leq i \leq N)$ are distinct elements of $\mathcal{E}(G)$. Put

$$V(P, \omega) = \sum_{1 \leq i \leq N} C(G/\Gamma, \Omega_i).$$

Lemma 7. *$\phi \to \boldsymbol{E}(P:\phi)$ $(\phi \in \mathfrak{D}(P, \omega))$ is a surjective mapping of $\mathfrak{D}(P, \omega)$ onto $V(P, \omega)$.*

Let $D(P, \omega)$ denote the space of all functions $f \in C(G)$ such that:

1) $f(xu) = f(x)$ $(x \in G, u \in U)$.

2) For every $x \in G$, the function $m \to f(xm)$ $(m \in M)$ lies in $C(M, \omega)$.

Fix $y \in G$ and for any $f \in D(P, \omega)$, define the function f_y by $f_y(x) = f(xy)$ $(x \in G)$. Then the mapping $f \to f_y$ is a bijection of $D(P, \omega)$ onto $D(P^y, \omega^y)$. Since

$$G = \bigcup_{1 \leq i \leq q} \Gamma y_i P,$$

it follows easily that the space

$$W = \sum_{1 \leq i \leq q} \sum_{\gamma \in \Gamma} D(P^{\gamma y_i}, \omega^{\gamma y_i})$$

is stable under both left and right translations of G. Therefore it is clear that

$$W = \sum_{1 \leq i \leq N} C(G, \Omega_i).$$

For any $f \in W$, define

$$F_f(x) = \sum_{\gamma \in \Gamma} f(x\gamma) \qquad (x \in G).$$

Then $f \to F_f$ is a surjective mapping of W onto

$$\sum_{1 \leq i \leq N} C(G/\Gamma, \Omega_i) = V(P, \omega).$$

Moreover $F_{f_\gamma} = F_f$ for $\gamma \in \Gamma$. Therefore every element of $V(P, \omega)$ can be written in the form F_f where

$$f \in \sum_{1 \leq i \leq q} D(P^{y_i}, \omega^{y_i}).$$

The assertion of Lemma 7 is now obvious.

Lemma 8. *Let ω, ω' be two distinct elements of ${}^0\mathcal{E}(M)$. Then the spaces $\mathfrak{D}(P, \omega)$ and $\mathfrak{D}(P, \omega')$ are mutually orthogonal in ${}^0\mathfrak{D}(P)$.*

We now return to the notation of §7. Fix $\omega \in {}^0\mathcal{E}(M_i)$ and $s \in \mathfrak{w}(\boldsymbol{A}_i, \boldsymbol{A}_j)$. Then $\Omega(P_i, \omega) = \Omega(P_j, \omega^s)$ from the corollary of Lemma 1. Hence

$$V(P_i, \omega) = V(P_j, \omega^s) = V \text{ (say)}.$$

Let $(\phi, \psi)_i$ $(\phi, \psi \in {}^0\mathfrak{D}(P_i))$ denote the scalar product in ${}^0\mathfrak{D}(P_i)$.

Lemma 9. *Fix $s \in \mathfrak{w}(\boldsymbol{A}_i, \boldsymbol{A}_j)$. Then*

$$(\psi, \boldsymbol{c}_{P_j|P_i}(s)\phi)_j = (\boldsymbol{c}_{P_i|P_j}(s^{-1})\psi, \phi)_i$$

for $\phi \in {}^0\mathfrak{D}(P_i)$ *and* $\psi \in {}^0\mathfrak{D}(P_j)$. *Moreover*

$$\boldsymbol{c}_{P_j|P_i}(s)\,\phi \in \mathfrak{D}(P_j, \omega^s)$$

for $\phi \in \mathfrak{D}(P_i, \omega)$. *Finally*

$$\boldsymbol{c}_{P_i|P_i}(1)\,\phi = \phi \qquad (\phi \in {}^0\mathfrak{D}(P_i)).$$

Here 1 *denotes the unit element of the group* $\mathfrak{w}(\boldsymbol{A}_i)$.

This follows from the results of § 5.

Now we come to the proof of Theorem 3. Fix $\phi \in \mathfrak{D}(P_i, \omega)$ and let

$$f = \boldsymbol{E}(P_i : \phi) - \boldsymbol{E}\left(P_j : \boldsymbol{c}_{P_j|P_i}(s)\,\phi\right) \in V.$$

We have to show that $f = 0$. Fix $\psi \in \mathfrak{D}(P_j, \omega^s)$. In view of Lemma 7, it is enough to verify that

$$(\boldsymbol{E}(P_j : \psi), f)_{G/\Gamma} = 0.$$

But

$$\begin{aligned}
(\boldsymbol{E}(P_j : \psi), f)_{G/\Gamma} &= \sum_{1 \leq l \leq q_j} (\psi_l, f_{P_{jl}})_{G/\Gamma \cap P_{jl}} \\
&= \sum_{t' \in \mathfrak{w}(\boldsymbol{A}_i, \boldsymbol{A}_j)} (\psi, \boldsymbol{c}_{P_j|P_i}(t')\,\phi)_j - \sum_{t \in \mathfrak{w}(\boldsymbol{A}_j)} (\psi, \boldsymbol{c}_{P_j|P_j}(t)\,\boldsymbol{c}_{P_j|P_i}(s)\,\phi)_j \\
&= \sum_{t \in \mathfrak{w}(\boldsymbol{A}_j)} (\psi, \boldsymbol{c}_{P_j|P_i}(ts)\,\phi - \boldsymbol{c}_{P_j|P_j}(t)\,\boldsymbol{c}_{P_j|P_i}(s)\,\phi)_j
\end{aligned}$$

from Lemmas 5 and 6. Now by Lemma 9, $\boldsymbol{c}_{P_j|P_i}(ts)\,\phi$ and $\boldsymbol{c}_{P_j|P_j}(t)\,\boldsymbol{c}_{P_j|P_i}(s)\,\phi$ are both in $\mathfrak{D}(P_j, \omega^{ts})$. Since ω is unramified, $\omega^{ts} \neq \omega^s$ unless $t = 1$. Therefore we conclude from Lemmas 8 and 9 that

$$(\boldsymbol{E}(P_j : \psi), f)_{G/\Gamma} = 0.$$

This proves that $f = 0$.

Now fix s and t as in Theorem 3. Then it follows from the above result that

$$\boldsymbol{E}(P_i : \phi) = \boldsymbol{E}\left(P_j : \boldsymbol{c}_{P_j|P_i}(s)\,\phi\right) = \boldsymbol{E}\left(P_k : \boldsymbol{c}_{P_k|P_j}(t)\,\boldsymbol{c}_{P_j|P_i}(s)\,\phi\right).$$

Hence

$$f = \boldsymbol{E}(P_i : \phi) - \boldsymbol{E}\left(P_k : \boldsymbol{c}_{P_k|P_j}(t)\,\boldsymbol{c}_{P_j|P_i}(s)\,\phi\right) = 0.$$

Now fix $\psi \in \mathfrak{D}(P_k : \omega^{ts})$. Then

$$\begin{aligned}
0 = (\boldsymbol{E}(P_k : \psi), f)_{G/\Gamma} &= \sum_{1 \leq l \leq q_k} (\psi_l, f_{P_{kl}})_{G/\Gamma \cap P_{kl}} \\
&= \sum_{u \in \mathfrak{w}(\boldsymbol{A}_i, \boldsymbol{A}_k)} (\psi, \boldsymbol{c}_{P_k|P_i}(u)\,\phi)_k - \sum_{v \in \mathfrak{w}(\boldsymbol{A}_k)} (\psi, \boldsymbol{c}_{P_k|P_k}(v)\,\boldsymbol{c}_{P_k|P_j}(t)\,\boldsymbol{c}_{P_j|P_i}(s)\,\phi)_k \\
&= \sum_{v \in \mathfrak{w}(\boldsymbol{A}_k)} (\psi, \boldsymbol{c}_{P_k|P_i}(vts)\,\phi - \boldsymbol{c}_{P_k|P_k}(v)\,\boldsymbol{c}_{P_k|P_j}(t)\,\boldsymbol{c}_{P_j|P_i}(s)\,\phi)_k
\end{aligned}$$

from Lemmas 5 and 6. Since ω is unramified, we again conclude from Lemmas 8 and 9 that the right side is equal to

$$(\psi, \boldsymbol{c}_{P_k|P_i}(ts)\phi - \boldsymbol{c}_{P_k|P_j}(t)\,\boldsymbol{c}_{P_j|P_i}(s)\phi)_k.$$

This implies that

$$\boldsymbol{c}_{P_k|P_i}(ts)\phi - \boldsymbol{c}_{P_k|P_j}(t)\,\boldsymbol{c}_{P_j|P_i}(s)\phi = 0.$$

On the other hand from Lemma 9,

$$(\boldsymbol{c}_{P_j|P_i}(s))^* = \boldsymbol{c}_{P_i|P_j}(s^{-1}) \qquad (s \in \mathfrak{w}(\boldsymbol{A}_i, \boldsymbol{A}_j)),$$

where the star denotes the adjoint. Therefore if ω is unramified it follows from the above proof that

$$(\boldsymbol{c}_{P_j|P_i}(s))^*\boldsymbol{c}_{P_j|P_i}(s)\phi = \boldsymbol{c}_{P_i|P_j}(s^{-1})\,\boldsymbol{c}_{P_j|P_i}(s)\phi = \boldsymbol{c}_{P_i|P_i}(1)\phi = \phi$$

for $\phi \in \mathfrak{D}(P_i, \omega)$. Since the situation is symmetrical in i and j, it is clear that $\boldsymbol{c}_{P_j|P_i}(s)$ defines a unitary transformation of $\mathfrak{D}(P_i, \omega)$ onto $\mathfrak{D}(P_j, \omega^s)$. This completes the proof of Theorem 3.

9. A counterexample

Let $\boldsymbol{G} = SL(2)$ and $\boldsymbol{P}$ the group of all upper triangular matrices in $\boldsymbol{G}$. Then $\boldsymbol{P}$ is a K-parabolic subgroup of $\boldsymbol{G}$. Let $\boldsymbol{M}$ be the group of all diagonal matrices in $\boldsymbol{G}$ and $\boldsymbol{U}$ the group of all unipotent matrices in $\boldsymbol{P}$. Then $\boldsymbol{P} = \boldsymbol{M}\boldsymbol{U}$ is a Levi K-decomposition of $\boldsymbol{P}$ and $\boldsymbol{A} = \boldsymbol{M}$ the corresponding split component of $\boldsymbol{P}$.

Now $G = SL(2, K)$, the pair $(P, \boldsymbol{A})$ is mincuspidal and prk $P = 1$. Put $\Gamma = U = \boldsymbol{U} \cap G$ and

$$y = \begin{pmatrix} 0 & 1 \\ -1 & 0 \end{pmatrix}.$$

Then $\{1, y\}$ is a complete set of representatives of $\Gamma \backslash G/P$. Put $P_1 = P$, $P_2 = P^y$. Then $P_2 = \bar{P}$ where $\bar{P}$ is the subgroup of all lower triangular matrices in G. The group $\mathfrak{w} = \mathfrak{w}(\boldsymbol{A})$ consists of just two elements $\{1, s\}$ and y is a representative of s in G.

We observe that

$${}^0\mathfrak{D}(P) = {}^0\mathscr{D}(P_1) \times {}^0\mathscr{D}(P_2)$$

and $\boldsymbol{c}_{P|P}(1) = 1$. Put $C = \boldsymbol{c}_{P|P}(s)$. Then if $\phi \in {}^0\mathfrak{D}(P)$, we have

$$(C\phi)_1 = c_{P|P}(s)\phi_1 + c_{P|\bar{P}}(1)\phi_2$$
$$(C\phi)_2 = c_{\bar{P}|P}(1)\phi_1 + c_{\bar{P}|\bar{P}}(s)\phi_2.$$

Moreover $c_{P|P}(s)=0$ since $\Gamma\cap(PyP)=\emptyset$. Hence

$$C=\begin{pmatrix}0 & , c_{P|\bar{P}}(1)\\ c_{\bar{P}|P}(1), & c_{\bar{P}|\bar{P}}(s)\end{pmatrix}.$$

On the other hand $\Gamma\cap(P\,\bar{P})=\Gamma\cap(\bar{P}\,P)=U=\Gamma$. Hence

$$\left(c_{P|\bar{P}}(1)\,\phi_2\right)(x)=\int_U \phi_2(xu)\,du$$

and

$$(c_{\bar{P}|P}(1)\,\phi_1)(x)=[\bar{U}]^{-1}\int_{\bar{U}}\phi_2(x\bar{u})\,d\bar{u} \qquad (x\in G),$$

where du and $d\bar{u}$ are the standard measures on U and $\bar{U}$ respectively. ($\bar{U}=U^y$ is the unipotent radical of $\bar{P}$.) Finally

$$\boldsymbol{E}(P:\phi:x)=\phi_1(x)+\int_U \phi_2(xu)\,du$$

and

$$\boldsymbol{E}(P:C\phi:x)=\int_U \phi_2(xu)\,du+[\bar{U}]^{-1}\int_{U\times\bar{U}}\phi_1(xu\bar{u})\,du\,d\bar{u}$$

for $\phi\in{}^0\mathcal{D}(P)$ and $x\in G$. Hence

$$\boldsymbol{E}(P:\phi)=\boldsymbol{E}(P:C\phi)$$

if and only if

$$\phi_1(x)=[\bar{U}]^{-1}\int_{U\times\bar{U}}\phi_1(xu\bar{u})\,du\,d\bar{u} \qquad (x\in G).$$

Similarly the second assertion of Theorem 3 may be written in the form

$$C^2\phi=\phi \qquad (\phi\in\mathcal{D}(P,\omega)).$$

But a simple calculation shows that the above condition implies that $c_{\bar{P}|\bar{P}}(s)\,\phi_2=0$ for $\phi_2\in\mathcal{D}(\bar{P},\omega^s)$.

Since P is mincuspidal, $\mathcal{E}(M)={}^0\mathcal{E}(M)$. Moreover $M=A$ where A is the group of all diagonal matrices in G. Hence $\mathcal{E}(M)$ may be identified with the set A^* of all characters of the finite abelian group A. Fix $\chi\in A^*$ and suppose χ is ramified. Then there are two cases.

1) $\chi=1$.

2) $\chi\neq 1$, $\chi^2=1$.

If $\chi=1$, we can take $\phi_2=1$. Then $\phi_2\in\mathcal{D}(\bar{P},\chi^s)$ and

$$\left(c_{\bar{P}|\bar{P}}(s)\,\phi_2\right)(x)=[\bar{U}]^{-1}\int_{\bar{U}}\sum_{\gamma\in U\cap(\bar{P}y\bar{P})}\phi_2(x\bar{u}\gamma)\,d\bar{u}=[U\cap(\bar{P}y\bar{P})] \qquad (x\in G).$$

By Bruhat's lemma G is the disjoint union of $\bar{P}$ and $\bar{P}y\bar{P}$. Since $U\cap\bar{P}=\{1\}$, we conclude that

$$\begin{pmatrix}1 & 1\\ 0 & 1\end{pmatrix}\in U\cap(\bar{P}y\bar{P})$$

and therefore

$$c_{\bar{P}|\bar{P}}(s)\,\phi_2 \neq 0.$$

This means that Theorem 3 does not hold in this case. On the other hand it can be shown that the statements of Theorem 3 remain valid for case 2) above.

10. Concluding remarks

It is clear from the above discussion that the ramified case requires further study. However, in my opinion, a more serious problem is to find an effective method of obtaining the elements of ${}^0\mathscr{E}(G)$. This is entirely similar to the problem of determining the discrete series in the real or the $\mathfrak{p}$-adic case.

Let $\boldsymbol{B}$ be a Cartan subgroup of $\boldsymbol{G}$ defined over K and let $X_K(\boldsymbol{B})$ denote the group of all K-morphisms of $\boldsymbol{B}$ in $GL(1)$. We say that $\boldsymbol{B}$ is (K-)anisotropic if $X_K(\boldsymbol{B}) = \{1\}$. Put $B = \boldsymbol{B} \cap G$ and let B^* be the group of all (complex) characters of the finite abelian group B. There seem to be some indications (see TANAKA [6, p. 83]) that in case $\operatorname{prk} G = 0$, ${}^0\mathscr{E}(G)$ is „parameterized" by B^* for the various anisotropic Cartan K-subgroups $\boldsymbol{B}$ of G. This is in fact so for real semisimple Lie groups and the same is believed to be true for the $\mathfrak{p}$-adic groups. Thus the construction of the „discrete series" appears to be the central problem in all cases.

References

1. BOREL, A., and J. TITS: Groupes rèductifs. Pub. Math. I.H.E.S., No. 27 (1965).
2. GELFAND, I. M.: Automorphic functions and the theory of representations. Proc. Intern. Congr. Math., 1962, p. 74—85.
3. HARISH-CHANDRA: Automorphic forms on semisimple Lie groups. Notes by J. G. M. MARS, Lecture notes in mathematics, vol. 62. Berlin-Heidelberg-New York: Springer 1968.
4. LANGLANDS, R. P.: (a) On the functional equations satisfied by Eisenstein series. Mimeographed manuscript, 1965 (unpublished). (b) Eisenstein series. Algebraic groups and discontinuous groups (1966). Am. Math. Soc., p. 235—252.
5. SELBERG, A.: Discontinuous groups and harmonic analysis. Proc. Intern. Congr. Math., 1962, p. 177—189.
6. TANAKA, S.: Construction and classification of irreducible representations of special linear group of the second order over a finite field, Osaka J. Math. 4, 65—84 (1967).

$\mathfrak{L}_p$ Transforms on Compact Groups

By Edwin Hewitt

University of Washington, Seattle

It is an honor and a privilege to address this conference, honoring Marshall H. Stone on the occasion of his retirement from the University of Chicago. I offer my sincere thanks to the program committee for the opportunity of doing so.

Let me first follow Professor Mackey's good example by indulging in a few minutes of retrospection. Harmonic analysis on locally compact Abelian groups and on compact groups has evolved at a dazzling pace during the last two decades. The roots of this development are clearly seen in the work of Weyl, Peter, and Bochner in the 1920's, especially in the use of integral equations (convolutions actually) to establish the main theorem on almost periodic functions on the line and the completeness of the irreducible unitary representations of a compact Lie group. The work of Wiener in the early 1930's, culminating in his famous Tauberian theorem, set the stage for the study of the $\mathfrak{L}_1$ algebra of a locally compact Abelian group, a study which is far advanced but far from complete at the present day.

In 1933, Alfred Haar published the famous paper in which Haar measure on locally compact groups was constructed. It is a sad fact that Haar died in the same year, so that he never saw the flowering of the discipline which his discovery made possible. A second indispensable tool for abstract harmonic analysis is the Pontryagin-van Kampen duality theorem, published in 1934 and 1935.

A milestone in the development of the theory was the publication in 1940 of André Weil's fundamental book [27], in which many basic facts about Fourier series and Fourier integrals were extended to arbitrary compact and locally compact Abelian groups. Among these facts are Plancherel's theorem, Bochner's theorem, the Weyl-Peter theorem, and the Hausdorff-Young inequality. It is true that Weil's style is exceedingly condensed, and that a conscientious reader has many details to fill in for himself. But the book is there, it is a genuine milestone, and its importance in the subsequent evolution of abstract harmonic analysis is enormous.

Also in the early 1940's other writers discovered parts of abstract harmonic analysis, independent of WEIL's work. Thus M. G. KREĬN [14] proved PLANCHEREL's theorem for locally compact Abelian groups, by a method quite unlike WEIL's. KREĬN's proof, by the way, is the one that appears in most textbooks today. RAĬKOV [22] and POVZNER [21] also gave proofs in 1940 of BOCHNER's theorem on arbitrary locally compact Abelian groups.

Another important line of thought was initiated in the 1930's by several analysts, who took up the algebraic side of a number of analytic constructs. Algebraic notions were already present implicitly in WIENER's classic book *The Fourier integral and certain of its applications* [28] and also in the paper [29] of WIENER and PITT. A decisively algebraic point of view was introduced by M. H. STONE in his long 1937 memoir in the Transactions of the American Mathematical Society [24]. In this paper, he studied, among many other things, the algebra $\mathfrak{C}(X, R)$ of all real-valued continuous functions on a compact Hausdorff space X. He obtained a complete description of the closed ideals in this algebra, classified linear isometries between two of them, and in passing proved the Stone-Weierstrass theorem, which today is part of the equipment of every analyst. ARNE BEURLING, in a paper presented to the IX. Congress of Scandinavian Mathematicians at Helsingfors (1938) [1], studied the algebra $\boldsymbol{M}(R)$ of all complex Borel measures on the line, under convolution, and stated for the first time the spectral radius formula

$$\lim_{n\to\infty} \|\mu^n\|^{\frac{1}{n}} = \sup_{t\in R} \{|\hat{\mu}(t)|\},$$

which he proved for all measures μ having singular component zero. These papers were the forerunners of GEL'FAND'S fundamental discovery of Banach algebras [*normed rings* in his terminology] in 1941 [5]. The Banach algebra point of view toward algebras of functions and measures on groups has of course completely transformed the theory. It forms the *mise en scène* of standard texts (see for example [16], [19], [9], and [23]), and it dictates the questions one asks. The recent textbook on Fourier series by R. E. EDWARDS [3] sheds much light on classical matters by adopting the Banach algebra point of view.

Later work in the 1940's and 1950's by SEGAL, GODEMENT, CARTAN and GODEMENT, MACKEY, LOOMIS, AMBROSE, KREĬN, RAĬKOV, GEL'FAND, NAĬMARK, and many other writers have set the stage for the present efflorescence of abstract harmonic analysis. A useful survey of the state of the art in 1952, together with some original viewpoints and constructions, appears in M. H. STONE's 1952 memoir [25]. We now have a generation of younger workers, who are carrying the field forward at an ever accelerating tempo.

What is the goal of abstract harmonic analysis? One may say that it is to rewrite ZYGMUND [30] for every locally compact Abelian group and every compact non-Abelian group. This is not strictly true, of course: but a major aim is to provide the sort of detailed knowledge about each locally compact Abelian or compact group that we have for the circle T and the line R. Unquestionably some of this can be done. The p-adic numbers, for example, are just as good a group as R, and there is no reason why Hilbert transforms, conjugate functions, CARLESON's theorem, SALEM's singular measures with small Fourier-Stieltjes transforms and Cantor set supports, et cetera, should not be studied on this group. The same is true of other neo-classical groups, such as the character group of the discrete additive rationals. The classical compact non-Abelian groups are also wide open for detailed analysis. Extremely refined studies of $\mathfrak{S}\mathfrak{U}(2)$ are being carried on by R. A. MAYER [17] and DAVID RAGOZIN. PAUL SALLY, MITCHELL TAIBLESON, KEITH PHILLIPS, to name three others of a large group, are concerned with detailed analysis on one or another group. The higher-dimensional unitary groups $\mathfrak{S}\mathfrak{U}(n)$ for $n \geqq 3$ are of great interest to physicists. Here difficulties arise in obtaining explicitly the irreducible unitary representations, although WEYL's work gives an algorithm for computing them all. No reasonable formula is known for decomposing the tensor product of two irreducible representations into its irreducible components. But the future looks bright, and for the older generation of harmonic analysts, the main problem is to grasp the new work as quickly as the younger people write it.

After this perhaps overlengthy preamble, let me turn to the main topic of my address. This is joint work with KENNETH A. ROSS of the University of Oregon, and will appear in a book [10], which we hope will be in print in 1970. A preliminary announcement of some of our results is given in [11]. We are concerned with certain algebras and function spaces on general compact groups. The level of abstraction is about like that of the general theory of $\mathfrak{L}_1(G)$ for a locally compact Abelian group G, although some surprisingly explicit details appear.

Throughout the remainder of this talk, the symbol G will denote an arbitrary compact group, and Σ will denote the dual object of G. Let us explain exactly what Σ is. We consider a homomorphism $x \rightarrow U_x$ of G into the group of unitary operators on a Hilbert space H with the property that the function $x \rightarrow \langle U_x \xi, \eta \rangle$ is continuous on G for every pair of vectors ξ, η in H. If the representation U is irreducible, that is, if the set of operators $\{U_x : x \in G\}$ leaves no proper subspace of H invariant, then H must be finite-dimensional: a very simple proof of this theorem is due to L. NACHBIN [18]. For a given irreducible U, consider the class σ of all representations of G unitarily equivalent with U. Such a class σ

is an element of the dual object Σ, and, obviously, Σ consists of all the classes σ. Let d_σ be the dimension of the Hilbert space H_σ on which a certain $U^{(\sigma)} \in \sigma$ acts. Plainly d_σ depends only upon σ, and as noted above, d_σ is a positive integer.

For $1 \leqq p < \infty$, the symbol $\mathfrak{L}_p(G)$ means the space of Borel measurable complex-valued functions f on G such that $|f|^p$ has a finite integral with respect to Haar measure λ on G, which we normalize by the convention that $\lambda(G) = 1$. For a function $f \in \mathfrak{L}_1(G)$, we define the Fourier transform f as a certain function on the dual object Σ. We choose a fixed $U^{(\sigma)} \in \sigma$ for each $\sigma \in \Sigma$. The entity $\hat{f}(\sigma)$ is defined as the operator on the representation space H_σ of $U^{(\sigma)}$ such that

$$\langle \hat{f}(\sigma)\, \xi, \eta \rangle = \int_G \overline{\langle U_x^{(\sigma)}\, \xi, \eta \rangle}\, f(x)\, dx$$

for all $\xi, \eta \in H_\sigma$. Obviously $\hat{f}(\sigma)$ depends upon the particular choice of $U^{(\sigma)} \in \sigma$, but upon changing from $U^{(\sigma)}$ to a unitarily equivalent representation, $\hat{f}(\sigma)$ changes into a unitarily equivalent representation, as a trivial computation shows. In what follows, we will choose a fixed $U^{(\sigma)} \in \sigma$ for each $\sigma \in \Sigma$. The representation space of $U^{(\sigma)}$ will be denoted by H_σ.

We are concerned with certain norms for the operators $\hat{f}(\sigma)$, which were discovered by J. v. NEUMANN, and described in a 1937 paper in the Tomsk University Izvestiya [20]. Given a linear operator A on a d-dimensional Hilbert space H (d is a positive integer), consider the adjoint operator $A^\sim$, and the self-adjoint positive-definite operator $A A^\sim$. This operator has nonnegative eigenvalues, let us say $\alpha_1, \alpha_2, \ldots, \alpha_d$. For $1 \leqq p < \infty$, we define $\|A\|_{\varphi_p}$ by

$$\|A\|_{\varphi_p} = \left[\alpha_1^{\frac{p}{2}} + \alpha_2^{\frac{p}{2}} + \cdots + \alpha_d^{\frac{p}{2}}\right]^{\frac{1}{p}}.$$

We define $\|A\|_{\varphi_\infty}$ by

$$\|A\|_{\varphi_\infty} = \max\{\alpha_1^{\frac{1}{2}}, \alpha_2^{\frac{1}{2}}, \ldots, \alpha_d^{\frac{1}{2}}\}.$$

It is a remarkable fact that each $\|\ \|_{\varphi_p}$ is a norm on $\mathfrak{B}(H)$, the linear space of all linear operators on H. (These norms are defined and discussed in a more general situation in [2], pp. 1088 *et. seq.*) The φ_p norms have a number of remarkable properties. For example:

$$\|U A V\|_{\varphi_p} = \|A\|_{\varphi_p} \text{ if } U \text{ and } V \text{ are unitary;}$$

$$\|A B\|_{\varphi_p} \leqq \|A\|_{\varphi_p} \|B\|_{\varphi_p};$$

$$\|A B\|_{\varphi_1} \leqq \|A\|_{\varphi_p} \|B\|_{\varphi_{p'}}, \quad \text{where} \quad p' = \frac{p}{p-1};$$

and so on. The v. NEUMANN norm $\|A\|_{\varphi_\infty}$ is just the operator norm of A.

v. NEUMANN's norms received no attention that we know of until the late 1940's, when M. G. KREĬN used the φ_1 and φ_∞ norms in his study of continuous positive-definite functions on compact groups. KREĬN's paper [13] also has lingered in undeserved obscurity, perhaps because of the obscure journal where it appeared. My own attention to it was drawn by the detailed and thoughtful review written by G. W. MACKEY (MR 12, p. 719).

In 1958, R. A. KUNZE [15] published a construction of Fourier transforms on unimodular locally compact groups, which contains KREĬN's work on compact groups as a special case. There is no evidence that KUNZE knew of KREĬN's work. Let us sketch the special case of KUNZE's construction that we will use. Consider the Cartesian product $\mathop{\mathsf{P}}_{\sigma\in\Sigma}\mathfrak{B}(H_\sigma)$, which we denote by the symbol $\mathfrak{E}$. That is, $\mathfrak{E}$ is the space of *all* $\mathfrak{B}(H_\sigma)$-valued operator functions on Σ. If G is Abelian so that every $d_\sigma=1$, $\mathfrak{E}$ is the space of *all* complex-valued functions on the character group Σ of G. The space $\mathfrak{E}$ is far too large to be useful for analytical purposes. We single out certain subspaces, as follows:

$$\mathfrak{E}_\infty=\{(E_\sigma)\in\mathfrak{E}:\|E\|_\infty=\sup_{\sigma\in\Sigma}\|E_\sigma\|_{\varphi_\infty}<\infty\};$$

$\mathfrak{E}_0=\{(E_\sigma)\in\mathfrak{E}$: for every $\varepsilon>0$, there are only finitely many σ such that $\|E_\sigma\|_{\varphi_\infty}>\varepsilon\}$, and we give $\mathfrak{E}_0$ the $\|\ \|_\infty$ norm.

The Weyl-Peter formula

$$\|f\|_2^2=\sum_{\sigma\in\Sigma}d_\sigma\|\hat f(\sigma)\|_{\varphi_2}^2\qquad (f\in\mathfrak{L}_2(G))$$

compels us to proceed differently in defining the spaces $\mathfrak{E}_q$ for $1\leqq q<\infty$:

$$\mathfrak{E}_q=\Big\{(E_\sigma)\in\mathfrak{E}:\sum_{\sigma\in\Sigma}d_\sigma\|E_\sigma\|_{\varphi_q}^q<\infty\Big\}.$$

For $E\in\mathfrak{E}_q$, we define $\|E\|_q$ as $\Big[\sum_{\sigma\in\Sigma}d_\sigma\|E_\sigma\|_{\varphi_q}^q\Big]^{\frac{1}{q}}$. All of these normed spaces are Banach spaces. Again, the Abelian case produces the familiar function spaces $l_q(\Sigma)$.

With this notation, two familiar facts appear in a succinct form:

$\mathfrak{L}_1(G)^\frown\subset\mathfrak{E}_0$, and $\|\hat f\|_\infty\leqq\|f\|_1$ (this is the Riemann-Lebesgue lemma);

$\mathfrak{L}_2(G)^\frown=\mathfrak{E}_2$, and $\|\hat f\|_2=\|f\|_2$ (this is the Weyl-Peter formula).

As KUNZE showed, we can apply the M. Riesz convexity theorem to the two foregoing facts and obtain:

for $1<p<2$, we have $\mathfrak{L}_p(G)^\frown\subset\mathfrak{E}_{p'}$, and $\|\hat f\|_{p'}\leqq\|f\|_p$. (HY)

Here p' is the usual conjugate exponent $\frac{p}{p-1}$. The inequality (HY) is of course the Hausdorff-Young inequality for Fourier series, generalized to all compact groups. I remark here J. E. LITTLEWOOD's observation that his principal contribution to Fourier analysis is the profound fact that in (HY), the big number stands on the left. From here on, p will denote an arbitrary number such that $1<p<2$.

ROSS and I (referred to as "we" hereafter) asked ourselves for what functions in $\mathfrak{L}_p(G)$ does equality occur in (HY), and — a companion question — for what $f \in \mathfrak{L}_{p'}(G)$ is $\hat{f}$ in $\mathfrak{E}_p$ and $\|f\|_{p'} = \|\hat{f}\|_p$? [Functions in $\mathfrak{L}_p(G)$ with equality in (HY) are called *$\mathfrak{L}_p$-maximal*, and functions in $\mathfrak{L}_{p'}(G)$ with $\|\hat{f}\|_p = \|f\|_{p'}$ are called *$\mathfrak{L}_{p'}$-maximal*.] The above questions have a long history. In 1926, HARDY and LITTLEWOOD [6] answered the first for the circle group T, *i.e.*, for ordinary Fourier series. The only $\mathfrak{L}_p$-maximal functions on T are those which are trivially $\mathfrak{L}_p$-maximal, namely, functions $t \to \alpha \exp[int]$, where α is a complex constant and n is an integer.

The question was taken up many years later (1954) by I. I. HIRSCHMAN, JR., and me [7]. We considered an arbitrary locally compact *Abelian* group H [not necessarily compact]. Let A be a compact open subgroup of H, let χ be a continuous character of A, and let χ_A be the function on G such that

$$\chi_A(x) = \begin{cases} \chi(x) & \text{for } x \in A, \\ 0 & \text{for } x \notin A. \end{cases}$$

Such a function is called a *subcharacter of G*. It is simple to show that every subcharacter is $\mathfrak{L}_p$-maximal, and obvious that a constant times a translate of an $\mathfrak{L}_p$-maximal function is also $\mathfrak{L}_p$-maximal. The Hewitt-Hirschman theorem states that these are *all* of the $\mathfrak{L}_p$-maximal functions. On R, for example, the inequality $\|\hat{f}\|_{p'} < \|f\|_p$ holds for all nonzero f in $\mathfrak{L}_p(R)$, since R lacks compact open subgroups.

In 1959, HIRSCHMAN [12] took up compact non-Abelian groups G, introducing the following norm in part of $\mathfrak{E}$:

$$\|E\|_q = \Big[\sum_{\sigma \in \Sigma} d_\sigma^{2-\frac{1}{2}q} \|E_\sigma\|_{\varphi_2}^q\Big]^{\frac{1}{q}},$$

for $1 \leqq q < \infty$. He proved that for a function f in $\mathfrak{L}_p(G)$, the equality $\|f\|_p = \|\hat{f}\|_{p'}$ holds if and only if f is a constant times a translate of a subcharacter χ_{G_0}, where G_0 is an open *normal* subgroup of G. He also proved that these are exactly the functions that are $\mathfrak{L}_{p'}$-maximal for his norm $\|\hat{f}\|_p$.

Looking at KUNZE's norms $\|E\|_q$ on $\mathfrak{E}_q$ and noting the inequality (HY), one is quickly convinced that KUNZE's norm, rather than HIRSCHMAN's, is the appropriate one to use on $\hat{f}$ for $f \in \mathfrak{L}_p(G)$. ROSS and I were able to show that $\mathfrak{L}_p$-maximal and $\mathfrak{L}_{p'}$-maximal functions under KUNZE's norm can also be characterized. They are exactly the functions that are constants times translates of subcharacters χ_{G_0}, where now G_0 is an arbitrary open subgroup of G. HIRSCHMAN's maximal functions are translates of central functions that are maximal in our sense.

It is easy and also instructive to check that χ_{G_0} is $\mathfrak{L}_p$- and $\mathfrak{L}_{p'}$-maximal. For typographical convenience, we write φ for χ_{G_0}. One sees at once that $\varphi(xy^{-1})\,\varphi(y) = \varphi(x)$ for all $x, y \in G$, and so

$$\varphi * \varphi(x) = \varphi(x) \int_{G_0} 1\, d\lambda(y) = \varphi(x)\, \lambda(G_0) = \frac{1}{n}\, \varphi(x),$$

where n is the cardinal number of G/G_0. Thus $(n\varphi) * (n\varphi)$ is $n\varphi$. Also we have $(n\varphi)^{\sim}(x) = \overline{n\varphi(x^{-1})}$ by the definition of $\tilde{f}$ for all functions f on G, and clearly $\overline{\varphi(x^{-1})} = \varphi(x)$. Thus $(n\varphi)^{\sim} = n\varphi$. Under Fourier transforms, we have

$$(f * g)^{\wedge}(\sigma) = \hat{f}(\sigma)\, \hat{g}(\sigma) \qquad \text{(operator multiplication)}$$

and

$$(\tilde{f})^{\wedge}(\sigma) = \hat{f}(\sigma)^{\sim} \qquad \text{(the last } {}^{\sim} \text{ means adjoint)}.$$

Therefore $(n\varphi)^{\wedge}(\sigma)$ is idempotent and Hermitian, *i.e.*, a projection. Thus there is an integer $l_\sigma \in \{0, 1, \ldots, d_\sigma\}$ such that $\hat{\varphi}(\sigma)^{\sim}\, \hat{\varphi}(\sigma)$ has eigenvalue $\frac{1}{n}$ with multiplicity l_σ and eigenvalue 0 with multiplicity $d_\sigma - l_\sigma$. It follows that

$$\|\hat{\varphi}\|_{p'}^{p'} = \sum_{\sigma \in \Sigma} d_\sigma \|\hat{\varphi}(\sigma)\|_{\varphi_{p'}}^{p'} = \sum_{\sigma \in \Sigma} d_\sigma\, l_\sigma\, n^{-p'} = n^{-p'} \sum_{\sigma \in \Sigma} d_\sigma\, l_\sigma.$$

For $p = p' = 2$, we have $\|\varphi\|_2^2 = \|\hat{\varphi}\|_2^2$, and so

$$\frac{1}{n} = n^{-2} \sum_{\sigma \in \Sigma} d_\sigma\, l_\sigma,$$

so that

$$\|\hat{\varphi}\|_{p'}^{p'} = n^{1-p'}$$

and

$$\|\hat{\varphi}\|_{p'} = n^{\frac{1}{p'} - 1} = n^{-\frac{1}{p}} = \|\varphi\|_p.$$

The proof that φ is $\mathfrak{L}_{p'}$-maximal is just the same.

The converse is much harder to prove. The proof is long, and uses many delicate facts about v. NEUMANN's norms. In the end, it comes down to a careful application of the maximum modulus principle. It is so messy in fact that only its authors, one feels, could love it. Rather than go into its details, I would prefer to spend my remaining time discussing an algebra of functions on G that KREĬN [13] first discussed.

Recall that $\mathfrak{E}_1$ is the space of all $(E_\sigma) \in \mathfrak{E}$ for which $\sum_{\sigma \in \Sigma} d_\sigma \| E_\sigma \|_{\varphi_1} = \| E \|_1$ is finite. KREĬN proved the following two theorems [13].

I. A function f on G is continuous and positive-definite [we write $f \in \mathfrak{P}(G)$] if and only if $\hat{f}(\sigma)$ is a positive-definite operator for all $\sigma \in \Sigma$ and $\hat{f} \in \mathfrak{E}_1$.

II. A function f on G is a linear combination of functions in $\mathfrak{P}(G)$ [we write $f \in \mathfrak{K}(G)$] if and only if $\hat{f} \in \mathfrak{E}_1$. For $f \in \mathfrak{K}(G)$, we have

$$f(x) = \sum_{\sigma \in \Sigma} d_\sigma \, tr(A_\sigma U_x^{(\sigma)}),$$

where $(A_\sigma) \in \mathfrak{E}_1$, and the series converges absolutely and uniformly on G.*

KREIN's first theorem is an exact analogue for compact non-Abelian groups of BOCHNER's theorem, at least in one direction. (There is no really satisfactory description of the Fourier-Stieltjes transforms $\hat{\mu} \in \mathfrak{E}$ for nonnegative measures on G — this would be BOCHNER's theorem in the other direction.)

Since the pointwise product of positive-definite functions is positive-definite, $\mathfrak{K}(G)$ is an algebra of functions on G. Under the norm $\|f\|_{\varphi_1} = \sum_{\sigma \in \Sigma} d_\sigma \|\hat{f}(\sigma)\|_{\varphi_1}$, $\mathfrak{K}(G)$ turns out to be a Banach algebra, which can very reasonably be called *the algebra of absolutely convergent Fourier series on G*. It is elementary although somewhat tedious to verify that $\|fg\|_{\varphi_1} \leqq \|f\|_{\varphi_1} \|g\|_{\varphi_1}$. If G is Abelian, $\mathfrak{K}(G)$ is of course the algebra of functions f on G of the form $\sum_{n=1}^{\infty} \alpha_n \chi_n$, where $\|f\| = \sum_{n=1}^{\infty} |\alpha_n| < \infty$ and each χ_n is a continuous character of G.

KREĬN [13] observed that every multiplicative linear functional on $\mathfrak{K}(G)$ has the form $f \to f(a)$ for some $a \in G$. (This fact has been generalized by EYMARD [4], who apparently was unaware of KREĬN's work.) KREĬN also showed that the algebra $\mathfrak{K}(G)$ is regular, so that the Gel'fand topology of G regarded as the space of multiplicative linear functionals of $\mathfrak{K}(G)$ is the original topology of G.

ROSS and I have investigated $\mathfrak{K}(G)$ as a Banach algebra and have found some interesting facts about it. Here, without any proofs, are some of these facts.

* We can express A_σ in terms of $\hat{f}(\sigma)$: $A_\sigma = D_\sigma \hat{f}(\sigma)^\sim D_\sigma$, where D_σ is a conjugate-linear mapping of H_σ onto itself such that $\langle D_\sigma \zeta, D_\sigma \eta \rangle = \langle \eta, \zeta \rangle$ for all $\zeta, \eta \in H_\sigma$, and $D_\sigma^2 = I$.

The algebra $\mathfrak{K}(G)$ is exactly the set $\mathfrak{L}_2(G) * \mathfrak{L}_2(G)$: and we know of no other way to show that $\mathfrak{L}_2(G) * \mathfrak{L}_2(G)$ is a linear space. (In [4], EYMARD has proved a generalization of this fact.) There seems no simple way to characterize $\mathfrak{K}(G) * \mathfrak{K}(G)$.

Next, $\mathfrak{K}(G)$ satisfies DITKIN's condition. If $f \in \mathfrak{K}(G)$ and $f(a) = 0$, one can find a sequence $\{h_n\}_{n=1}^{\infty}$ such that $h_n \in \mathfrak{K}(G)$, $\|h_n\|_{\varphi_1} \leqq 2$, each h_n vanishes in some neighborhood of a, and $\lim_{n\to\infty} \|h_n f - f\|_{\varphi_1} = 0$. Thus ŠILOV's theorem on closed ideals holds for $\mathfrak{K}(G)$ (see for example [16], p. 86).

Finally, spectral synthesis fails for $\mathfrak{K}(G)$ if G contains an infinite Abelian subgroup. That is, there is a closed subset F of G such that there are two distinct closed ideals $\mathfrak{J}_1$ and $\mathfrak{J}_2$ of $\mathfrak{K}(G)$ for which $F = \bigcap_{f \in \mathfrak{J}_1} \{x \in G : f(x) = 0\} = \bigcap_{f \in \mathfrak{J}_2} \{x \in G : f(x) = 0\}$. This result is not very satisfactory, for two reasons. First, we conjecture, but have been unable to prove, that every infinite compact group contains an infinite Abelian subgroup. Second, our proof that spectral synthesis fails for $\mathfrak{K}(G)$ in this case merely appeals to MALLIAVIN's theorem. A proper proof would construct "independent" sets in G like those constructed by KAKUTANI and myself [8] and would then follow VAROPOULOS's proof [26] of MALLIAVIN's theorem. But we have no idea as yet how this can be done.

References

1. BEURLING, A.: Sur les intégrales de Fourier absolument convergentes et leur applications à une transformation fonctionnelle. Nionde Skandinaviska Matematikerkongressen, Helsingfors, 1938, p. 345—366. Helsingfors: Mercator 1939.
2. DUNFORD, N., and J. T. SCHWARTZ: Linear operators. Part II. New York: Interscience Publishers 1963.
3. EDWARDS, R. E.: Fourier series, a modern interpretation, vols. I and II. New York: Holt, Rinehart, and Winston, Inc. 1967.
4. EYMARD, P.: L'algèbre de Fourier d'un groupe localement compact. Bull. Soc. Math. France **92**, 181—236 (1964).
5. GEL'FAND, I. M.: Normierte Ringe. Mat. Sb. (N.S.) **9 (51)**, 3—24 (1941).
6. HARDY, G. H., and J. E. LITTLEWOOD: Some new properties of Fourier constants. Math. Ann. **97**, 158—209 (1926).
7. HEWITT, E., and I. I. HIRSCHMAN, JR.: A maximal problem in harmonic analysis. Am. J. Math. **76**, 839—852 (1954).
8. —, and SHIZUO KAKUTANI: A class of multiplicative linear functionals on the measure algebra of a locally compact Abelian group. Illinois J. Math. **4**, 553—574 (1960).
9. —, and K. A. ROSS: Abstract harmonic analysis, vol. I. Göttingen-Berlin-Heidelberg: Springer 1963.
10. — — Abstract harmonic analysis, vol. II. Berlin-Heidelberg-New York: Springer 1970.
11. — — A maximal problem in harmonic analysis. III. Bull. Am. Math. Soc. **74**, 225—227 (1968).

12. Hirschman, I. I., Jr.: A maximal problem in harmonic analysis. II. Pacific J. Math. **9**, 525—540 (1959).
13. Kreĭn, M. G.: Ermitovo-položitel'nye yadra na odnorodnyh prostranstvah. I and II. Ukrain. Mat. Ž. **1**, 64—98 (1949); **2**, 10—59 (1950). [Am. Math. Soc. Translations (2) **34**, 69—108, 109—164 (1963).]
14. — Sur une généralisation du théorème de Plancherel au cas des intégrales de Fourier sur les groupes topologiques commutatifs. Dokl. Akad. Nauk SSSR (N. S.) **30**, 484—488 (1941).
15. Kunze, R. A.: L_p Fourier transforms on locally compact unimodular groups. Trans. Am. Math. Soc. **89**, 519—540 (1958).
16. Loomis, L. H.: An introduction to abstract harmonic analysis. New York: D. van Nostrand Co. 1953.
17. Mayer, R. A.: Summation of Fourier series on compact groups. Am. J. Math. **89**, 661—692 (1967).
18. Nachbin, L.: On the finite dimensionality of every irreducible representation of a compact group. Proc. Am. Math. Soc. **12**, 11—12 (1961).
19. Naĭmark, M. A.: Normirovannye kol'ca. Moscow: Gostehizdat 1956. Revised German translation: Normierte Algebren. Berlin: VEB Deutscher Verlag der Wiss. 1959. English translation: Normed rings. Groningen: P. Noordhoff N. V. 1959.
20. Neumann, J. v.: Some matrix-inequalities and metrization of matricspace. Tomsk Univ. Inst. Mat. i Meh. Izvest. **1**, 286—300 (1937). Also in Collected works, vol. IV, p. 205—219. New York: The MacMillan Co. 1962.
21. Povzner, A.: Über positive Funktionen auf einer abelschen Gruppe. Dokl. Akad. Nauk SSSR (N. S.) **28**, 294—295 (1940).
22. Raĭkov, D. A.: Positive definite functions on commutative groups with an invariant measure. Dokl. Akad. Nauk SSSR (N. S.) **28**, 296—300 (1940).
23. Rudin, W.: Fourier analysis on groups. New York: Interscience Publ. Co. 1962.
24. Stone, M. H.: Applications of the theory of Boolean rings to general topology. Trans. Am. Math. Soc. **41**, 375—481 (1937).
25. — On the foundations of harmonic analysis. Kgl. Fysiogr. Sällsk. Lund Förh. **21**, 21 pp. (1952), also in Medd. Lunds. Univ. Mat. Sem., Tome suppl., 207—227 (1952).
26. Varopoulos, N. Th.: Tensor algebras and harmonic analysis. Acta Math. **119**, 51—112 (1967).
27. Weil, A.: L'intégration dans les groupes topologiques et ses applications. Actualités Sci. et Ind., No. 869, 1145. Paris: Hermann & Cie. 1941 and 1951.
28. Wiener, N.: The Fourier integral and certain of its applications. Cambridge, England: Cambridge University Press 1933. Reprinted, New York: Dover Publications, Inc.
29. —, and H. R. Pitt: On absolutely convergent Fourier-Stieltjes transforms. Duke Math. J. **4**, 420—438 (1938).
30. Zygmund, A.: Trigonometric series, 2nd edit. 2 vols. Cambridge, England: Cambridge University Press 1959.

Theory of Simple Scattering and Eigenfunction Expansions

By Tosio Kato and S. T. Kuroda

University of California, Berkeley,
and University of Tokyo, and Yale University

Introduction

In the theory of *simple scattering systems* we consider the *wave operators* $W^{\pm} = \underset{t\to\pm\infty}{\text{s-lim}}\, e^{itH_2} e^{-itH_1}$ and the *scattering operator* $S = (W^+)^* W^-$, where H_1, H_2 are selfadjoint operators in a Hilbert space $\mathfrak{H}$ describing the unperturbed and perturbed systems (see Jauch [12]). $W^{\pm}$ are isometric and intertwine H_1 and H_2 ($H_2 W^{\pm} \supset W^{\pm} H_1$) whenever they exist. S is unitary if and only if the ranges of $W^{\pm}$ are identical. For technical reasons, it was found convenient to define the (generalized) wave operators by $W^{\pm} = \underset{t\to\pm\infty}{\text{s-lim}}\, e^{itH_2} e^{-itH_1} P_{1,ac}$, where $P_{1,ac}$ is the projection of $\mathfrak{H}$ onto the *subspace* $\mathfrak{H}_{1,ac}$ *of absolute continuity* with respect to H_1 (see Kato [17], Kuroda [18]). Then $W^{\pm}$ are, if they exist, partial isometries with initial sets $\mathfrak{H}_{1,ac}$ and final sets (ranges) in $\mathfrak{H}_{2,ac}$. S is unitary on $\mathfrak{H}_{1,ac}$ if and only if the ranges of $W^{\pm}$ coincide. In particular we say that W^+ (W^-) is *complete* if its range is $\mathfrak{H}_{2,ac}$.

Construction of the wave operators directly based on the above definition is customarily referred to as the *time-dependent* (or *non-stationary*) theory. Without going into details of this theory, we simply note that most of sufficient conditions so far obtained for the existence and completeness of the wave operators assume some *trace condition*, which states that $H_2 - H_1$ (or some related quantity) belongs to the trace class of compact operators. We note also that the *invariance principle* has been proved under rather mild assumptions, to the effect that $W^{\pm} = W^{\pm}(H_2, H_1)$ is equal to $W^{\pm}(\gamma(H_2), \gamma(H_1))$ for any piecewise monotone increasing function γ satisfying some continuity condition (see Kato [15, 17]). For details of the time-dependent theory see the references mentioned above.

Other methods for constructing $W^{\pm}$ are known as the *stationary theory* and date back to Friedrichs [4, 5]. Here $W^{\pm}$ are constructed by solving certain singular integral equations, which are given in terms of a spectral representation for the unperturbed operator H_1. $W^{\pm}$ are then shown to be equal to the time limits used in the definition of $W^{\pm}$ in the time-dependent theory. Further developments in this direction (theory of *gentle perturbations*) were given by FRIEDRICHS [6], REJTO [24, 25], FADDEEV [3], and others.

A stationary method of somewhat different kind was given by KATO [13], in which $H_2 - H_1$ is assumed to be of rank one. Here certain spectral representations for H_1, H_2 are constructed, through which $W^{\pm}$ are defined in such a way that their partial isometry and intertwining property are obvious. This method was later generalized to more general cases in KURODA [19, 20, 21, 22]. Its final form [22], in which use is made of a spectral representation by Hilbert spaces constructed in terms of certain operator-valued measures, is rather general and has wide range of application. It can not only be applied to cases with trace conditions but it also contains the main results of IKEBE [11] concerning the existence and completeness of $W^{\pm}$ for the Schrödinger operators $H_1 = -\Delta, H_2 = -\Delta + V(x)\,\cdot$ in $\mathfrak{H} = L^2(R^3)$, for example. Moreover, it shows that the invariance principle holds in this case, a result not proved in [11]. But it is still restricted in that the perturbation is assumed to be factorizable into the product of two operators acting in $\mathfrak{H}$.

Stationary methods of another type were proposed by REJTO [26, 27] and HOWLAND [9] independently. These are characterized by the use of an auxiliary Banach space $\mathfrak{X}$, part of which is densely embedded in $\mathfrak{H}$. Roughly speaking, $\mathfrak{X}$ is assumed to be such that the resolvents $R_j(\zeta) = (H_j - \zeta)^{-1}$ have weak boundary values on the two edges of the real axis in an operator topology associated with the topology of $\mathfrak{X}$ and that $Q_j(\zeta) = (H_2 - H_1) R_j(\zeta)$ map $\mathfrak{X}$ into $\mathfrak{X}$ continuously and have boundary values on the real axis in the strong operator topology. $W^{\pm}$ are then defined in a direct way not necessarily using spectral representations. Their methods were applied successfully to many interesting cases, but it appears that the cases with trace conditions or the invariance principle were not covered by their results.

There are still other stationary methods of "direct" type, due to BIRMAN-ENTINA [2], BIRMAN [1], KATO [16]. But these are somewhat special, being related to trace conditions or to "smooth" perturbations.

In the present work, the authors intend to construct a unified theory which hopefully comprises most of the stationary methods mentioned above. In one direction it simplifies the method of [22] by constructing the spectral representations in a more direct and simple way. At the same time it generalizes the latter by eliminating the assumption that the perturbation be factorizable. As it turns out, our basic assumptions resemble those of [26] and [9] inasmuch as an auxiliary subspace $\mathfrak{X}$ of $\mathfrak{H}$ is used. But it differs from [26] and [9] in that $\mathfrak{X}$ need not be a normed space nor a dense subset of $\mathfrak{H}$ and that we do not assume continuous dependence of the $Q_j(\zeta)$ on ζ. This makes it possible to apply the results not only to gentle and smooth perturbations but also to perturbations with the trace conditions. Also we are able to prove the invariance principle in a general setting, with the single but essential restriction that $\mathfrak{X}$ has the structure of a pre-Hilbert space. In the case of an arbitrary $\mathfrak{X}$, we have not been able to prove the time-limit formula for $W^{\pm}$, but we do have a slightly weakened result that the time-limit formula is true in the sense of the *Abel limit* (this was proved in [9] under somewhat stronger assumptions).

Actually the theory of the wave operators are here presented in the case of two *unitary* operators U_1, U_2 rather than of two selfadjoint operators H_1, H_2. In the time-dependent theory, this corresponds to considering the discrete groups $\{U_j^k\}$ rather than the continuous groups $\{e^{itH_j}\}$. We plan to discuss the selfadjoint case in detail in a later publication, together with applications. But it should be noted that some results on the perturbation of selfadjoint operators follow directly from the ones for the unitary operators through the Cayley transformation.

The composition of the present paper is as follows. We start with an abstract theorem on the perturbation of spectral systems. By a *spectral system* we mean a spectral measure E, with values in the set of all orthogonal projections in a Hilbert space $\mathfrak{H}$, on a measurable space $(\Gamma_0, \boldsymbol{B})$, together with a numerical measure on $\boldsymbol{B}$ with completion m. By a standard process, E is then decomposed into the absolutely continuous part E_{ac} and the singular part E_s. A *spectral form* for such a system is a complex-valued function $f(\theta; x, y)$ defined on $\Gamma \times \mathfrak{X} \times \mathfrak{X}$, where $\Gamma \in \boldsymbol{B}$ and $\mathfrak{X}$ is a linear manifold in $\mathfrak{H}$, such that $f(\theta; x, y)$ is a nonnegative Hermitian form in x, y and m-measurable in θ, and that $\int_\Delta f(\theta; x, y)\, m\,(d\theta) = (E_{ac}(\Delta)\, x, y)$ for $\Delta \in \boldsymbol{B}$. Under these assumptions we construct a spectral representation for E_{ac} on the subspace $E_{ac}(\Gamma)\,\mathfrak{L}$, where $\mathfrak{L}$ is the reducing subspace for E generated by $\mathfrak{X}$. It is essentially a unitary map of $E_{ac}(\Gamma)\,\mathfrak{L}$ onto a direct integral of Hilbert spaces $\widetilde{\mathfrak{X}}(\theta)$, each obtained by completing $\mathfrak{X}$ with respect to the norm induced by $f(\theta; \cdot, \cdot)$ (see Section 1).

If we have two such spectral systems 1, 2 and if there is an isometric map $\widetilde{G}(\theta)$ of $\widetilde{X}_1(\theta)$ into $\widetilde{X}_2(\theta)$ for each $\theta \in \Gamma$, then it is easy to construct an isometric map W of $E_{1,ac}(\Gamma)\,\mathfrak{L}_1$ into $E_{2,ac}(\Gamma)\,\mathfrak{L}_2$ that intertwines E_1 and E_2. We call W the *general wave operator*; it is the prototype of the wave operators $W^{\pm}$ between two unitary or selfadjoint operators (see Section 2).

So far we need no topologies in $\mathfrak{X}_1$ or $\mathfrak{X}_2$. In applications, however, the $\widetilde{G}(\theta)$ are usually constructed as limits of mappings from $\mathfrak{X}_1$ to $\mathfrak{X}_2$;

then we need a topology at least in $\mathfrak{X}_2$. But it is found that the topology plays a minor role and it suffices to assume that $\mathfrak{X}_2$ is a linear topological space*. We need a topology in $\mathfrak{X}_1$ too if we want W to be complete. The construction of the $\tilde{G}(\theta)$ is carried out in Section 3 along the lines indicated. In Section 5 we consider a more specific situation involving two unitary operators, constructing two wave operators $W^{\pm}$. In Section 6 it is shown that many new as well as known results for wave operators follow as applications of the general results of Section 5. Section 7 is devoted to the proof of the invariance principle, under the additional assumption that $\mathfrak{X}$ is a pre-Hilbert space.

Another problem considered in this paper is the theory of *eigenfunction expansions* as a perturbation problem (see Section 4). A preliminary report of this theory based on [22] was given by KURODA [23], but here we give a detailed proof in a more general setting. The eigenfunction expansion is formulated in an abstract way, analogous to that given by GELFAND-SHILOV [7] and GELFAND-VILENKIN [7a]. Suppose that the unitary operator U_1 has "eigenfunctions" $\varphi_1(\xi)$, with the parameter ξ varying on a measure space $(\Omega, \Sigma, \varrho)$, which form a "complete set" in $E_{1,ac}(\Gamma)\mathfrak{H}_1$. We assume that the parameter $\xi \in \Omega$ corresponds to a more refined spectral representation than the one constructed above, so that in particular $\theta \in \Gamma$ is a function $\omega(\xi)$ of ξ. Furthermore, we regard $\varphi_1(\xi)$ as belonging to $\mathfrak{X}_1^*$, the adjoint space of $\mathfrak{X}_1$. Under certain conditions similar to the ones used for the construction of the $\tilde{G}(\theta)$, we construct a map $G(\omega(\xi))^*$, depending on ξ, from $\mathfrak{X}_2^*$ to $\mathfrak{X}_1^*$ and show that $\varphi_2(\xi) = G(\omega(\xi))^{*-1}\varphi_1(\xi)$ gives the "eigenfunctions" for U_2. This relationship between $\varphi_1(\xi)$ and $\varphi_2(\xi)$ is an abstract analog of the Lippmann-Schwinger equation. It should be noted that HOWLAND [10] develops a similar theory of eigenfunction expansions on the basis of his stationary theory of wave operators.

The authors are indebted to Professors S. KAKUTANI and C. E. RICKART for stimulating discussions, which, among others, led to great improvement of the basic theorem. This work was partly supported by NSF GP 6838.

1. Spectral systems and spectral representations

Definition 1.1. A *spectral system* $(\Gamma_0, \boldsymbol{B}, m, \mathfrak{H}, E)$ is a spectral measure E on a measurable space $(\Gamma_0, \boldsymbol{B})$, with values in the set of all orthogonal projections in a complex Hilbert space $\mathfrak{H}$, together with a σ-finite, nonnegative scalar measure on $\boldsymbol{B}$, the completion of which is denoted by m.

* Throughout this paper, linear topological spaces are always assumed to be Hausdorff.

We denote the norm and inner product in $\mathfrak{H}$ by $\| \|$ and $(,)$. Sometimes we write $\| \|_{\mathfrak{H}}$ and $(,)_{\mathfrak{H}}$ to indicate that these symbols refer to $\mathfrak{H}$. We denote by $[\boldsymbol{B}]$ the set of all $\boldsymbol{B}$-measurable, bounded scalar functions on Γ_0. We define the m-measurability of a scalar function on Γ_0 in the usual way. (In applications Γ_0 is usually the real line or the unit circle on the plane, $\boldsymbol{B}$ is the set of all Borel subsets of Γ_0, and m is the Lebesgue measure.)

Let $\mathfrak{H}_{ac}$ be the set of all $u \in \mathfrak{H}$ such that the numerical measure $\|E(\cdot)u\|^2$ is m-absolutely continuous; $\mathfrak{H}_{ac}$ is a closed subspace of $\mathfrak{H}$ reducing all $E(\Delta)$, $\Delta \in \boldsymbol{B}$. Let P_{ac} be the projection of $\mathfrak{H}$ onto $\mathfrak{H}_{ac}$, so that P_{ac} commutes with all $E(\Delta)$. Set $E_{ac}(\Delta) = E(\Delta) P_{ac}$. E_{ac} is the *absolutely continuous part* of E; it is also a spectral measure on $\boldsymbol{B}$, except that $E_{ac}(\Gamma_0)$ is in general not the identity operator. E_{ac} is absolutely continuous in the sense that $m(\Delta) = 0$ implies $E_{ac}(\Delta) = 0$. We note that for each $u \in \mathfrak{H}$ there is $\Delta_u \in \boldsymbol{B}$ such that $m(\Gamma_0 - \Delta_u) = 0$ and $P_{ac} u = E(\Delta_u) u$ (cf. [17]).

Definition 1.2. Let $(\Gamma_0, \boldsymbol{B}, m, \mathfrak{H}, E)$ be a spectral system. A *spectral form* for this system is a complex-valued function f on $\Gamma \times \mathfrak{X} \times \mathfrak{X}$, where $\Gamma \subset \Gamma_0$, $\Gamma \in \boldsymbol{B}$, and $\mathfrak{X}$ is a linear manifold in $\mathfrak{H}$, with the following properties. (a) For each $x, y \in \mathfrak{X}$, $\theta \mapsto f(\theta; x, y)$ is m-integrable on Γ and its integral on each $\Delta \subset \Gamma$, $\Delta \in \boldsymbol{B}$, is equal to $(E_{ac}(\Delta) x, y)$. (b) For each $\theta \in \Gamma$, $x, y \mapsto f(\theta; x, y)$ is a nonnegative Hermitian form on $\mathfrak{X} \times \mathfrak{X}$. (We write $f(\theta; x)$ for $f(\theta; x, x)$.)

Remark 1.3. Since $(E_{ac}(\cdot) u, v)$ is a complex-valued, m-absolutely continuous measure for each $u, v \in \mathfrak{H}$, it is an indefinite integral of a complex-valued, m-integrable function $f(\cdot; u, v)$ on Γ_0. Since such f is not unique for a given pair u, v, however, it would be in general difficult to choose $f(\theta; u, v)$ as a Hermitian form in $u, v \in \mathfrak{H}$ for each $\theta \in \Gamma_0$. But it can be done easily if u, v, and θ are suitably restricted. This is what Definition 1.2 is concerned with.

Example 1.4. If $\mathfrak{X}$ is finite-dimensional, there always exists a spectral form f on $\Gamma_0 \times \mathfrak{X} \times \mathfrak{X}$. To see this, choose a basis $\{e_k\}$ of $\mathfrak{X}$ and a density function $f_{jk}(\cdot)$ for the measure $(E_{ac}(\cdot) e_j, e_k)$ for each pair e_j, e_k (Remark 1.3). Then $f(\theta; x, y) = \sum \alpha_j \bar{\beta}_k f_{jk}(\theta)$ for $x = \sum \alpha_k e_k$, $y = \sum \beta_k e_k$ defines a spectral form on $\Gamma_0 \times \mathfrak{X} \times \mathfrak{X}$. This example is not altogether trivial. If E has finite multiplicity, a finite-dimensional $\mathfrak{X}$ can generate the whole space with respect to E. As is seen below, this is sufficient for the purpose of the spectral representation of E_{ac}.

Example 1.5. Let $\mathfrak{H} = L^2(-\infty, \infty)$, $\Gamma_0 = R^1$ with the Lebesgue measure m, and let $E(\Delta)$ be the operator of multiplication by the characteristic function of Δ, defined for all Lebesgue-measurable set $\Delta \subset R^1$. If $\mathfrak{X}$ is the set of all *continuous* functions in $L^2(-\infty, \infty)$, then $f(\theta; x, y) = x(\theta)\overline{y(\theta)}$ defines a spectral form on $R^1 \times \mathfrak{X} \times \mathfrak{X}$. Note that $E_{ac} = E$ here. In this example $\mathfrak{X}$ is dense in $\mathfrak{H}$, but this is not necessary for the spectral representation introduced below.

Suppose f is a spectral form on $\Gamma \times \mathfrak{X} \times \mathfrak{X}$. For each $\theta \in \Gamma$, $f(\theta; \cdot, \cdot)$ defines a semi-inner product in $\mathfrak{X}$. Let $\mathfrak{N}(\theta) \subset \mathfrak{X}$ be the set of all x with $f(\theta; x) = 0$. Then the quotient space $\mathfrak{X}/\mathfrak{N}(\theta)$ is a pre-Hilbert space with the inner product induced by $f(\theta; \cdot, \cdot)$. We denote by $\tilde{\mathfrak{X}}(\theta)$ its completion, by $(\cdot, \cdot)_\theta$ and $\|\cdot\|_\theta$ the inner product and norm in $\tilde{\mathfrak{X}}(\theta)$, and by $J(\theta)$ the canonical map of $\mathfrak{X}$ onto $\mathfrak{X}/\mathfrak{N}(\theta) \subset \tilde{\mathfrak{X}}(\theta)$. Thus $(J(\theta)x, J(\theta)y)_\theta = f(\theta; x, y)$ for $x, y \in \mathfrak{X}$, and $J(\theta)x = 0$ if and only if $f(\theta; x) = 0$.

Consider the product vector space $\tilde{\tilde{\mathfrak{X}}} = \prod_{\theta \in \Gamma} \tilde{\mathfrak{X}}(\theta)$ consisting of all vector fields $g = \{g(\theta)\}_{\theta \in \Gamma}$ with $g(\theta) \in \tilde{\mathfrak{X}}(\theta)$. We say two elements $g_1, g_2 \in \tilde{\tilde{\mathfrak{X}}}$ are equivalent, in symbol $g_1 \sim g_2$, if $g_1(\theta) = g_2(\theta)$ for m-a.e. $\theta \in \Gamma$. Clearly it is an equivalence relation compatible with linear operations in $\tilde{\tilde{\mathfrak{X}}}$ and with the operation of multiplication $g \mapsto \alpha g = \{\alpha(\theta) g(\theta)\}$ with a scalar function α on Γ.

Definition 1.6. $g \in \tilde{\tilde{\mathfrak{X}}}$ is said to be *f-measurable* if there is a sequence $\{h_n\}$ of quasi-simple functions on Γ to $\mathfrak{X}$ such that

$$\lim_{n\to\infty} \|g(\theta) - J(\theta) h_n(\theta)\|_\theta = 0 \quad \text{for } m\text{-a.e.} \quad \theta \in \Gamma. \tag{1.1}$$

Here we mean by a quasi-simple function h a function of the form (finite sum)

$$h(\theta) = \sum \alpha_k(\theta) x_k, \qquad \alpha_k \in [\boldsymbol{B}], \qquad x_k \in \mathfrak{X}. \tag{1.2}$$

(We use quasi-simple functions rather than simple functions for convenience.)

The following are easy consequences of the definition. If $g_1 \sim g_2$, g_1 is f-measurable if and only if g_2 is. If g is f-measurable and α is an m-measurable scalar function, then αg is also f-measurable.

Lemma 1.7. *If $\{g_n\}$ is a sequence of f-measurable elements of $\tilde{\tilde{\mathfrak{X}}}$ such that $\lim_{n\to\infty} g_n(\theta) = g(\theta)$ in $\tilde{\mathfrak{X}}(\theta)$ for m-a.e. $\theta \in \Gamma$, then $g = \{g(\theta)\}$ is f-measurable.*

Proof. This follows immediately from Corollary B of Appendix.

Proposition 1.8. If $g_1, g_2 \in \tilde{\tilde{\mathfrak{X}}}$ are f-measurable, then $\theta \mapsto (g_1(\theta), g_2(\theta))_\theta$ is m-measurable on Γ.

Proof. Let h_{jn} be quasi-simple functions on Γ to $\mathfrak{X}$ such that $\lim_{n\to\infty} \|g_j(\theta) - J(\theta) h_{jn}(\theta)\|_\theta = 0$ m-a.e., $j = 1, 2$. Then $f(\theta; h_{1n}(\theta), h_{2n}(\theta)) = (J(\theta) h_{1n}(\theta), J(\theta) h_{2n}(\theta))_\theta \to (g_1(\theta), g_2(\theta))_\theta$ m-a.e. Since $f(\cdot\,; h_{1n}(\cdot), h_{2n}(\cdot))$ is obviously m-measurable, the desired result follows.

We denote by $\mathfrak{M}$ the set of all f-measurable elements $g \in \tilde{\mathfrak{X}}$ such that $\|g\|_{\mathfrak{M}}^2 = \int_\Gamma \|g(\theta)\|_\theta^2\, m(d\theta) < \infty$, where equivalent elements are to be identified. Thus $\mathfrak{M}$ is a pre-Hilbert space with the inner product

$$(g_1, g_2)_{\mathfrak{M}} = \int_\Gamma (g_1(\theta), g_2(\theta))_\theta\, m(d\theta). \tag{1.3}$$

Proposition 1.9. $\mathfrak{M}$ is a Hilbert space.

Proof. It suffices to show that $\mathfrak{M}$ is complete. The proof is essentially the same as the proof of the completeness of L^2-spaces and may be omitted.

Proposition 1.10. Quasi-simple functions on Γ to $\mathfrak{X}$ are densely embedded in $\mathfrak{M}$ in the following sense. (a) For any quasi-simple function h, we have $Jh \equiv \{J(\theta) h(\theta)\} \in \mathfrak{M}$. (b) For each $g \in \mathfrak{M}$ and $\varepsilon > 0$, there is a quasi-simple function h such that $\|g - Jh\|_{\mathfrak{M}} < \varepsilon$.

Proof. Let h be given by (1.2). Then $\|J(\theta) h(\theta)\|_\theta^2 = f(\theta; h(\theta)) = \sum \alpha_j(\theta) \overline{\alpha_k(\theta)} f(\theta; x_j, x_k)$. But since $f(\cdot\,; x, y)$ is equal to a density function for $(E_{ac}(\cdot) x, y)$ by definition, it follows that

$$\begin{aligned} &\int_\Gamma \|J(\theta) h(\theta)\|_\theta^2\, m(d\theta) \\ &\qquad = \sum \int_\Gamma \alpha_j(\theta) \overline{\alpha_k(\theta)}\, (E_{ac}(d\theta) x_j, x_k) = \|v\|_{\mathfrak{H}}^2 < \infty, \end{aligned} \tag{1.4}$$

where

$$v = \sum \alpha_k(E) E_{ac}(\Gamma) x_k. \tag{1.5}$$

Here we use the notation $\alpha(E) = \int_{\Gamma_0} \alpha(\theta) E(d\theta)$ for any $\alpha \in [\boldsymbol{B}]$. (1.4) shows that $Jh \in \mathfrak{M}$ with $\|Jh\|_{\mathfrak{M}} = \|v\|_{\mathfrak{H}}$.

To prove (b), let $g \in \mathfrak{M}$. Since $\theta \mapsto \|g(\theta)\|_\theta^2$ is m-integrable on Γ, there is $\Gamma' \subset \Gamma$, $\Gamma' \in \boldsymbol{B}$, with $m(\Gamma') < \infty$, such that $\int_{\Gamma - \Gamma'} \|g(\theta)\|_\theta^2 m(d\theta) < \varepsilon$. Also there is $\delta > 0$ such that $m(\Delta) < \delta$ implies $\int_\Delta \|g(\theta)\|_\theta^2\, m(d\theta) < \varepsilon$.

Since g is f-measurable, there are quasi-simple functions h_n, $n = 1, 2, \ldots,$ on Γ to $\mathfrak{X}$ such that $\|g(\theta) - J(\theta) h_n(\theta)\|_\theta \to 0$, $n \to \infty$, for m-a.e. $\theta \in \Gamma$. Since $m(\Gamma') < \infty$, there is $\Gamma'' \subset \Gamma'$, $\Gamma'' \in \boldsymbol{B}$, such that $m(\Gamma' - \Gamma'') < \delta$ and $\|g(\theta) - J(\theta) h_n(\theta)\|_\theta \to 0$, $n \to \infty$, uniformly on Γ'' (EGOROFF's theorem). Hence $\int_{\Gamma''} \|g(\theta) - J(\theta) h_n(\theta)\|_\theta^2\, m(d\theta) < \varepsilon$ for sufficiently large n. Fix one such n and set $h = \chi_{\Gamma''} h_n$, where $\chi_{\Gamma''}$ is the characteristic function of Γ''. Then

$$\int_{\Gamma} \|g(\theta) - J(\theta)h(\theta)\|_{\theta}^{2} m(d\theta) = \left[\int_{\Gamma-\Gamma'} + \int_{\Gamma'-\Gamma''}\right] \|g(\theta)\|_{\theta}^{2} m(d\theta) + \int_{\Gamma''} \|g(\theta) - J(\theta)h_n(\theta)\|_{\theta}^{2} m(d\theta) < 3\varepsilon.$$

But h is quasi-simple, so that $Jh \in \mathfrak{M}$ by (a) and the last inequality can be written $\|g - Jh\|_{\mathfrak{M}}^{2} < 3\varepsilon$. This proves (b).

We can now prove the main result of this section on the spectral representation for E. We denote by $\mathfrak{L}$ the smallest closed subspace of $\mathfrak{H}$ containing $\mathfrak{X}$ and reducing E (i.e. reducing the $\alpha(E)$ for all $\alpha \in [\boldsymbol{B}]$). $\mathfrak{L}$ is the closed span of the set of all vectors of the form $\alpha(E)x$ with $\alpha \in [\boldsymbol{B}]$ and $x \in \mathfrak{X}$. Hence $\mathfrak{L}_{ac}(\Gamma) \equiv E_{ac}(\Gamma)\mathfrak{L}$ is the closure of the set of all vectors of the form (1.5). $\mathfrak{L}_{ac}(\Gamma)$ also reduces E. $\mathfrak{L}_{ac}(\Gamma) = E_{ac}(\Gamma)\mathfrak{H}$ if $\mathfrak{X}$ generates $\mathfrak{H}$ (i.e. if $\mathfrak{L} = \mathfrak{H}$). For this it is sufficient, but not necessary, that $\mathfrak{X}$ be dense in $\mathfrak{H}$.

Theorem 1.11. *There is a unitary map Π on $\mathfrak{L}_{ac}(\Gamma)$ to $\mathfrak{M}$ such that (a) $\Pi\alpha(E)u = \alpha\Pi u = \{\alpha(\theta)\,(\Pi u)\,(\theta)\}$ for each $\alpha \in [\boldsymbol{B}]$ and $u \in \mathfrak{L}_{ac}(\Gamma)$, and (b) $\Pi E_{ac}(\Gamma)x = \{J(\theta)x\}$ for each $x \in \mathfrak{X}$.*

Proof. Consider all $v \in \mathfrak{L}_{ac}(\Gamma)$ of the form (1.5) and all h of the form (1.2). (1.4) shows that the map $v \mapsto Jh \in \mathfrak{M}$ is well-defined and is isometric. Since the v are dense in $\mathfrak{L}_{ac}(\Gamma)$ by the remark given above and since the Jh are dense in $\mathfrak{M}$ by Proposition 1.10, the isometry can be extended to a unitary operator Π on $\mathfrak{L}_{ac}(\Gamma)$ to $\mathfrak{M}$. It is clear that (a) is true for $u = v$ of the above form, and it is extended to all $u \in \mathfrak{L}_{ac}(\Gamma)$ by continuity. (b) is a special case of $v \mapsto Jh$ when $h(\theta) = x$.

2. The general wave operator

In this section we consider two spectral systems $(\Gamma_0, \boldsymbol{B}, m, \mathfrak{H}_j, E_j)$ with spectral forms $(\Gamma, \mathfrak{X}_j, f_j)$, $j = 1, 2$, where the measure space $(\Gamma_0, \boldsymbol{B}, m)$ and $\Gamma \subset \boldsymbol{B}$ are common to the two systems (see Definitions 1.1 and 1.2). We use the obvious notations such as $\mathfrak{H}_{j,ac}$, $E_{j,ac}$, $\widetilde{\mathfrak{X}}_j(\theta)$, $J_j(\theta)$, $\widetilde{\mathfrak{X}}_j$, $(,)_{j\theta}$, $\| \|_{j\theta}$, $\mathfrak{M}_j$, Π_j, $\mathfrak{L}_j$, $\mathfrak{L}_{j,ac}(\Gamma)$, etc.

As a link between the two systems, we introduce the following assumption.

(**2A**) For each $\theta \in \Gamma$ there is an isometric operator $\widetilde{G}(\theta)$ on $\widetilde{\mathfrak{X}}_1(\theta)$ to $\widetilde{\mathfrak{X}}_2(\theta)$, such that $\theta \mapsto \widetilde{G}(\theta)J_1(\theta)x$ is f_2-measurable on Γ for each $x \in \mathfrak{X}_1$.

For each $g = \{g(\theta)\} \in \widetilde{\mathfrak{X}}_1$, we set $\widetilde{G}g = \{\widetilde{G}(\theta)g(\theta)\} \in \widetilde{\mathfrak{X}}_2$. It is clear that $g \sim g'$ implies $\widetilde{G}g \sim \widetilde{G}g'$. The following propositions are easily proved.

Proposition 2.1. If $g \in \tilde{\mathfrak{X}}_1$ is f_1-measurable, then $\tilde{G} g \in \tilde{\mathfrak{X}}_2$ is f_2-measurable.

Proposition 2.2. $\tilde{G}$ restricted on $\mathfrak{M}_1$ is isometric on $\mathfrak{M}_1$ to $\mathfrak{M}_2$.

We now define a linear operator W on $\mathfrak{H}_1$ to $\mathfrak{H}_2$ by

$$\begin{aligned} W u &= \Pi_2^{-1} \tilde{G} \Pi_1 u \quad \text{for} \quad u \in \mathfrak{L}_{1,\,ac}(\Gamma), \\ W u &= 0 \quad \text{for} \quad u \perp \mathfrak{L}_{1,\,ac}(\Gamma). \end{aligned} \tag{2.1}$$

Recall that Π_j is a unitary operator on $\mathfrak{L}_{j,\,ac}(\Gamma)$ to $\mathfrak{M}_j$.

Theorem 2.3. *W is a partial isometry with initial set $\mathfrak{L}_{1,\,ac}(\Gamma)$ and final set contained in $\mathfrak{L}_{2,\,ac}(\Gamma)$. W has the intertwining property $\alpha(E_2) W = W \alpha(E_1)$ for $\alpha \in [\boldsymbol{B}]$.*

Proof. The first assertion follows directly from Proposition 2.2. The intertwining property follows from the fact that the operator $\alpha(E_j)$ acting in $\mathfrak{L}_{j,\,ac}(\Gamma)$ is transformed into the operator of multiplication by α, acting in $\mathfrak{M}_j$, under the unitary transformation Π_j.

Theorem 2.4. *The final set of W is exactly $\mathfrak{L}_{2,\,ac}(\Gamma)$ if the following additional condition is satisfied:*

(**2B**) *For each $\theta \in \Gamma$ $\tilde{G}(\theta)$ is unitary on $\tilde{\mathfrak{X}}_1(\theta)$ to $\tilde{\mathfrak{X}}_2(\theta)$, and $\theta \mapsto \tilde{G}(\theta)^{-1} J_2(\theta) y$ is f_1-measurable for each $y \in \mathfrak{X}_2$.*

Proof. The $\tilde{G}(\theta)^{-1}$ satisfy (**2A**) with $j = 1, 2$ exchanged, so that $\tilde{G}^{-1}$ is isometric on $\mathfrak{M}_2$ to $\mathfrak{M}_1$. Hence $\tilde{G}$ is unitary on $\mathfrak{M}_1$ to $\mathfrak{M}_2$, and the range of W is $\mathfrak{L}_{2,\,ac}(\Gamma)$.

Theorems 2.3 and 2.4 are the basis for constructing the wave operators between two unitary or selfadjoint operators in this paper. We call W the *general wave operator* associated with the two systems. We say W is *complete* if the range of W is equal to $\mathfrak{L}_{2,\,ac}(\Gamma)$. W is "local" since Γ is in general a subset of Γ_0. We are naturally most interested in the case in which $\mathfrak{L}_{1,\,ac}(\Gamma)$ and $\mathfrak{L}_{2,\,ac}(\Gamma)$ are large, e.g. $\mathfrak{L}_1 = \mathfrak{H}_1$ and $\mathfrak{L}_2 = \mathfrak{H}_2$, so that $\mathfrak{L}_{1,\,ac}(\Gamma) = E_{1,ac}(\Gamma)\,\mathfrak{H}_1$ and $\mathfrak{L}_{2,\,ac}(\Gamma) = E_{2,\,ac}(\Gamma)\,\mathfrak{H}_2$, and in which Γ is as large as possible.

Although the theorems given above are quite general, they are rather formal and have no direct application. The real problem is how to construct the $\tilde{G}(\theta)$ satisfying (**2A**) and (**2B**). The following sections are devoted to this question. Before closing this section, we prove a preliminary lemma.

Lemma 2.5. *Condition* (**2A**) *is equivalent to*

(**2A′**) *For each $\theta \in \Gamma$ there is a linear operator $G'(\theta)$ on $\mathfrak{X}_1$ to $\tilde{\mathfrak{X}}_2(\theta)$, such that (a) $\|G'(\theta)\, x\|_{2\theta}^2 = f_1(\theta;\, x)$ for each $x \in \mathfrak{X}_1$, and (b) $G'(\cdot)\, x$ is f_2-measurable for each $x \in \mathfrak{X}_1$.*

Proof. If (**2A**) is satisfied, we obtain the $G'(\theta)$ satisfying (**2A'**) by setting $G'(\theta)\,x = \widetilde{G}(\theta)\,J_1(\theta)\,x$ for $x \in \mathfrak{X}_1$. Conversely, suppose (**2A'**) is satisfied. By (a), $G'(\theta)$ induces an isometric operator on $\mathfrak{X}_1/\mathfrak{N}_1(\theta)$ to $\widetilde{\mathfrak{X}}_2(\theta)$, which can be extended by continuity to an isometric operator $\widetilde{G}(\theta)$ on $\widetilde{\mathfrak{X}}_1(\theta)$ to $\widetilde{\mathfrak{X}}_2(\theta)$. Then (b) implies that $\widetilde{G}(\cdot)\,J_1(\cdot)\,x = G'(\cdot)\,x$ is f_2-measurable.

3. Construction of the $\widetilde{G}(\theta)$

We continue to consider two spectral systems as in the previous section.

We want to construct $\widetilde{G}(\theta)$ satisfying (**2A**) or, equivalently, $G'(\theta)$ satisfying (**2A'**). The problem would be rather simple if we constructed $G'(\theta)$ from an operator on $\mathfrak{X}_1$ to $\mathfrak{X}_2$, but this is too restrictive to be useful in applications. What we are going to do is, roughly speaking, to derive $G'(\theta)$ from a *converging sequence* of operators* on $\mathfrak{X}_1$ to $\mathfrak{X}_2$. This requires that we introduce a topology into $\mathfrak{X}_2$. With this in mind we consider the following conditions.

(**3A**$_j$) There is a sequence of approximating spectral forms f_{jn}, $n = 1, 2, \ldots$, on $\Gamma \times \mathfrak{X}_j \times \mathfrak{X}_j$, in the following sense. For each $\theta \in \Gamma$, (a) $f_{jn}(\theta; \cdot, \cdot)$ is a nonnegative Hermitian form on $\mathfrak{X}_j \times \mathfrak{X}_j$, and (b) $f_{jn}(\theta; x, y) \to f_j(\theta; x, y)$, $n \to \infty$, for each $x, y \in \mathfrak{X}_j$. (Again we write $f_{jn}(\theta; x) = f_{jn}(\theta; x, x)$.)

(**3B**$_j$) $\mathfrak{X}_j$ is a linear topological space with its own topology. For each $\theta \in \Gamma$, $\{f_{jn}(\theta; \cdot, \cdot)\}_n$ is equicontinuous on $\mathfrak{X}_j \times \mathfrak{X}_j$.

(**3C**$_1$) For each $\theta \in \Gamma$ there is a sequence $\{G_n(\theta)\}$ of linear operators on $\mathfrak{X}_1$ to $\mathfrak{X}_2$ such that $f_{2n}(\theta; G_n(\theta)\,x) = f_{1n}(\theta; x)$ for $x \in \mathfrak{X}_1$. Furthermore, the following conditions are satisfied with respect to the topology of $\mathfrak{X}_2$ given in (**3B**$_2$): (a) For each $\theta \in \Gamma$ and $x \in \mathfrak{X}_1$, $\{G_n(\theta)\,x\}_n$ is a Cauchy sequence in $\mathfrak{X}_2$. (b) For each $x \in \mathfrak{X}_1$ and n, $G_n(\cdot)\,x$ is strongly measurable on Γ (i.e. it is the limit m-a.e. of a sequence of quasi-simple functions).

We note that the *existence* of $\lim f_{1n}(\theta; x, y)$ is a consequence of (**3A**$_2$), (**3B**$_2$), and (**3C**$_1$); see Lemma D of Appendix.

* Another method for constructing the $G'(\theta)$ is to construct first mappings $\overline{G}(\theta)$ on $\mathfrak{X}_1$ to $\overline{\mathfrak{X}}_2$, the completion of $\mathfrak{X}_2$, assuming that $\mathfrak{X}_2$ is a linear topological space and the $f_2(\theta; \cdot, \cdot)$ are continuous on $\mathfrak{X}_2 \times \mathfrak{X}_2$ so that they can be extended continuously to $\overline{\mathfrak{X}}_2 \times \overline{\mathfrak{X}}_2$. Then $J_2(\theta)$ can be extended continuously to $\bar{J}_2(\theta)$ on $\overline{\mathfrak{X}}_2$ to $\widetilde{\mathfrak{X}}_2(\theta)$, and we set $G'(\theta) = \bar{J}_2(\theta)\overline{G}(\theta)$. Now $\overline{G}(\theta)$ may be constructed as the limit of a sequence $G_n(\theta)$ of operators on $\mathfrak{X}_1$ to $\mathfrak{X}_2$ under conditions given below. This method is not essentially different from the one given in the text. But sometimes one may find it convenient to use operators $G_n(\theta)$ on $\mathfrak{X}_1$ to $\overline{\mathfrak{X}}_2$. In this paper we do not use $\overline{\mathfrak{X}}_2$, working exclusively within $\mathfrak{X}_2$ and $\mathfrak{X}_1$.

Proposition 3.1. Assume $(\mathbf{3A}_j)$ and $(\mathbf{3B}_j)$. Then for each $\theta \in \Gamma$, (a) the Hermitian form $f_j(\theta; \cdot, \cdot)$ is continuous on $\mathfrak{X}_j \times \mathfrak{X}_j$, and (b) the map $J_j(\theta): \mathfrak{X}_j \to \widetilde{\mathfrak{X}}_j(\theta)$ is continuous.

Proof. f_j is continuous since it is the limit of an equicontinuous sequence. The continuity of $J_j(\theta)$ follows from this immediately.

Proposition 3.2. $(\mathbf{3A}_1)$, $(\mathbf{3A}_2)$, $(\mathbf{3B}_2)$, and $(\mathbf{3C}_1)$ together imply $(\mathbf{2A}')$ and hence $(\mathbf{2A})$.

Proof. We shall construct $G'(\theta)$ satisfying $(\mathbf{2A}')$; then $(\mathbf{2A})$ follows by Lemma 2.5.

Let $\theta \in \Gamma$ and $x \in \mathfrak{X}_1$. Since $\{G_n(\theta)x\}$ is Cauchy in $\mathfrak{X}_2$, $\{J_2(\theta)G_n(\theta)x\}$ is Cauchy in $\widetilde{\mathfrak{X}}_2(\theta)$ by the continuity of $J_2(\theta)$ (Proposition 3.1). Thus $G'(\theta)x = \lim J_2(\theta)G_n(\theta)x$ exists. $G'(\theta)$ is a linear operator on $\mathfrak{X}_1$ to $\widetilde{\mathfrak{X}}_2(\theta)$.

In particular we have $f_2(\theta; G_n(\theta)x) = \|J_2(\theta)G_n(\theta)x\|_{2\theta}^2 \to \|G'(\theta)x\|_{2\theta}^2$, $n \to \infty$. On the other hand

$$f_2(\theta; G_n(\theta)x) - f_{2n}(\theta; G_n(\theta)x) \to 0, \; n \to \infty,$$

because $f_{2n}(\theta; y) \to f_2(\theta; y)$ for each $y \in \mathfrak{X}_2$, $\{G_n(\theta)x\}$ is Cauchy in $\mathfrak{X}_2$, and $\{f_{2n}\}$ is equicontinuous (see Lemma D). Since $f_{2n}(\theta; G_n(\theta)x) = f_{1n}(\theta; x) \to f_1(\theta; x)$, we obtain $\|G'(\theta)x\|_{2\theta}^2 = f_1(\theta; x)$. This proves (a) of $(\mathbf{2A}')$.

Since $G_n(\cdot)x$ is strongly measurable, it can be approximated in $\mathfrak{X}_2$ m-a.e. by a sequence of quasi-simple functions. Since $J_2(\theta)$ is continuous, it follows that $J_2(\cdot)G_n(\cdot)x$ is f_2-measurable. Hence $G'(\cdot)x = \lim J_2(\cdot)G_n(\cdot)x$ is also f_2-measurable (see Lemma 1.7). This proves (b) of $(\mathbf{2A}')$.

In order to satisfy $(\mathbf{2B})$ we need additional conditions.

$(\mathbf{3C}_2)$ The $G_n(\theta)$ introduced in $(\mathbf{3C}_1)$ are onto $\mathfrak{X}_2$, and the $G_n(\theta)^{-1}: \mathfrak{X}_2 \to \mathfrak{X}_1$ satisfy condition $(\mathbf{3C}_1)$ with subscripts 1, 2 exchanged (so that the topology of $\mathfrak{X}_1$ is involved).

$(\mathbf{3D})$ With the topologies in $\mathfrak{X}_j$ as in $(\mathbf{3B}_j)$, $\{G_n(\theta)\}_n$ and $\{G_n(\theta)^{-1}\}_n$ are equicontinuous for each $\theta \in \Gamma$.

Proposition 3.3. $(\mathbf{3A}_j)$, $(\mathbf{3B}_j)$, $(\mathbf{3C}_j)$ for both $j = 1, 2$ and $(\mathbf{3D})$ together imply $(\mathbf{2A})$ and $(\mathbf{2B})$.

Proof. We write $G_n(\theta)^{-1} = H_n(\theta)$. As in Proposition 3.2, we can construct $H'(\theta): \mathfrak{X}_2 \to \widetilde{\mathfrak{X}}_1(\theta)$ and hence an isometric operator $\widetilde{H}(\theta): \widetilde{\mathfrak{X}}_2(\theta) \to \widetilde{\mathfrak{X}}_1(\theta)$ such that $\widetilde{H}(\theta)J_2(\theta)y$ is f_1-measurable for $y \in \mathfrak{X}_2$. What remains is to show that $\widetilde{H}(\theta) = \widetilde{G}(\theta)^{-1}$.

The construction of $G'(\theta)$ and $\widetilde{G}(\theta)$ shows that

$$J_2(\theta)G_n(\theta)x \to G'(\theta)x = \widetilde{G}(\theta)J_1(\theta)x \quad \text{in } \widetilde{\mathfrak{X}}_2(\theta), \quad x \in \mathfrak{X}_1. \tag{3.1}$$

Similarly

$$[J_1(\theta) H_n(\theta) - \tilde{H}(\theta) J_2(\theta)] y \to 0 \quad \text{in} \quad \tilde{\mathfrak{X}}_1(\theta), \qquad y \in \mathfrak{X}_2. \tag{3.2}$$

The operators in [] are equicontinuous on $\mathfrak{X}_2$ to $\tilde{\mathfrak{X}}_1(\theta)$ for $n = 1, 2, \ldots$. Since $\{G_n(\theta) x\}$ is Cauchy in $\mathfrak{X}_2$, it follows from Lemma C (Appendix) that

$$[J_1(\theta) H_n(\theta) - \tilde{H}(\theta) J_2(\theta)] G_n(\theta) x \to 0 \quad \text{in} \quad \tilde{\mathfrak{X}}_1(\theta). \tag{3.3}$$

Using (3.1) and the fact that $H_n(\theta) = G_n(\theta)^{-1}$ and noting that $\tilde{H}(\theta)$ is continuous (being isometric), we obtain

$$\tilde{H}(\theta) \tilde{G}(\theta) J_1(\theta) x = J_1(\theta) x, \qquad x \in \mathfrak{X}_1. \tag{3.4}$$

Since $J_1(\theta)\mathfrak{X}_1$ is dense in $\tilde{\mathfrak{X}}_1(\theta)$, it follows that $\tilde{H}(\theta)\tilde{G}(\theta) = 1$ (the identity on $\tilde{\mathfrak{X}}_1(\theta)$).

In the same way we can show that $\tilde{G}(\theta)\tilde{H}(\theta) = 1$. This proves that $\tilde{H}(\theta) = \tilde{G}(\theta)^{-1}$.

In the remainder of this section we consider the adjoint operators of the $G_n(\theta)$ and their convergence.

For a linear topological space $\mathfrak{X}$, we denote by $\mathfrak{X}^*$ its adjoint space; $\mathfrak{X}^*$ is the set of all continuous antilinear forms on $\mathfrak{X}$. The value of $x^* \in \mathfrak{X}^*$ at $x \in \mathfrak{X}$ is denoted by $\langle x^*, x\rangle$, and we write $\langle x, x^*\rangle = \overline{\langle x^*, x\rangle}$ so that $\langle x, x^*\rangle$ is linear in x and antilinear in x^*.

If $\mathfrak{X}$, $\mathfrak{Y}$ are linear topological spaces and $A: \mathfrak{X} \to \mathfrak{Y}$ is linear and continuous, there is a unique linear operator $A^*: \mathfrak{Y}^* \to \mathfrak{X}^*$ such that $\langle A x, y^*\rangle = \langle x, A^* y^*\rangle$ for $x \in \mathfrak{X}$, $y^* \in \mathfrak{Y}^*$. (We do not consider topologies in $\mathfrak{X}^*$, $\mathfrak{Y}^*$ nor the continuity of A^*.)

Under the assumptions of Proposition 3.3, $G_n(\theta): \mathfrak{X}_1 \to \mathfrak{X}_2$ is linear and continuous so that $G_n(\theta)^*: \mathfrak{X}_2^* \to \mathfrak{X}_1^*$ exists. Since $\{G_n(\theta) x\}$ is Cauchy for each $x \in \mathfrak{X}_1$, $\lim \langle x, G_n(\theta)^* y^*\rangle = \lim \langle G_n(\theta) x, y^*\rangle$ exists for each $x \in \mathfrak{X}_1$ and $y^* \in \mathfrak{X}_2^*$. Since $\{G_n(\theta)\}_n$ is equicontinuous, the sequence $\{\langle \cdot\,, G_n(\theta)^* y^*\rangle\}_n$ is equicontinuous and hence the limit is continuous. Thus the limit can be written as $\langle \cdot\,, x^*\rangle$ with a unique $x^* \in \mathfrak{X}_1^*$. Setting $x^* = G(\theta)^* y^*$ defines a linear operator $G(\theta)^*: \mathfrak{X}_2^* \to \mathfrak{X}_1^*$. We may express this result by writing

$$G_n(\theta)^* \rightharpoonup G(\theta)^*, \qquad n \to \infty \quad \text{(weak* convergence)}. \tag{3.5}$$

Note that $G(\theta)$ does not in general make sense (unless one introduces the completion of $\mathfrak{X}_2$), only $G(\theta)^*$ does.

Similarly, for $H_n(\theta) = G_n(\theta)^{-1}$ we have

$$H_n(\theta)^* \rightharpoonup H(\theta)^*, \qquad n \to \infty. \tag{3.6}$$

Proposition 3.4. $G(\theta)^*: \mathfrak{X}_2^* \to \mathfrak{X}_1^*$ and $H(\theta)^*: \mathfrak{X}_1^* \to \mathfrak{X}_2^*$ are inverse to each other.

Proof. For simplicity we omit the argument θ. Let $y \in \mathfrak{X}_2$ and $y^* \in \mathfrak{X}_2^*$. Since $\{\langle \cdot , G_n^* y^* \rangle\}$ is equicontinuous and has limit $\langle \cdot , G^* y^* \rangle$ and since $\{H_n y\}$ is Cauchy in $\mathfrak{X}_1$, it follows from Lemma C that $\langle H_n y, G_n^* y^* \rangle - \langle H_n y, G^* y^* \rangle \to 0$. Here the first term is equal to $\langle G_n H_n y, y^* \rangle = \langle y, y^* \rangle$, while the second term is equal to $\langle y, H_n^* G^* y^* \rangle$ and tends to $\langle y, H^* G^* y^* \rangle$. Hence $\langle y, y^* \rangle = \langle y, H^* G^* y^* \rangle$ and so $H^* G^* = 1$. Similarly we prove $G^* H^* = 1$.

4. Eigenfunction expansions

In Section 1 we considered a spectral representation for a given spectral system. In this section we consider a sort of eigenfunction expansion for the system.

Let $(\Gamma_0, \boldsymbol{B}, m, \mathfrak{H}, E)$ be a spectral system with a spectral form $(\Gamma, \mathfrak{X}, f)$. Then there is a unitary operator Π on $\mathfrak{L}_{ac}(\Gamma)$ to $\mathfrak{M}$ with the properties stated in Theorem 1.11. We now introduce the following conditions, which imply that the system has a representation in a somewhat more refined sense.

(4A) There exist a σ-finite measure space $(\Omega, \Sigma, \varrho)$, a partial isometry Φ of $\mathfrak{H}$ onto $L^2(\varrho)$ with initial set $\mathfrak{L}_{ac}(\Gamma)$, and a measurable function ω on Ω to Γ_0 such that

$$(\Phi \alpha(E) u)(\xi) = \alpha(\omega(\xi))(\Phi u)(\xi), \quad \varrho\text{-a.e. } \xi \in \Omega, \tag{4.1}$$

for each $u \in \mathfrak{H}$ and $\alpha \in [\boldsymbol{B}]$. (The measurability of ω means that $\omega^{-1}(\Delta) \in \Sigma$ whenever $\Delta \in \boldsymbol{B}$. Thus $\alpha \circ \omega$ is ϱ-measurable on Ω if α is $\boldsymbol{B}$-measurable on Γ_0.)

(4B) There is a complex-valued function φ on $\Omega \times \mathfrak{X}$ such that for each fixed $\xi \in \Omega$, $x \mapsto \varphi(\xi; x)$ is linear and for each fixed $x \in \mathfrak{X}$,

$$\varphi(\xi; x) = (\Phi x)(\xi) \quad \text{for} \quad \varrho\text{-a.e. } \xi \in \Omega. \tag{4.2}$$

(4C) $\mathfrak{X}$ is a linear topological space and $x \mapsto \varphi(\xi; x)$ is continuous on $\mathfrak{X}$ for each $\xi \in \Omega$. (In this case we write $\varphi(\xi; x) = \langle x, \varphi(\xi) \rangle$, where $\varphi(\xi) \in \mathfrak{X}^*$; each $\varphi(\xi)$ will be called an *eigenfunction* of E. For the definition of $\mathfrak{X}^*$ see Section 3.)

Example 4.1. Let $\mathfrak{H} = L^2(R^3)$, Φ the Fourier-Plancherel transformation of $\mathfrak{H}$ onto $\hat{\mathfrak{H}} = L^2(\Omega)$, where Ω is another copy of R^3, Σ the set of all Borel subsets of Ω, ϱ the Lebesgue measure on Ω, ω the map of Ω into $\Gamma = \Gamma_0 = R^+$ given by $\omega(\xi) = |\xi|^2 = \xi_1^2 + \xi_2^2 + \xi_3^2$, $\boldsymbol{B}$ the set of all Borel subsets of Γ_0, m the Lebesgue measure on Γ_0, $E(\Delta) = \Phi^{-1} \hat{E}(\Delta) \Phi$, where $\hat{E}(\Delta)$ is the operator of multiplication by $\chi_{\omega^{-1}(\Delta)}$,

$\mathfrak{X} = L^1(R^3) \cap \mathfrak{H}$ with the L^1-topology, $\varphi(\xi) \in \mathfrak{X}^*$ given by the function $(2\pi)^{-\frac{3}{2}} e^{i\xi \cdot x}$ in the sense that

$$\langle u, \varphi(\xi) \rangle = (2\pi)^{-\frac{3}{2}} \int_{R^3} u(x) e^{-i\xi \cdot x} dx, \quad u \in \mathfrak{X}, \tag{4.3}$$

and

$$f(\theta; u, v) = (4\pi^2)^{-1} \int_{R^3 \times R^3} \frac{\sin(\theta^{\frac{1}{2}} |x-y|)}{|x-y|} u(x) \overline{v(y)} \, dx \, dy, \tag{4.4}$$

$$\theta \in \Gamma, \quad u, v \in \mathfrak{X}.$$

The spectral measure E is the one associated with the selfadjoint operator $-\Delta$ in $\mathfrak{H}$, and the $\varphi(\xi)$ are the eigenfunctions of $-\Delta$ in the usual sense. Note that $f(\theta; u, v)$ is well-defined by virtue of the Sobolev inequality because $\mathfrak{X} \subset L^{\frac{6}{5}}(R^3)$. $f(\theta; u, v)$ depends on θ continuously, though this is not required in the general theory. It is even continuous in u, v jointly if the topology of $\mathfrak{X}$ is strengthened to the $L^1 \cap L^{\frac{6}{5}}$-topology.

Proposition 4.2. Assume (**4A**). We have $\omega(\xi) \in \Gamma$ for ϱ-a.e. ξ. If $\Delta \in \boldsymbol{B}$ with $m(\Delta) = 0$, then $\varrho(\omega^{-1}(\Delta)) = 0$.

Proof. Set $S = \omega^{-1}(\Gamma_0 - \Gamma)$ and $S' \subset S$, $S' \in \boldsymbol{\Sigma}$, $\varrho(S') < \infty$. (**4A**) implies that there is $u \in \mathfrak{L}_{ac}(\Gamma)$ such that $\Phi u = \chi_{S'}$. If we set $\alpha = \chi_\Gamma$ in (4.1) and note that $\alpha(E) u = E(\Gamma) u = u$, we obtain $(\Phi u)(\xi) = 0$ ϱ-a.e. Hence $\varrho(S') = 0$. Since ϱ is σ-finite, we have $\varrho(S) = 0$. In other words, $\omega(\xi) \in \Gamma$ for ϱ-a.e. ξ.

Next let $\Delta \in \boldsymbol{B}$, $m(\Delta) = 0$, $S = \omega^{-1}(\Delta)$. Let $S' \subset S$ and $u \in \mathfrak{L}_{ac}(\Gamma)$ be as above. If we set $\alpha = \chi_\Delta$ in (4.1), the left member is zero because $\alpha(E) u = E(\Delta) u = E_{ac}(\Delta) u = 0$. But the right member is equal to $(\Phi u)(\xi)$ on S because $\chi_\Delta(\omega(\xi)) = \chi_S(\xi)$. Thus $\chi_{S'} = \Phi u = 0$, so that $\varrho(S') = 0$. It follows as above that $\varrho(S) = 0$.

Proposition 4.3. Assume (**4A**) and (**4B**). If h is an $\mathfrak{X}$-valued quasi-simple function on Γ and if $u = \Pi^{-1}(Jh) \in \mathfrak{L}_{ac}(\Gamma)$, then

$$(\Phi u)(\xi) = \varphi(\xi; h(\omega(\xi))) \quad \varrho\text{-a.e.} \tag{4.5}$$

Proof. Let h be given by (1.2). Then u is given by (1.5), so that

$$\begin{aligned}(\Phi u)(\xi) &= \sum \alpha_k(\omega(\xi)) (\Phi x_k)(\xi) && \varrho\text{-a.e.}\\ &= \sum \alpha_k(\omega(\xi)) \varphi(\xi; x_k) && \varrho\text{-a.e.}\\ &= \varphi(\xi; h(\omega(\xi))).\end{aligned}$$

Proposition 4.4. Assume (**4A**), (**4B**), and (**4C**). Let $J(\theta): \mathfrak{X} \to \tilde{\mathfrak{X}}(\theta)$ be continuous for each θ. Let $u \in \mathfrak{L}_{ac}(\Gamma)$, and let $\{h_n\}$ be a sequence of strongly measurable functions on Γ to $\mathfrak{X}$ such that for m-a.e. $\theta \in \Gamma$, $\{h_n(\theta)\}$ is Cauchy and $J(\theta)\, h_n(\theta) \to (\Pi u)(\theta)$, $n \to \infty$. Then

$$(\Phi u)\,(\xi) = \lim_{n\to\infty} \langle h_n(\omega(\xi)),\, \varphi(\xi)\rangle \qquad \varrho\text{-a.e.} \tag{4.6}$$

Proof. Since $\{h_n(\theta)\}$ is Cauchy for m-a.e. $\theta \in \Gamma$, $\{h_n(\omega(\xi))\}$ is Cauchy for ϱ-a.e. $\xi \in \Omega$ by Proposition 4.2. Hence

$$\psi(\xi) = \lim \langle h_n(\omega(\xi)),\, \varphi(\xi)\rangle \tag{4.7}$$

exists ϱ-a.e. We have to show that $\psi(\xi) = (\Phi u)\,(\xi)$ ϱ-a.e.

Each h_n is by definition an m-a.e. limit of a sequence $\{h_{nk}\}_k$ of quasi-simple functions. Hence $\langle h_{nk}(\omega(\xi)),\, \varphi(\xi)\rangle \to \langle h_n(\omega(\xi)),\, \varphi(\xi)\rangle$ ϱ-a.e. as $k\to\infty$. On the other hand $J(\theta)\, h_{nk}(\theta) \to J(\theta)\, h_n(\theta)$ m-a.e. as $k \to \infty$. It follows from Corollary B of Appendix that the repeated limits $\lim\limits_n \lim\limits_k$ of these two double sequences can be replaced by simple limits involving $h_{n_p k_p}$ with certain subsequences $\{n_p\}$, $\{k_p\}$ of positive integers. Writing $h'_p = h_{n_p k_p}$, we see that

$$\psi(\xi) = \lim_{p\to\infty} \langle h'_p(\omega(\xi)),\, \varphi(\xi)\rangle \qquad \varrho\text{-a.e.}$$

$$(\Pi u)\,(\theta) = \lim_{p\to\infty} J(\theta)\, h'_p(\theta) \qquad m\text{-a.e.},$$

where each h'_p is quasi-simple and therefore of the form

$$h'_p(\theta) = \sum_q \alpha_{pq}(\theta)\, x_{pq}, \qquad \alpha_{pq} \in [\boldsymbol{B}], \qquad x_{pq} \in \mathfrak{X}.$$

Recalling the proof of Proposition 1.10, we now see easily that there exist a sequence $\{\Gamma_r\}$, $\Gamma_r \in \boldsymbol{B}$, such that $\Gamma_1 \subset \Gamma_2 \subset \ldots\ldots$ and $\bigcup \Gamma_r = \Gamma$, and a subsequence $\{p_r\}$ of $\{p\}$ such that $\|\chi_{\Gamma_r} J\, h'_{p_r} - \Pi u\|_{\mathfrak{M}} \to 0$ as $r \to \infty$. Here $h''_r \equiv \chi_{\Gamma_r} h'_{p_r}$ is again a quasi-simple function. Let $u_r = \Pi^{-1} J\, h''_r \in \mathfrak{L}_{ac}(\Gamma)$. Then $\|u_r - u\|_{\mathfrak{H}} = \|\Pi u_r - \Pi u\|_{\mathfrak{M}} \to 0$ and hence $\Phi u_r \to \Phi u$ in $L^2(\varrho)$.

$h''_r = \chi_{\Gamma_r} h'_{p_r}$ implies that

$$\begin{aligned} \langle h''_r(\omega(\xi)),\, \varphi(\xi)\rangle &= \chi_{\Gamma_r}(\omega(\xi))\, \langle h'_{p_r}(\omega(\xi)),\, \varphi(\xi)\rangle \\ &\to \chi_\Gamma(\omega(\xi))\, \psi(\xi) = \psi(\xi) \qquad \varrho\text{-a.e.}; \end{aligned} \tag{4.8}$$

note that $\omega(\xi) \in \Gamma$ ϱ-a.e. by Proposition 4.2. On the other hand, the convergence $\Phi u_r \to \Phi u$ proved above implies that $(\Phi u_r)\,(\xi) \to (\Phi u)\,(\xi)$ ϱ-a.e. along a suitable subsequence of $\{r\}$. Since the left member of (4.8) is equal to $(\Phi u_r)\,(\xi)$ by Proposition 4.3, it follows that $\psi(\xi) = (\Phi u)\,(\xi)$ ϱ-a.e. as we wished to show.

We can now prove the main theorem on the relationship between eigenfunction expansions for two spectral systems.

Theorem 4.5. *Suppose we have two spectral systems, with indices 1 and 2, as in Section 2. Assume that conditions* (**3A**$_j$), (**3B**$_j$), (**3C**$_j$) *for both* $j=1,2$, *and* (**3D**) *are satisfied, so that the general wave operator* W *with the properties in Theorems 2.3, 2.4 exists. Assume further that the system 1 satisfies* (**4A**), (**4B**), *and* (**4C**), *with the partial isometry* $\Phi_1:\mathfrak{H}_1\to L^2(\varrho)$ *and the eigenfunctions* $\varphi_1(\xi)\in\mathfrak{X}_1^*$. *Then the system 2 also satisfies these conditions with the same measure space* $(\Omega, \Sigma, \varrho)$ *and with the partial isometry* $\Phi_2=\Phi_1 W^*:\mathfrak{H}_2\to L^2(\varrho)$ *and the eigenfunctions* $\varphi_2(\xi)=[G(\omega(\xi))^*]^{-1}\varphi_1(\xi)\in\mathfrak{X}_2^*$. *Here the operators* $G(\theta)^*:\mathfrak{X}_2^*\to\mathfrak{X}_1^*$ *and their inverses were defined in Proposition 3.4.*

Remark 4.6. The relation between the two sets of eigenfunctions may be written

$$G(\omega(\xi))^*\varphi_2(\xi)=\varphi_1(\xi). \tag{4.9}$$

This is the Lippmann-Schwinger equation in an abstract form. Recall that $G(\theta)^*$ is a linear operator on $\mathfrak{X}_1^*$ to $\mathfrak{X}_2^*$, but that $G(\theta)$ was not defined in general.

Proof of Theorem 4.5. First we note that $J_1(\theta):\mathfrak{X}_1\to\widetilde{\mathfrak{X}}_1(\theta)$ are continuous by Proposition 3.1, so that Proposition 4.4 holds for system 1.

W^* is a partial isometry on $\mathfrak{H}_2$ to $\mathfrak{H}_1$ with initial set $\mathfrak{L}_{2,ac}(\Gamma)$ and final set $\mathfrak{L}_{1,ac}(\Gamma)$ (see Theorems 2.3, 2.4 and Proposition 3.3). Since by hypothesis Φ_1 is a partial isometry on $\mathfrak{H}_1$ onto $L^2(\varrho)$ with initial set $\mathfrak{L}_{1,ac}(\Gamma)$, $\Phi_2=\Phi_1W^*$ is a partial isometry on $\mathfrak{H}_2$ onto $L^2(\varrho)$ with initial set $\mathfrak{L}_{2,ac}(\Gamma)$. Furthermore, $\Phi_2\alpha(E_2)u=\Phi_1W^*\alpha(E_2)u= =\Phi_1\alpha(E_1)W^*u$ by the intertwining property of W^* resulting from that of W. Hence $(\Phi_2\alpha(E_2)u)(\xi)=\alpha(\omega(\xi))(\Phi_1W^*u)(\xi)=\alpha(\omega(\xi))\cdot \cdot(\Phi_2u)(\xi)$ ϱ-a.e. by (4.1) for system 1. This proves (**4A**) for system 2.

To prove (**4B**), we note that $W^*y=\Pi_1^{-1}\widetilde{H}\Pi_2y$ by (2.1), where $\widetilde{H}=\widetilde{G}^{-1}$ and $y\in\mathfrak{X}_2$. Hence $(\Pi_1W^*y)(\theta)=\widetilde{H}(\theta)J_2(\theta)y$ for $\theta\in\Gamma$. According to the construction of $\widetilde{H}(\theta)=\widetilde{G}(\theta)^{-1}$ given in the proof of Proposition 3.3, we have $\widetilde{H}(\theta)J_2(\theta)y=\lim J_1(\theta)H_n(\theta)y$. It follows from Proposition 4.4 for system 1 (where we set $h_n(\theta)=H_n(\theta)y$) that

$$(\Phi_2y)(\xi)=(\Phi_1W^*y)(\xi)=\lim\langle H_n(\omega(\xi))y, \varphi_1(\xi)\rangle.$$

On the other hand

$$\begin{aligned}\langle H_n(\omega(\xi))y, \varphi_1(\xi)\rangle&=\langle y, H_n(\omega(\xi))^*\varphi_1(\xi)\rangle\\ &\to\langle y, H(\omega(\xi))^*\varphi_1(\xi)\rangle=\langle y, \varphi_2(\xi)\rangle\end{aligned}$$

by (3.6) and $H(\theta)^*=G(\theta)^{*-1}$ (see Proposition 3.4). This proves (4.2) as well as (**4C**) for system 2.

5. Perturbation of unitary operators

In this section we apply the foregoing general results to construct wave operators between two unitary operators in a Hilbert space $\mathfrak{H}$.

We begin with preliminary results concerning the resolvent of a unitary operator U in $\mathfrak{H}$. Let $U=\int_0^{2\pi} e^{i\theta} E(d\theta)$ be the spectral decomposition of U, where E is a spectral measure on the set $\boldsymbol{B}$ of all Borel subsets of the unit circle Γ_0, which we identify with the interval $[0,2\pi)$. Thus we have a spectral system $(\Gamma_0, \boldsymbol{B}, m, \mathfrak{H}, E)$, where m is the Lebesgue measure.

Let

$$R(\zeta)=U(U-\zeta)^{-1}=(1-\zeta U^*)^{-1}, \quad |\zeta| \neq 1, \tag{5.1}$$

be the "resolvent" of U. If we write $\zeta=re^{i\theta}$, $\zeta'=1/\bar{\zeta}=r^{-1}e^{i\theta}$, we have by an easy computation

$$R(\zeta)-R(\zeta')=(1-|\zeta|^2)R(\zeta)^*R(\zeta)=2\pi\delta_r(E;\theta), \tag{5.2}$$

where

$$\delta_r(\theta';\theta)=\frac{1}{2\pi}\cdot\frac{1-r^2}{1-2r\cos(\theta'-\theta)+r^2} \gtrless 0, \quad r \lessgtr 1, \tag{5.3}$$

is the Poisson kernel and where we used the general notation $\alpha(E)=\int \alpha(\theta)E(d\theta)$. Note that $\delta_{r^{-1}}=-\delta_r$.

According to Fatou's theorem, we have

$$\lim_{r\uparrow 1}(\delta_r(E;\theta)u,v)=(d/d\theta)(E_{ac}(\theta)u,v) \tag{5.4}$$

for a.e. $\theta\in[0,2\pi)$ for each fixed pair $u, v\in\mathfrak{H}$.

Suppose now that we have two unitary operators U_j, $j=1, 2$, in the same Hilbert space $\mathfrak{H}$. We use the obvious notations E_j, $E_{j,ac}$, $R_j(\zeta)$, etc. Set

$$\begin{aligned} G(\zeta)=R_2(\zeta)^{-1}R_1(\zeta)&=(1-\zeta U_2^*)(1-\zeta U_1^*)^{-1}\\ &=1+\zeta(U_1^*-U_2^*)R_1(\zeta). \end{aligned} \tag{5.5}$$

Then

$$G(\zeta)^{-1}=R_1(\zeta)^{-1}R_2(\zeta)=1+\zeta(U_2^*-U_1^*)R_2(\zeta). \tag{5.6}$$

Since $R_1(\zeta)=R_2(\zeta)G(\zeta)$, we have by (5.2)

$$\delta_r(E_1;\theta)=G(\zeta)^*\delta_r(E_2;\theta)G(\zeta), \quad \zeta=re^{i\theta}, \quad r\neq 1. \tag{5.7}$$

Let $\{r_n\}$ be a sequence of positive numbers such that $r_n<1$ and $r_n\to 1$, $n\to\infty$, and set

$$f_{jn}(\theta;u,v)=(\delta_{r_n}(E_j;\theta)u,v). \tag{5.8}$$

We see from (5.7) that

$$f_{1n}(\theta; u, v) = f_{2n}(\theta; G_n^{\pm}(\theta)\, u, G_n^{\pm}(\theta)\, v), \quad G_n^{\pm}(\theta) = G(r_n^{\pm 1} e^{i\theta}). \quad (5.9)$$

It is now clear that we have all the *formal* tools necessary for the application of the results of section 3. It is thus natural to assume that there is a linear manifold $\mathfrak{X} \subset \mathfrak{H}$ and a Borel set $\Gamma \subset [0, 2\pi)$ satisfying (some of) the following conditions, which roughly correspond to the ones given in section 3, with $\mathfrak{H}_1 = \mathfrak{H}_2 = \mathfrak{H}$ and $\mathfrak{X}_1 = \mathfrak{X}_2 = \mathfrak{X}$.

($\mathbf{5A}_j$) For each $\theta \in \Gamma$, $f_j(\theta; x, y) = \lim_{n\to\infty} (\delta_{r_n}(E_j; \theta)\, x, y)$ exists for all $x, y \in \mathfrak{X}$. (Then the limit is necessarily a nonnegative Hermitian form in x, y and is a density function for $(E_{j,ac}(\cdot)\, x, y)$ by (5.4). Thus f_j is a spectral form on $\Gamma \times \mathfrak{X} \times \mathfrak{X}$ for the spectral system j.)

($\mathbf{5B}_j$) $\mathfrak{X}$ is a linear topological space with its own topology. For each $\theta \in \Gamma$, $x, y \mapsto (\delta_{r_n}(E_j; \theta)\, x, y)$ is equicontinuous on $\mathfrak{X} \times \mathfrak{X}$ for varying n.

($\mathbf{5C}_j^{\pm}$) For each $\theta \in \Gamma$ and $n = 1, 2, \ldots$, $Q_{jn}^{\pm}(\theta) \equiv (U_2^* - U_1^*)\, R_j(r_n^{\pm 1} e^{i\theta})$ maps $\mathfrak{X}$ into itself, with the following properties. (a) For each $\theta \in \Gamma$ and $x \in \mathfrak{X}$, $\{Q_{jn}^{\pm}(\theta)\, x\}_n$ is a Cauchy sequence in $\mathfrak{X}$. (b) For each $x \in \mathfrak{X}$ and $n = 1, 2, \ldots$, $Q_{jn}^{\pm}(\cdot)\, x$ is strongly measurable on Γ as an $\mathfrak{X}$-valued function.

($\mathbf{5D}_j^{\pm}$) For each $\theta \in \Gamma$, $Q_{jn}^{\pm}(\theta) : \mathfrak{X} \to \mathfrak{X}$ is equicontinuous for varying n.

We can now state the main theorems of this section.

Theorem 5.1. *Let* $U_j = \int_0^{2\pi} e^{i\theta} E_j(d\theta)$, $j = 1, 2$, *be unitary operators in* $\mathfrak{H}$. *Assume that there is a linear manifold* $\mathfrak{X} \subset \mathfrak{H}$, *a Borel set* $\Gamma \subset [0, 2\pi)$, *and a sequence* $r_n \uparrow 1$ *of positive numbers satisfying conditions* ($\mathbf{5A}_2$), ($\mathbf{5B}_2$), *and* ($\mathbf{5C}_1^{+}$). *Then there is a partial isometry* W^+ *in* $\mathfrak{H}$ *with the following properties.*

(a) The initial set of W^+ *is* $\mathfrak{L}_{1,ac}(\Gamma)$ *and the final set is contained in* $\mathfrak{L}_{2,ac}(\Gamma)$. *Here* $\mathfrak{L}_{j,ac}(\Gamma) = E_{j,ac}(\Gamma)\, \mathfrak{L}_j$ *and* $\mathfrak{L}_j$ *is the smallest closed subspace of* $\mathfrak{H}$ *containing* $\mathfrak{X}$ *and reducing* E_j.

(b) W^+ *has the intertwining property* $\alpha(E_2)\, W^+ = W^+ \alpha(E_1)$ *for* $\alpha \in [\boldsymbol{B}]$. *In particular* $U_2 W^+ = W^+ U_1$.

(c) For each $u \in \mathfrak{L}_{1,ac}(\Gamma)$, $W^+ u = \lim_{n\to\infty} W_{r_n}^+ u$ *where*

$$W_r^+ = W_r^+(U_2, U_1) = (1 - r^2) \sum_{k=0}^{\infty} r^{2k} U_2^k U_1^{-k}, \quad 0 < r < 1. \quad (5.10)$$

Similarly, one obtains a partial isometry W^- *assuming* ($\mathbf{5C}_1^{-}$) *instead of* ($\mathbf{5C}_1^{+}$). *In (c) we have* $W^- u = \lim W_{r_n}^- u$, *where* W_r^- *is given by* (5.10) *with the sum taken over* $k = -\infty$ *to* 0.

Theorem 5.2. *In Theorem* 5.1 *assume further* ($\mathbf{5C}_2^+$), *and* ($\mathbf{5D}_j^+$) *for both* $j=1, 2$. *Then the final set of* W^+ *is equal to* $\mathfrak{L}_{2,ac}(\Gamma)$. *For* $v \in \mathfrak{L}_{2,ac}(\Gamma)$, *we have* $(W^+)^* v = \lim_{n\to\infty} W_{r_n}^+(U_1, U_2)$. *Similar results hold for* W^-.

Remark 5.3. Theorem 5.1, (c) shows that W^+ is on $\mathfrak{L}_{1,ac}(\Gamma)$ equal to a sort of *strong Abel limit* of the sequence $\{U_2^k U_1^{-k}\}$ as $k\to\infty$ (taken along a particular sequence $r_n \uparrow 1$). It would be desirable to show that $W^+ =$ s-lim $U_2^k U_1^{-k}$, which is the definition of the wave operator in the "time-dependent theory". But we have not been able to do this without introducing further restrictions (see section 7). In any case (c) implies that W^+ is independent of the auxiliary space $\mathfrak{X}$, in the sense that two W^+'s constructed with two different $\mathfrak{X}$'s (but with a common $\{r_n\}$) coincide on the intersection of their initial sets.

Remark 5.4. In most cases in application one need not take a special sequence $\{r_n\}$ but any sequence $r_n \uparrow 1$ will do, with the same f_j. In such a case $W^+ u$ is the strong Abel limit of $\{U_2^k U_1^{-k} u\}$ in the usual sense.

Remark 5.5. The assumptions in the theorems given above involve both U_1 and U_2. This entails some difficulty in applications, for the property of the "perturbed operator" U_2 is not well known in advance. This is a defect common to all theorems of a similar type. The difficulty can be overcome in some special cases, see Examples 6.1, 6.2; cf. also [22] and [9].

We shall now prove Theorems 5.1 and 5.2. It suffices to consider W^+. Except for (c) of Theorem 5.1 and the corresponding assertion in Theorem 5.2, the results follow directly from Theorems 2.3 and 2.4. Note that the assumptions in Theorem 5.1 imply not only ($\mathbf{3A}_2$), ($\mathbf{3B}_2$), and ($\mathbf{3C}_1$) but also ($\mathbf{3A}_1$); see (5.4) and the remark just after ($\mathbf{3C}_1$). It follows from Proposition 3.2 that ($\mathbf{2A}$) is true, so that Theorem 2.3 is applicable. Similarly, the assumptions of Theorem 5.2 imply ($\mathbf{3A}_j$), ($\mathbf{3B}_j$), ($\mathbf{3C}_j$), for $j=1, 2$, and ($\mathbf{3D}$), so that Theorem 2.4 is applicable.

The remainder of this section is devoted to the proof of (c) of Theorem 5.1.

Proposition 5.6. For any integer k, we have

$$\|U_2^k W_r^+ - W_r^+ U_1^k\| \to 0, \quad r\uparrow 1. \tag{5.11}$$

Proof. A simple calculation gives $U_2 W_r^+ U_1^{-1} = r^{-2} W_r^+ + 1 - r^{-2}$. Hence $\|U_2 W_r^+ - W_r^+ U_1\| \to 1$ as $r\uparrow 1$; note that $\|W_r^+\| \leqq 1$. (5.11) follows easily from this.

Proposition 5.7. To prove (c), it suffices to show that

$$(W_{r_n}^+ x, v) \to (W^+ x, v), \quad n\to\infty, \tag{5.12}$$

for every $x \in \mathfrak{X}$ and $v = \beta(E_2)\, E_{2,ac}(\Gamma) y$, where $y \in \mathfrak{X}$ and $\beta \in [\boldsymbol{B}]$ with support on Γ.

Proof. First we show that (5.12) implies

$$(W_{r_n}^+ u, w) \to (W^+ u, w), \qquad u \in \mathfrak{L}_1, \qquad w \in \mathfrak{L}_{2,ac}(\Gamma). \tag{5.13}$$

To this end it suffices to consider the case in which $u = U_1^k x$, $x \in \mathfrak{X}$, $k =$ integer, and w is of the form v stated above, for such vectors are fundamental in $\mathfrak{L}_1$ and $\mathfrak{L}_{2,ac}(\Gamma)$, respectively. But $\lim_{n\to\infty} (W_{r_n}^+ U_1^k x, v) = \lim_{n\to\infty} (U_2^k W_{r_n}^+ x, v) = \lim_{n\to\infty} (W_{r_n}^+ x, U_2^{-k} v) = (W^+ x, U_2^{-k} v)$ by Proposition 5.6 and (5.12); note that $U_2^{-k} v$ is of the same form as v.

Suppose now that $u \in \mathfrak{L}_{1,ac}(\Gamma) \subset \mathfrak{L}_1$. Since $\|W_r^+\| \leq 1$ and $\|W^+ u\| = \|u\|$, we have $\|W_{r_n}^+ u - W^+ u\|^2 \leq 2\, Re(W^+ u - W_{r_n}^+ u, W^+ u)$. The right-hand side of this inequality tends to zero as $n \to \infty$ by (5.13) because $W^+ u \in \mathfrak{L}_{2,ac}(\Gamma)$. Thus $W_{r_n}^+ u \to W^+ u$, as we wished to show.

In what follows we fix x, v, y, β as in Proposition 5.7.

Proposition 5.8. We have the following integral expressions:

$$(W_{r_n}^+ x, v) = \int_0^{2\pi} \phi_n(\theta)\, d\theta, \qquad (W^+ x, v) = \int_0^{2\pi} \phi(\theta)\, d\theta, \tag{5.14}$$

where

$$\begin{aligned} \phi_n(\theta) &= (G_n^+(\theta)\, x, \delta_{r_n}(E_2; \theta) v) = f_{2n}(\theta; G_n^+(\theta)\, x, v), \\ \phi(\theta) &= \overline{\beta(\theta)}\, (\tilde{G}^+(\theta) J_1(\theta)\, x, J_2(\theta)\, y)_{2\theta}. \end{aligned} \tag{5.15}$$

Proof. The expression for $(W_{r_n}^+ x, v)$ comes from

$$\begin{aligned} W_r^+ &= (1 - r^2) \sum_{k=0}^{\infty} r^{2k} U_2^k U_1^{-k} \\ &= \frac{1 - r^2}{2\pi} \int_0^{2\pi} (1 - r e^{-i\theta} U_2)^{-1} (1 - r e^{i\theta} U_1^*)^{-1}\, d\theta \\ &= \frac{1 - r^2}{2\pi} \int_0^{2\pi} R_2(r e^{i\theta})^* R_1(r e^{i\theta})\, d\theta \\ &= \int_0^{2\pi} \delta_r(E_2; \theta) G(r e^{i\theta})\, d\theta, \end{aligned}$$

for $R_1(\zeta) = R_2(\zeta) G(\zeta)$ and $R_2(\zeta)^* R_2(\zeta) = 2\pi (1 - r^2)^{-1} \delta_r(E_2; \theta)$, $\zeta = r e^{i\theta}$, by (5.2).

The expression for $(W^+ x, v)$ is obtained by noting that it is equal to $(W^+ E_{1,ac}(\Gamma)\, x, \beta(E_2)\, E_{2,ac}(\Gamma)\, y)$, where $W^+ = \Pi_2^{-1} \tilde{G}^+ \Pi_1$ and $\Pi_1 E_{1,ac}(\Gamma)\, x = J_1 x$, $\Pi_2\, \beta(E_2)\, E_{2,ac}(\Gamma)\, y = \beta J_2\, y$ (see (2.1) and Theorem 1.11).

Proposition 5.9. There is a subsequence $\{n_k\}$ of $\{n\}$ such that for a.e. fixed $\theta \in \Gamma$, the linear form $f_{2n_k}(\theta; \cdot, v)$ is equicontinuous on $\mathfrak{X}$ for varying $k = 1, 2, \ldots$.

Proof. Note that the assertion does *not* follow directly from $(5\mathbf{B}_2)$ since v need not be in $\mathfrak{X}$. For the proof we use the Schwarz inequality

$$|f_{2n}(\theta; z', v) - f_{2n}(\theta; z'', v)| \leqq f_{2n}(\theta; z' - z'')^{\frac{1}{2}} f_{2n}(\theta; v)^{\frac{1}{2}}.$$

Since $\{f_{2n}(\theta; \cdot)\}$ is equicontinuous on $\mathfrak{X}$ by $(5\mathbf{B}_2)$, it suffices to show that there is a subsequence $\{n_k\}$ such that $q_n(\theta) \equiv f_{2n}(\theta; v)$ is bounded for $n = n_k$, $k = 1, 2, \ldots$, for a.e. $\theta \in \Gamma$.

$q_n(\theta)$ admits the Poisson integral expression

$$q_n(\theta) = \frac{1 - r_n^2}{2\pi} \int_0^{2\pi} \frac{q(\theta')\, d\theta'}{1 - 2r_n \cos(\theta' - \theta) + r_n^2},$$

where $q(\theta) = (d/d\theta)\,\big(E_2(\theta)v, v\big)$; note that $\big(E_2(\theta)v, v\big)$ is absolutely continuous because $v \in \mathfrak{H}_{2,ac}$. Since $q \in L^1(0,2\pi)$, it follows that $q_n \to q$ in L^1 as $n \to \infty$. Hence there is a subsequence $\{n_k\}$ such that $q_{n_k}(\theta) \to q(\theta)$ a.e. In particular $\{q_{n_k}(\theta)\}$ is bounded for a.e. θ.

Proposition 5.10. $\phi_{n_k}(\theta) \to \phi(\theta)$, $k \to \infty$, for a.e. $\theta \in \Gamma$.

Proof. $\{f_{2n_k}(\theta; \cdot, v)\}$ is equicontinuous on $\mathfrak{X}$ by Proposition 5.9. On the other hand, for each $z \in \mathfrak{X}$,

$$\begin{aligned}
\lim_{k\to\infty} f_{2n_k}(\theta; z, v) &= (d/d\theta)\,(E_{2,ac}(\theta)\, z, v) && \text{a.e.} && \text{by (5.4)}\\
&= \overline{\beta(\theta)}\,(d/d\theta)\,(E_{2,ac}(\theta)\, z, y) && \text{a.e.} &&\\
&= \overline{\beta(\theta)}\, f_2(\theta; z, y) && \text{a.e.} \quad \theta \in \Gamma &&\\
&= \overline{\beta(\theta)}\,(J_2(\theta)\, z, J_2(\theta)\, y)_{2\theta} && \text{a.e.} \quad \theta \in \Gamma. &&
\end{aligned}$$

Furthermore, $\{G_n^+(\theta)\, x\}$ is Cauchy in $\mathfrak{X}$ for fixed θ and $J_2(\theta)\, G_n^+(\theta)\, x \to \tilde{G}^+(\theta)\, J_1(\theta)\, x$ in $\tilde{\mathfrak{X}}_2(\theta)$; see $(5\mathbf{C}_1^+)$ and (3.1). Hence $\lim f_{2n_k}(\theta; G_n^+(\theta)\, x, v) = \overline{\beta(\theta)}\,\big(\tilde{G}^+(\theta)\, J_1(\theta)\, x, J_2(\theta)\, y\big)_{2\theta}$ by Lemma E of Appendix.

Proposition 5.11. The ϕ_n have uniformly equicontinuous integrals. In other words, for any $\varepsilon > 0$ there is $\delta > 0$ such that $m(\Delta) < \delta$ implies $\int_\Delta |\phi_n(\theta)|\, d\theta < \varepsilon$ for all n. Furthermore, $\int_{\Gamma'} |\phi_n(\theta)|\, d\theta \to 0$, $n \to \infty$, where $\Gamma' = (0, 2\pi) - \Gamma$.

Proof. Since $q_n \to q$ in $L^1(0, 2\pi)$ as was shown in the proof of Proposition 5.9, $\{q_n\}$ has uniformly equicontinuous integrals (Vitali-Hahn-Saks's theorem). Thus for any $\varepsilon > 0$ there is $\delta > 0$ such that $m(\Delta) < \delta$ implies $\int_\Delta q_n(\theta)\, d\theta < \varepsilon$.

By (5.15) we have $\phi_n(\theta)=f_{2n}(\theta; G_n^+(\theta)x, v)$, so that $|\phi_n(\theta)|^2 \leqq f_{2n}(\theta; G_n^+(\theta)x)\, f_{2n}(\theta; v)=f_{1n}(\theta, x)\, q_n(\theta)$ by (5.9). Hence

$$\begin{aligned}\int_\Delta |\phi_n(\theta)|\, d\theta &\leqq \Big[\int_\Delta f_{1n}(\theta; x)\, d\theta\Big]^{\frac{1}{2}} \Big[\int_\Delta q_n(\theta)\, d\theta\Big]^{\frac{1}{2}} \\ &\leqq \|x\| \varepsilon^{\frac{1}{2}} \quad \text{if} \quad m(\Delta)<\delta.\end{aligned} \tag{5.16}$$

This proves the first assertion of the proposition.

The second assertion follows from (5.16) with $\Delta=\Gamma'=(0,2\pi)-\Gamma$, for then $\int_{\Gamma'} q_n(\theta)\, d\theta=\int_{\Gamma'} (q_n(\theta)-q(\theta))\, d\theta \leqq \|q_n-q\|_{L^1}\to 0$; note that $q(\theta)=(d/d\theta)\,(E_2(\theta)v, v)=0$ on Γ' because β is supported on Γ.

Completion of the proof of (c). Since $\phi_{nk}(\theta)\to\phi(\theta)$ a.e. on Γ (Proposition 5.10) and since the ϕ_n have uniformly equicontinuous integrals (Proposition 5.11), it follows from the Vitali convergence theorem that $\int_\Gamma \phi_{nk}(\theta)\, d\theta \to \int_\Gamma \phi(\theta)\, d\theta$. Since $\int_{\Gamma'} \phi_{nk}(\theta)\, d\theta\to 0=\int_{\Gamma'} \phi(\theta)\, d\theta$ (Proposition 5.11), we have $(W_{r_n}^+ x, v)\to(W^+ x, v)$ along the subsequence $\{n_k\}$ [see (5.14)]. But the whole argument could have started with $\{n\}$ replaced by any of its subsequences. Hence the convergence must take place for the original sequence $\{n\}$. This proves (5.12) and completes the proof of (c).

6. Some applications

In what follows we denote by $\mathscr{B}(\mathfrak{X}, \mathfrak{Y})$ the set of all continuous linear operators on a linear topological space $\mathfrak{X}$ to another one $\mathfrak{Y}$. In most cases $\mathfrak{X}$, $\mathfrak{Y}$ will be Banach spaces. In such a case we denote by $\|T\|_{\mathfrak{X}\to\mathfrak{Y}}$ the norm of $T\in\mathscr{B}(\mathfrak{X}, \mathfrak{Y})$.

I. Let $\mathfrak{X}$ be a linear manifold in a Hilbert space $\mathfrak{H}$. We assume that $\mathfrak{X}$ is a normed space with its own norm. We denote by $\overline{\mathfrak{X}}$ the completion of $\mathfrak{X}$. (We are not interested in the question whether some of the ideal elements of $\overline{\mathfrak{X}}$ can be identified with elements of $\mathfrak{H}$.)

Let Γ be a subset of the unit circle, identified with the interval $[0,2\pi)$. Let $D^\pm=\{\zeta=re^{i\theta};\, a<r^{\pm 1}\leqq 1,\, \theta\in\Gamma\}$, where a is a positive number <1, so that $\Gamma\subset D^\pm$. Let U_j, $j=1, 2$, be unitary operators in $\mathfrak{H}$, with the resolvents R_j defined by (5.1). Consider the following conditions.

(6 A$_j$) The function $\zeta, x, y\to((R_j(\zeta)-R_j(\zeta'))x, y)$, defined for $|\zeta|<1$, $x, y\in\mathfrak{X}$, has a continuation which is continuous on $D^+\times\overline{\mathfrak{X}}\times\overline{\mathfrak{X}}$.

(6 B$_j^\pm$) For $|\zeta|\neq 1$, $Q_j(\zeta)=(U_2^*-U_1^*)\,R_j(\zeta)$ maps $\mathfrak{X}$ into $\mathfrak{X}$ continuously, so that it can be extended to an operator $\overline{Q}_j(\zeta)\in\mathscr{B}(\overline{\mathfrak{X}})\equiv\mathscr{B}(\overline{\mathfrak{X}}, \overline{\mathfrak{X}})$. Furthermore, $\overline{Q}_j(\cdot)$ can be continued on $D^\pm$ as strongly continuous functions with values in $\mathscr{B}(\overline{\mathfrak{X}})$.

It is easy to see that ($6\mathbf{A}_j$) implies ($5\mathbf{A}_j$) and ($5\mathbf{B}_j$), and that ($6\mathbf{B}_j^{\pm}$) implies ($5\mathbf{C}_j^{\pm}$) and ($5\mathbf{D}_j^{\pm}$) for any sequence $r_n \uparrow 1$ (principle of uniform boundedness). It follows, for example, that ($6\mathbf{A}_2$) and ($6\mathbf{B}_1^{+}$) imply the existence of W^+ with properties stated in Theorem 5.1, with (c) strengthened to $W^+ u = \lim_{r \uparrow 1} W_r u$, $u \in \mathfrak{L}_{1,ac}(\Gamma)$ (see Remark 5.4).

Thus we have analogs of the results of [9], [26], and [27], which are concerned with two selfadjoint rather than unitary operators. The selfadjoint case can be treated in a similar way.

II. Let U_1, U_2 be unitary operators such that

$$U_2^* - U_1^* = A\,B, \qquad A \in \mathscr{B}(\mathfrak{K}, \mathfrak{H}), \qquad B \in \mathscr{B}(\mathfrak{H}, \mathfrak{K}), \tag{6.1}$$

where $\mathfrak{K}$ is a Banach space, which may coincide with $\mathfrak{H}$. Let $\Gamma \subset [0, 2\pi)$, $r_n \uparrow 1$, and consider the following conditions.

($6\mathbf{C}_j$) For each $\theta \in \Gamma$, $\lim_{n \to \infty} \big(\delta_{r_n}(E_j; \theta) A z, A u\big)$ exists for every $z, u \in \mathfrak{K}$.

($6\mathbf{D}_j^{\pm}$) For each $\theta \in \Gamma$, the sequence $B R_j(r_n^{\pm 1} e^{i\theta}) A \in \mathscr{B}(\mathfrak{K})$ has a strong limit in $B(\mathfrak{K})$ as $n \to \infty$.

We shall show that by an appropriate choice of $\mathfrak{X}$, ($6\mathbf{C}_j$) implies ($5\mathbf{A}_j$) and ($5\mathbf{B}_j$), and that ($6\mathbf{D}_j^{\pm}$) implies ($5\mathbf{C}_j^{\pm}$) and ($5\mathbf{D}_j^{\pm}$). Thus we can construct wave operators $W^{\pm}$ with properties given by Theorem 5.1 or 5.2.

We choose $\mathfrak{X} = \mathfrak{R}(A)$ (range of A) and make it into a normed space with the norm

$$\|x\|_{\mathfrak{X}} = \inf_{Az = x} \|z\|_{\mathfrak{K}}. \tag{6.2}$$

Since $A \in \mathscr{B}(\mathfrak{K}, \mathfrak{H})$, $\mathfrak{X}$ is a Banach space isometrically isomorphic with $\mathfrak{K}/\mathfrak{N}(A)$, where $\mathfrak{N}(A)$ is the null space of A. We note that $A \in \mathscr{B}(\mathfrak{K}, \mathfrak{X})$ too, with $\|A\|_{\mathfrak{K} \to \mathfrak{X}} \leqq 1$.

Assume now ($6\mathbf{C}_j$). If $x = Az \in \mathfrak{X}$, $y = Au \in \mathfrak{X}$, where $z, u \in \mathfrak{K}$, then $\big(\delta_{r_n}(E_j; \theta) x, y\big) = \big(\delta_{r_n}(E_j; \theta) A z, A u\big)$ is convergent as $n \to \infty$; thus ($5\mathbf{A}_j$) is true. Also we have, by the principle of uniform boundedness, $|\big(\delta_{r_n}(E_j; \theta) x, y\big)| \leqq M(\theta) \|x\|_{\mathfrak{X}} \|y\|_{\mathfrak{X}}$, where $M(\theta) < \infty$ does not depend on n. Hence the Hermitian form $\big(\delta_{r_n}(E_j; \theta) x, y\big)$ is equicontinuous on $\mathfrak{X} \times \mathfrak{X}$ for varying n. This proves ($5\mathbf{B}_j$).

$Q_{jn}(\theta) = (U_2^* - U_1^*) R_j(r_n e^{i\theta}) = A B R_j(r_n e^{i\theta})$ obviously maps $\mathfrak{X}$ into itself. Assume now ($6\mathbf{D}_j^{+}$) and let $x = Az \in \mathfrak{X}$, $z \in \mathfrak{K}$. Then $Q_{jn}(\theta)\, x = A u_n(\theta)$ with $u_n(\theta) = B R_j(r_n e^{i\theta}) A z$, and $\{A u_n(\theta)\}$ converges in $\mathfrak{X}$ because $\{u_n(\theta)\}$ converges in $\mathfrak{K}$ and $A \in \mathscr{B}(\mathfrak{K}, \mathfrak{X})$. Moreover, since $u_n(\cdot)$ is continuous in $\mathfrak{K}$-norm (because $R_j(\cdot)$ is analytic in $\mathfrak{H}$-norm and $B \in \mathscr{B}(\mathfrak{H}, \mathfrak{K})$) $A u_n(\cdot)$ is continuous in $\mathfrak{X}$-norm and hence strongly measurable. This proves ($5\mathbf{C}_j^{+}$). Again the principle of uniform boundedness implies that $\|Q_{jn}(\theta)\|_{\mathfrak{X} \to \mathfrak{X}} \leqq M'(\theta) < \infty$ with $M'(\theta)$ independent of n. Thus $\{Q_{jn}(\theta)\}_n$ is equicontinuous on $\mathfrak{X}$ to $\mathfrak{X}$, which proves ($5\mathbf{D}_j^{+}$).

Similarly it can be proved that ($6\mathbf{D}_j^-$) implies ($5\mathbf{C}_j^-$) and ($5\mathbf{D}_j^-$). Moreover, we have in this case

$$\mathfrak{L}_1 = \mathfrak{L}_2 \equiv \mathfrak{L}.$$

In fact, $U_2^* \mathfrak{L}_1 = (U_1^* + A B) \mathfrak{L}_1 \subset U_1^* \mathfrak{L}_1 + A B \mathfrak{L}_1 \subset \mathfrak{L}_1$ since $\mathfrak{L}_1$ is invariant under U_1^* and $\mathfrak{R}(A) = \mathfrak{X} \subset \mathfrak{L}_1$. Furthermore, $A B = U_2^* - U_1^* = U_2^{-1} - U_1^{-1}$ so that $U_1 A B U_2 = U_1 - U_2$, and

$$U_2 \mathfrak{L}_1 = (U_1 - U_1 A B U_2) \mathfrak{L}_1 \subset U_1 \mathfrak{L}_1 + U_1 A B U_2 \mathfrak{L}_1 \subset \mathfrak{L}_1 + U_1 \mathfrak{X} \subset \mathfrak{L}_1$$

because $\mathfrak{L}_1$ is invariant under U_1. Thus $\mathfrak{L}_1$ is reduced by E_2, being invariant under U_2^* and U_2. Since $\mathfrak{L}_1 \supset \mathfrak{X}$, it follows that $\mathfrak{L}_1 \supset \mathfrak{L}_2$. Similarly one proves $\mathfrak{L}_1 \subset \mathfrak{L}_2$.

Finally we show that $x \perp \mathfrak{L}$ implies $U_1 x = U_2 x$ and $U_1^* x = U_2^* x$. Let $x \perp \mathfrak{L}$. Then $x \perp \mathfrak{X} = \mathfrak{R}(A)$ and so $A^* x = 0$. Hence $(U_2 - U_1) x = B^* A^* x = 0$. Also $U_1^* x \perp \mathfrak{L}$, so that we may replace x by $U_1^* x$ in the result just obtained. Thus $x = U_2 U_1^* x$ or $U_1^* x = U_2^* x$.

We may disregard the subspace $\mathfrak{L}^\perp$ of $\mathfrak{H}$, in which nothing happens. In the subspace $\mathfrak{L}$, we have the local wave operators $W^\pm$ which implement the unitary equivalence of U_1 and U_2 restricted on $\mathfrak{L}_{1,ac}(\Gamma) = E_{1,ac}(\Gamma)\mathfrak{L}$ and $\mathfrak{L}_{2,ac}(\Gamma) = E_{2,ac}(\Gamma)\mathfrak{L}$, respectively, and which are given as the Abel limits of $U_2^k U_1^{-k}$ as $k \to \pm\infty$. As will be shown in the following section, we have $W^\pm = \operatorname*{s-lim}_{k\to\pm\infty} U_2^k U_1^{-k}$ on $\mathfrak{L}_{1,ac}(\Gamma)$ if $\mathfrak{K}$ is a Hilbert space, in which case $\mathfrak{X}$ is also a Hilbert space.

These results generalize those of [22], where $\mathfrak{K} = \mathfrak{H}$ is assumed. Thus it follows, for example, that $W^\pm$ exist and are complete, with $\Gamma = [0, 2\pi)$, provided $U_1 - U_2$ belongs to the trace class. Also it can be proved that $W^\pm$ exist and are complete if $U_2 - U_1$ is "smooth" with respect to U_1 on Γ. Actually the corresponding results in [22] are mostly stated in the case of two selfadjoint operators, but it is not difficult to give their unitary versions.

Example 6.1. *Let $\mathfrak{H} = L^2(R^3)$. Let H_1 be the selfadjoint realization in $\mathfrak{H}$ of $-\Delta$ (negative Laplacian). Let q be a real function such that*

$$q \in L^{\frac{3}{2}}(R^3). \tag{6.3}$$

Under these conditions it can be shown (see [16]) that there exists a unique selfadjoint operator H_2 in $\mathfrak{H}$ with the property that, for any nonreal complex number z,

$$\begin{aligned}(H_2 - z)^{-1} - (H_1 - z)^{-1} &= -[(H_2 - z)^{-1} A'] \, B' (H_1 - z)^{-1} \\ &= -[(H_1 - z)^{-1} A'] \, B' (H_2 - z)^{-1},\end{aligned} \tag{6.4}$$

where A', B' are the operators of multiplication by $|q|^{\frac{1}{2}}$, $|q|^{\frac{1}{2}}$ sign q, respectively, and where

$$B'(H_j-z)^{-1}\in\mathscr{B}(\mathfrak{H}), \qquad (H_j-z)^{-1}A'\subset[(H_j-z)^{-1}A']\in\mathscr{B}(\mathfrak{H}), \qquad j=1,2.$$

It should be noted that $\overline{Q}_1(z)\equiv B'[(H_1-z)^{-1}A']$ has finite Hilbert-Schmidt norm, which is smaller than 1 if $-Im\, z^{\frac{1}{2}}$ is sufficiently large. (This is an example of smooth perturbation.)

We note further that H_2 is a selfadjoint realization in $\mathfrak{H}$ of $-\Delta+q$ and coincides with the selfadjoint operator associated with the semibounded, closed quadratic form $\int_{R^3}[|\operatorname{grad} u|^2+q(x)|u|^2]\,dx$ (see [17], Chapter 6).

Let $U_j=(H_j-i)(H_j+i)^{-1}$ be the Cayley transform of H_j, $j=1,2$. From (6.4) we obtain

$$U_2^*-U_1^*=AB, \tag{6.5}$$

where

$$A=[(H_1-i)^{-1}A'], \qquad B=2iB'(H_2-i)^{-1}. \tag{6.6}$$

We can now verify conditions (**6** $\mathbf{C}_j$), (**6** $\mathbf{D}_j^{\pm}$), $j=1,2$, where $\mathfrak{K}=\mathfrak{H}$ and Γ is a certain open subset of $\Gamma_0=[0,2\pi)$ with $m(\Gamma_0-\Gamma)=0$. A straightforward calculation gives namely

$$\begin{aligned} B R_1(\zeta)A &=\zeta^{-1}\big((1-\overline{Q}_2(i))\,\overline{Q}_1(z)-\overline{Q}_2(i)\big),\\ B R_2(\zeta)A &=\zeta^{-1}\big(\overline{Q}_2(z)\,(1+\overline{Q}_1(i))-\overline{Q}_1(i)\big), \end{aligned} \tag{6.7}$$

where $\zeta=(z-i)(z+i)$, $\overline{Q}_1(z)$ is as given above, and $\overline{Q}_2(z)=1-(1+\overline{Q}_1(z))^{-1}$. $\overline{Q}_1(z)$ is compact and analytic in z in the complex plane cut along the nonnegative real axis and has continuous (in the Hilbert-Schmidt norm) boundary values on the two edges of the positive real axis. It follows (see [22]) that $\overline{Q}_2(z)$ is compact and meromorphic (with poles only on the negative real axis) in the cut plane and continuous up to the boundary, except for a closed subset of the real axis with measure zero. Removing this closed set, the poles, and the origin from the real axis R^1, we obtain an open subset Γ' of R^1. Let Γ be the image of Γ' under the map $z\to\zeta$. Then we see easily that (**6** $\mathbf{D}_j^{\pm}$), $j=1,2$, are satisfied. (**6** $\mathbf{C}_j$) can be verified in the same way noting that $A^*=A'(H_1+i)^{-1}$ has the same form as B.

It follows that the wave operators $W^{\pm}$ exist and satisfies the invariance principle. In particular $W^{\pm}=\operatorname*{s-lim}_{t\to\pm\infty} e^{itH_2}e^{-itH_1}$. $H_{2,ac}$ is unitarily equivalent to* H_1.

In this case we have $\mathfrak{X}=\mathfrak{R}(A)\subset L^{\frac{6}{5}}(R^3)$. The adjoint space $\mathfrak{X}^*$ is however not large enough to contain the "eigenfunctions" of U_1 or H_1

* It is not known whether H_2 has continuous singular spectrum, nor whether it has positive (point) eigenvalues. In any case we have shown that the singular spectrum is a closed subset of R^1 with measure zero.

given by the plane waves $e^{i\xi\cdot x}$. But $\mathfrak{X}^* \supset L^6(R^3)$ does contain all eigenfunctions of U_1 in the form of spherical waves (products of spherical harmonics and radial functions), for they are $O(r^{-1})$ as $r \to \infty$. According to the results of section 4 (cf. Example 4.1), it follows that U_2 admits an eigenfunction expansion, with the eigenfunctions in L^6 representing distorted spherical waves*.

Remark 6.2. In Example 6.1 we can also use conditions $(\mathbf{6A}_j)$, $(\mathbf{6B}_j^{\pm})$. Without going into details we note that we can choose $\mathfrak{X} = \mathfrak{H} \cap \overline{\mathfrak{X}}$, where $\overline{\mathfrak{X}} = L^p \cap L^{\frac{6}{5}}$ with its natural topology, with any p such that $\frac{6}{5} \geqq p > 1$. (For the use of $L^{\frac{6}{5}}$ cf. HØEGH-KROHN [8]). $p = \frac{6}{5}$ is the simplest choice, but a smaller p gives a smaller $\mathfrak{X}$ and so a larger $\mathfrak{X}^*$ and would be more convenient for eigenfunction expansions. Since $\mathfrak{X} \supset L^6$, we see again that U_2 admits eigenfunction expansion in distorted spherical waves in L^6. Since $p = 1$ is not permitted, however, the possibility of the eigenfunction expansion for U_2 in distorted plane waves cannot be proved in this way. But the fact that p can be arbitrarily close to 1 makes such an expansion possible** under a slightly stronger assumption on q, namely,

$$q \in L^{\frac{3}{2}-\varepsilon} \cap L^{\frac{3}{2}+\varepsilon} \quad \text{for some} \quad \varepsilon > 0. \tag{6.8}$$

Details of the proof of these results will be given in a subsequent publication.

7. Time limits and the invariance principle

The purpose of this section is to prove that the $W^{\pm}$ constructed in Section 5 are on $\mathfrak{L}_{1,\,ac}(\Gamma)$ equal to the strong limits of $U_2^k U_1^{-k}$ as $k \to \pm\infty$, under the following additional assumption.

$(\mathbf{7A})$ $\mathfrak{X}$ is a pre-Hilbert space (independent of the structure of the Hilbert space $\mathfrak{H}$).

The result is an analog of the formula $W^{\pm} u = \lim_{t\to\pm\infty} e^{itH_2} e^{-itH_1} u$ for the selfadjoint case. Actually we can prove a stronger result, the *invariance* of the wave operators.

Theorem 7.1. *Let U_1, U_2 be unitary operators in $\mathfrak{H}$ and assume that conditions* $(\mathbf{5A}_2)$, $(\mathbf{5B}_2)$, $(\mathbf{5C}_1^+)$, $(\mathbf{5C}_1^-)$, *and* $(\mathbf{7A})$ *are satisfied, so that the wave operators $W^{\pm}$ exist by Theorem 5.1. Let γ be a real-valued function*

* Eigenfunctions in our general theory are rather abstract objects. Thus it is not clear whether the eigenfunctions we deduced for H_2 satisfy differential equations of the form $(-\Delta + q)\,\varphi = \lambda\,\varphi$, although this can be proved easily under somewhat stronger assumptions on q. The general case will be discussed in a later publication.

** See the preceding footnote.

on $[0, 2\pi)$ *such that*

$$\lim_{t\to\infty}\sum_{k=0}^{\infty}\left|\int_0^{2\pi} e^{-ik\theta-it\gamma(\theta)}\omega(\theta)\,d\theta\right|^2=0 \tag{7.1}$$

for every $\omega\in L^2(0, 2\pi)$. *Then*

$$\lim_{t\to\pm\infty} e^{it\gamma(E_2)}e^{-it\gamma(E_1)}u=W^{\pm}u,\qquad u\in\mathfrak{L}_{1,ac}(\Gamma). \tag{7.2}$$

Remark 7.2. Obviously $\gamma(\theta)=\theta$ satisfies (7.1), so that (7.2) implies $W^{\pm}u=\lim U_2^k U_1^{-k}u$. More generally, it is known that (7.1) is satisfied if γ is piecewise smooth with the derivative γ' positive and of bounded variation (for more precise statement see [15, 17]).

We now prove Theorem 7.1 in several steps.

Let $\overline{\mathfrak{X}}$ be the completion of the pre-Hilbert space $\mathfrak{X}$; the ideal elements of $\overline{\mathfrak{X}}$ are supposed to be outside $\mathfrak{H}$. We denote by $(,)_{\mathfrak{X}}$ and $\|\;\|_{\mathfrak{X}}$ the inner product and norm in $\overline{\mathfrak{X}}$. We continue to use $(,)$ and $\|\;\|$ for the ones in $\mathfrak{H}$ without subscript.

Proposition 7.3. For each $\theta\in\Gamma$ there is a unique $F_2(\theta)\in\mathscr{B}(\mathfrak{X}, \overline{\mathfrak{X}})$ such that

$$f_2(\theta;x,y)=(F_2(\theta)x,y)_{\mathfrak{X}}=(x,F_2(\theta)y)_{\mathfrak{X}},\qquad x,y\in\mathfrak{X}. \tag{7.3}$$

For each $x\in\mathfrak{X}$, $F_2(\cdot)x$ is weakly measurable on Γ. If $\overline{P}$ is an orthogonal projection of $\overline{\mathfrak{X}}$ onto a separable subspace, $\overline{P}F_2(\cdot)x$ is strongly measurable.

Proof. Obvious since the Hermitian form $f(\theta;\cdot,\cdot)$ is continuous on $\mathfrak{X}\times\mathfrak{X}$ for each $\theta\in\Gamma$ (Proposition 3.1) and since $f_2(\cdot;x,y)$ is measurable on Γ for each $x,y\in\mathfrak{X}$. $\overline{P}F_2(\cdot)x$ is weakly measurable and separably-valued, hence strongly measurable.

Let $x\in\mathfrak{X}$ be arbitrary but fixed. By ($\mathbf{5C}_1^{\pm}$), $G_n^{\pm}(\theta)x\in\mathfrak{X}$ are strongly measurable in $\theta\in\Gamma$. Hence there exists a separable, closed subspace $\mathfrak{Z}\subset\mathfrak{X}$ such that $G_n^{\pm}(\theta)x\in\mathfrak{Z}$ for a.e. $\theta\in\Gamma$ and $n=1,2,\ldots$. Let $\overline{\mathfrak{Z}}$ be the closure of $\mathfrak{Z}$ in $\overline{\mathfrak{X}}$, and let $\overline{P}$ be the orthogonal projection of $\overline{\mathfrak{X}}$ onto $\overline{\mathfrak{Z}}$.

Since $\{G_n^{\pm}(\theta)x\}$ is a Cauchy sequence in $\mathfrak{X}$ for each $\theta\in\Gamma$,

$$G^0(\theta)x=\lim_{n\to\infty}[G_n^+(\theta)x-G_n^-(\theta)x]\in\overline{\mathfrak{Z}} \tag{7.4}$$

exists for a.e. $\theta\in\Gamma$. We set $G^0(\theta)x=0$ if the limit is not in $\overline{\mathfrak{Z}}$. $G^0(\theta)x\in\overline{\mathfrak{Z}}$ is strongly measurable in θ. In particular $\|G^0(\theta)x\|_{\mathfrak{X}}$ is measurable in θ. We introduce the measurable sets

$$\Gamma_m=\{\theta\in\Gamma;\ \|G^0(\theta)x\|_{\mathfrak{X}}\leq m\},\qquad m>0. \tag{7.5}$$

Let another $y\in\mathfrak{X}$ be fixed. $\overline{P}F_2(\theta)y\in\overline{\mathfrak{Z}}$ is strongly measurable in $\theta\in\Gamma$ by Proposition 7.3. Hence $\|\overline{P}F_2(\theta)y\|_{\mathfrak{X}}$ is measurable in θ. We introduce the measurable sets

$$\Gamma'_m = \{\theta \in \Gamma;\ \|\bar{P} F_2(\theta)\, y\|_{\mathfrak{X}} \leqq m\}, \qquad m > 0. \tag{7.6}$$

Proposition 7.4. Let $\beta \in [\mathbf{B}]$ and $v = \beta(E_2) E_{2,ac}(\Gamma)\, y$. Assume further that β is supported on Γ'_m. Then for each $z \in \mathfrak{Z}$, $|\zeta| \neq 1$, $\zeta' = 1/\bar{\zeta}$,

$$(R_2(\zeta') z, v) = (z, g(\zeta; v))_{\mathfrak{X}}, \tag{7.7}$$

where

$$g(\zeta; v) = \int_0^{2\pi} \frac{\beta(\theta)}{1 - \zeta^{-1} e^{i\theta}} \bar{P} F_2(\theta)\, y\, d\theta \in \bar{\mathfrak{Z}}. \tag{7.8}$$

Proof. Since $\bar{P} F_2(\theta)\, y \in \bar{\mathfrak{Z}}$ is strongly measurable and bounded on Γ'_m while β is supported on Γ'_m and bounded, the integral in (7.8) makes sense and $g(\cdot\,; v)$ is a $\bar{\mathfrak{Z}}$-valued analytic function for $|\zeta| \lessgtr 1$ belonging to the Hardy class H^2. To prove (7.7), therefore, it suffices to observe that

$$(R_2(\zeta') z, v) = \int_0^{2\pi} \frac{1}{1 - \zeta' e^{-i\theta}} (d/d\theta)\, (E_{2,ac}(\theta) z, v)\, d\theta$$

because $v \in \mathfrak{H}_{2,ac}$, and that $(d/d\theta)\, (E_{2,ac}(\theta) z, v) = \overline{\beta(\theta)}\, (d/d\theta)$ $(E_{2,ac}(\theta) z, y) = \overline{\beta(\theta)} f_2(\theta; z, y) = \overline{\beta(\theta)}\, (z, F_2(\theta)\, y)_{\mathfrak{X}} = \overline{\beta(\theta)}\, (z, \bar{P} F_2(\theta)\, y)_{\mathfrak{X}}$.

Note that $g(\cdot\,; v)$ is determined by v, independently of the particular expression $v = \beta(E_2) E_{2,ac}(\Gamma)\, y$, as is easily seen from (7.7), which is true for all $z \in \mathfrak{Z}$.

Since $g(\cdot\,; v)$ is in the Hardy class, the boundary values $g^{\pm}(\theta; v) = \lim_{r \uparrow\downarrow 1} g(r e^{i\theta}; v)$ exist, with $g^{\pm}(\cdot\,; v) \in L^2(0, 2\pi; \bar{\mathfrak{Z}})$, in the sense of pointwise convergence a.e. as well as in the sense of L^2-convergence.

Proposition 7.5. If y satisfies (7.1), then

$$\int_0^{2\pi} \|g^{\pm}(\theta; e^{-it\gamma(E_2)} v)\|^2_{\mathfrak{X}}\, d\theta \to 0, \qquad t \to \pm\infty. \tag{7.9}$$

Proof. Consider g^+. Expanding the integrand in (7.8) into the power series in ζ and going to the limit $r \uparrow 1$, we see easily that

$$\int_0^{2\pi} \|g^+(\theta; v)\|^2_{\mathfrak{X}}\, d\theta = 2\pi \sum_{k=1}^{\infty} \left\| \int_0^{2\pi} e^{-ik\theta} \beta(\theta) \bar{P} F_2(\theta)\, y\ d\theta \right\|^2. \tag{7.10}$$

Replacing v by $e^{-it\gamma(E_2)} v$ in (7.10) is equivalent to multiplying $\beta(\theta)$ by $e^{-it\gamma(\theta)}$. Thus the required result follows from Lemma 7.6 below. In the case of g^-, we have an expression similar to (7.10), with $e^{-ik\theta}$ replaced by $e^{ik\theta}$ and with summation ranging from $k = 0$ to ∞, from which the required result follows. Note that (7.1) is unchanged by replacing

$-i$ by i and letting $t \to -\infty$, for this amounts to taking the complex conjugate of the expression in $|\ |$.

Lemma 7.6. *Let γ satisfy (7.1). Then (7.1) is also true for ω replaced by any element of $L^2(0, 2\pi; \mathfrak{K})$, where $\mathfrak{K}$ is any Hilbert space, if $|\ |$ is replaced by $\|\ \|_{\mathfrak{K}}$.*

Proof. It is known (see [17]) that (7.1) is true for all $\omega \in L^2(0, 2\pi)$ if it is true for all ω in a fundamental subset of $L^2(0, 2\pi)$. The same result is seen to hold in the case of vector-valued functions. Thus it suffices to prove (7.1) for ω replaced by functions of the form ωw, where $\omega \in L^2(0, 2\pi)$ and w is a fixed vector. But this is obvious by (7.1) itself.

Proposition 7.7. Let v be as above. Let $u = \alpha(E_1) E_{1,ac}(\Gamma) x \in \mathfrak{L}_{1,ac}(\Gamma)$, where $\alpha \in [\mathbf{B}]$ is supported on Γ_m. Then

$$((W^{\pm} - 1) u, v) = -\frac{1}{2\pi} \int_0^{2\pi} \alpha(\theta) \left(G^0(\theta) x, g^{\pm}(\theta; v)\right) d\theta. \qquad (7.11)$$

Proof. Using the relation $R_1(\zeta) = R_2(\zeta) G(\zeta)$, we have for $\zeta = r e^{i\theta}$, $r \neq 1, \zeta' = 1/\bar{\zeta}$,

$$\begin{aligned} 2\pi\, \delta_r(E_1; \theta) &= R_1(\zeta) - R_1(\zeta') \\ &= [R_2(\zeta) - R_2(\zeta')] G(\zeta) + R_2(\zeta') [G(\zeta) - G(\zeta')] \\ &= 2\pi\, \delta_r(E_2; \theta) G(\zeta) + R_2(\zeta') [G(\zeta) - G(\zeta')]. \end{aligned}$$

Setting $r = r_n^{\pm 1}$ and using (7.7), we obtain

$$\begin{aligned} (\delta_{r_n}(E_1; \theta) x; v) &= (\delta_{r_n}(E_2; \theta) G_n^{\pm}(\theta) x, v) \\ &\quad + \frac{1}{2\pi} (G_n^{+}(\theta) x - G_n^{-}(\theta) x, g(r_n^{\pm 1} e^{i\theta}; v))_{\mathfrak{X}}; \end{aligned} \qquad (7.12)$$

note that $\delta_{r^{-1}} = -\delta_r$.

Let $n \to \infty$ in (7.12). The left member tends to $(d/d\theta)\left(E_{1,ac}(\theta) x, v\right)$ a.e. The first term on the right tends a.e. to $\phi^{\pm}(\theta)$, where ϕ^{+} is given by the ϕ of (5.15) and ϕ^{-} is defined in a similar way, at least along a certain subsequence of $\{n\}$ (Proposition 5.10). The third term tends to $\frac{1}{2\pi} \left(G^0(\theta) x, g^{\pm}(\theta; v)\right)_{\mathfrak{X}}$ a.e.

If we multiply the result by $\alpha(\theta)$ and integrate on $(0, 2\pi)$, we obtain (7.11). Note that $u \in \mathfrak{H}_{1,ac}$ and $(W^{\pm} u, v) = \left(W^{\pm} \alpha(E_1) x, v\right) = \left(W^{\pm} x, \alpha(E_2)^* v\right) = \left(W^{\pm} x, \alpha(E_2)^* \beta(E_2) y\right) = \int \alpha(\theta) \phi^{\pm}(\theta) d\theta$, as is seen from Proposition 5.8 applied with β replaced by $\bar{\alpha}\beta$.

Proposition 7.8. Let u, v be as above. Then

$$\left((W^{\pm} - 1) e^{-it\gamma(E_1)} u, e^{-it\gamma(E_2)} v\right) \to 0, \quad t \to \pm\infty. \qquad (7.13)$$

Proof. In (7.11) replace u, v by $e^{-it\gamma(E_1)}u$, $e^{-it\gamma(E_2)}v$, respectively. Then $\alpha(\theta), \beta(\theta)$ should be replaced by $e^{-it\gamma(\theta)}\alpha(\theta)$, $e^{-it\gamma(\theta)}\beta(\theta)$, respectively. In view of (7.9) and the Schwarz inequality, (7.13) follows easily if one notices that $\int|\alpha(\theta)|^2\|G^0(\theta)x\|_{\mathfrak{X}}^2 d\theta<\infty$ because $\|G^0(\theta)x\|_{\mathfrak{X}}$ is bounded on Γ_m while $\alpha(\theta)$ is bounded and supported on Γ_m.

Proposition 7.9. (7.13) is true for any $u\in\mathfrak{L}_{1,ac}(\Gamma)$ and $v\in\mathfrak{L}_{2,ac}(\Gamma)$.

Proof. Since all the operators in (7.13) are uniformly bounded, it suffices to prove it for all u, v of the form $u=\alpha(E_1)E_{1,ac}(\Gamma)x$ and $v=\beta(E_2)E_{2,ac}(\Gamma)y$, where $x, y\in\mathfrak{X}$ and $\alpha, \beta\in[\mathbf{B}]$. Here we may assume further that α, β are supported on Γ_m, Γ'_m, respectively, for some $m>0$, for the measures of $\Gamma-\Gamma_m$ and $\Gamma-\Gamma'_m$ tend to zero as $m\to\infty$. But then (7.13) is true by proposition 7.8.

Now we can complete the proof of the theorem. Set $W(t)=e^{it\gamma(E_2)}e^{-it\gamma(E_1)}$. Since $W^\pm$ is isometric on $\mathfrak{L}_{1,ac}(\Gamma)$ and $W(t)$ is unitary, we have for each $u\in\mathfrak{L}_{1,ac}(\Gamma)$

$$\begin{aligned}\|W^\pm u-W(t)u\|^2&=2\,Re\left(W^\pm u-W(t)u,\ W^\pm u\right)\\&=2\,Re\left((W^\pm-1)e^{-it\gamma(E_1)}u,\ e^{-it\gamma(E_2)}W^\pm u\right)\to 0\end{aligned}$$

as $t\to\pm\infty$, where the intertwining property of $W^\pm$ and Proposition 7.9 are used (note that $W^\pm u\in\mathfrak{L}_{2,ac}(\Gamma)$).

Appendix

We collect here some lemmas on measure theory and linear topological spaces, which were used in the text.

Lemma A. *For $j=1, 2, \ldots, m$, let $\{\phi^j_{nk}\}$, $n, k=1, 2, \ldots$, be a double sequence of nonnegative real functions on a σ-finite measure space S^j. If $\limsup_{k\to\infty}\phi^j_{nk}(s)=\phi^j_n(s)$ for a.e. $s\in S^j$ for each j and n and if $\lim_{n\to\infty}\phi^j_n(s)=0$ for a.e. $s\in S^j$ for each j, there exist sequences $\{n_p\}$, $\{k_p\}$ of positive integers such that $\lim_{p\to\infty}\phi^j_{n_pk_p}(s)=0$ for a.e. $s\in S^j$ for all $j=1, 2, \ldots, m$.*

The proof depends on elementary properties of the measure and may be omitted.

Corollary B. *For each $j=1, 2, \ldots, m$, let $\{f^j_{nk}\}$ be a double sequence of functions on a σ-finite measure space S^j to a metric space $\mathfrak{X}^j$. If $\lim_{k\to\infty}f^j_{nk}(s)=f^j_n(s)$ exists for a.e. $s\in S^j$ for each j and n and if $\lim_{n\to\infty}f^j_n(s)=f^j(s)$ exists for a.e. $s\in S^j$ for each j, then there are sequences $\{n_p\}$, $\{k_p\}$ of positive integers such that $\lim_{p\to\infty}f^j_{n_pk_p}(s)=f^j(s)$ for a.e. $s\in S^j$ for all j.*

Proof. Denoting by d^j the metric in $\mathfrak{X}^i$, we set $\phi^j_{nk}(s)=d^j(f^j_{nk}(s), f^j(s))$, $\phi^j_n(s)=d^j(f^j_n(s), f^j(s))$ and apply Lemma A.

Lemma C. *Let $\mathfrak{X}$, $\mathfrak{Y}$ be linear topological spaces. Let $\{A_n\}$ be an equicontinuous sequence of linear operators on $\mathfrak{X}$ to $\mathfrak{Y}$ such that* $\lim A_n x = 0$ *for each $x \in \mathfrak{X}$. If $\{x_n\}$ is a Cauchy sequence in $\mathfrak{X}$, then* $\lim A_n x_n = 0$.

Proof. Let $\mathfrak{V}$ be an arbitrary neighborhood of 0 in $\mathfrak{Y}$. Since the A_n are equicontinuous, there is a neighborhood $\mathfrak{U}$ of 0 in $\mathfrak{X}$ such that $A_n \mathfrak{U} \subset \mathfrak{V}$ for all n. Since $\{x_n\}$ is Cauchy, there is p such that $m, n \geqq p$ implies $x_n - x_m \in \mathfrak{U}$. Since $\lim A_n x_p = 0$ as $n \to \infty$, there is $q \geqq p$ such that $n \geqq q$ implies $A_n x_p \in \mathfrak{V}$. If $n \geqq q$, we have then $A_n x_n = A_n(x_n - x_p) + A_n x_p \in \mathfrak{V} + \mathfrak{V}$. Since for any neighborhood $\mathfrak{W}$ of 0 in $\mathfrak{Y}$ there is one $\mathfrak{V}$ such that $\mathfrak{V} + \mathfrak{V} \subset \mathfrak{W}$, the lemma was proved.

Lemma D. *Let $\mathfrak{X}$ be a linear topological space, and let $\{f_n\}$ be an equicontinuous sequence of nonnegative Hermitian forms on $\mathfrak{X} \times \mathfrak{X}$. Furthermore, let* $\lim f_n(x, y) = f(x, y)$ *exist for each $x, y \in \mathfrak{X}$. Then f is continuous on $\mathfrak{X} \times \mathfrak{X}$. If $\{x_n\}$, $\{y_n\}$ are Cauchy sequences in $\mathfrak{X}$, then $\{f_n(x_n, y_n)\}$ and $\{f(x_n, y_n)\}$ are Cauchy and $f_n(x_n, y_n) - f(x_n, y_n) \to 0$.*

Proof. It is obvious that f is a continuous, nonnegative Hermitian form on $\mathfrak{X} \times \mathfrak{X}$. We write $f(x) = f(x, x) \geqq 0$, $g(x) = f(x)^{\frac{1}{2}}$ and similarly define g_n. g is continuous on $\mathfrak{X}$ and, in virtue of the Minkowsky inequality $|g(x) - g(y)| \leqq g(x - y)$, g is uniformly continuous on $\mathfrak{X}$. It follows that if $\{x_n\}$ is a Cauchy sequence in $\mathfrak{X}$, $\{g(x_n)\}$ is a Cauchy sequence of nonnegative numbers.

Since $\{g_n\}$ is equicontinuous, there is for any $\varepsilon > 0$ a neighborhood $\mathfrak{U}$ of 0 in $\mathfrak{X}$ such that $x \in \mathfrak{U}$ implies $g_n(x) \leqq \varepsilon$ for all n. In view of the Minkowsky inequality for g_n, then, $x - y \in \mathfrak{U}$ implies $|g_n(x) - g_n(y)| \leqq \varepsilon$ for all n and hence $|g(x) - g(y)| \leqq \varepsilon$ too. Since $\{x_n\}$ is Cauchy, there is p such that $n \geqq p$ implies $x_n - x_p \in \mathfrak{U}$. Hence $|g_n(x_n) - g_n(x_p)| \leqq \varepsilon$ and $|g(x_n) - g(x_p)| \leqq \varepsilon$ for $n \geqq p$, so that $|g_n(x_n) - g(x_n)| \leqq 2\varepsilon + |g_n(x_p) - g(x_p)|$. Since $g_n(x_p) - g(x_p) \to 0$ as $n \to \infty$, it follows that $\limsup |g_n(x_n) - g(x_n)| \leqq 2\varepsilon$. This shows that $g_n(x_n) - g(x_n) \to 0$.

Since $\{g(x_n)\}$ is Cauchy as noted above, it follows that $\{g_n(x_n)\}$ is also Cauchy. In particular these sequences are bounded. Hence $f_n(x_n) - f(x_n) = g_n(x_n)^2 - g(x_n)^2 = [g_n(x_n) - g(x_n)]\,[g_n(x_n) + g(x_n)] \to 0$. Also $f_n(x_n) = g_n(x_n)^2$ and $f(x_n) = g(x_n)^2$ form Cauchy sequences. This proves the lemma when $y_n = x_n$. The general case can be dealt with by polarization.

Lemma E. *Let $\mathfrak{X}$, $\tilde{\mathfrak{X}}$ be linear topological spaces and let $J: \mathfrak{X} \to \tilde{\mathfrak{X}}$ be a continuous linear operator. Let $\{x_n\}$ be a Cauchy sequence in $\mathfrak{X}$ and let $\lim\limits_{n\to\infty} J x_n = \tilde{x} \in \tilde{\mathfrak{X}}$. Let $\{f_n\}$ be an equicontinuous sequence of linear forms on $\mathfrak{X}$ and $\tilde{f}$ a continuous linear form on $\tilde{\mathfrak{X}}$ such that $\lim\limits_{n\to\infty} f_n(x) = \tilde{f}(J x)$ for each $x \in \mathfrak{X}$. Then* $\lim f_n(x_n) = \tilde{f}(\tilde{x})$.

Proof. We have $\tilde{f}(\tilde{x}) - f_n(x_n) = \tilde{f}(\tilde{x} - J x_n) + g_n(x_n)$, where $g_n(x) = \tilde{f}(J x) - f_n(x)$, $x \in \mathfrak{X}$. But $\tilde{f}(\tilde{x} - J x_n) \to 0$ since $\tilde{x} - J x_n \to 0$ and $\tilde{f}$ is continuous. Also $g_n(x_n) \to 0$ since $\{g_n\}$ is equicontinuous, $g_n(x) \to 0$ for each $x \in \mathfrak{X}$, and $\{x_n\}$ is Cauchy (see Lemma C).

References

1. BIRMAN, M. SH.: A local test of the existence of wave operators. Dokl. Akad. Nauk SSSR **159**, 485—488 (1964) [Russian].
2. —, and S. B. ENTINA: Stationary approach in the abstract theory of scattering. Dokl. Akad. Nauk SSSR **155**, 506—508 (1964); Izv. Akad. Nauk SSSR **31**, 401—430 (1967) [Russian].
3. FADDEEV, L. D.: On the Friedrichs model in the perturbation theory of continuous spectrum. Trudy Mat. Inst. Steklov. **73**, 292—313 (1964) [Russian].
4. FRIEDRICHS, K. O.: Über die Spektralzerlegung eines Integraloperators. Math. Ann. **115**, 249—272 (1938).
5. — On the perturbation of continuous spectra, Comm. Pure Appl. Math. **1**, 361—406 (1948).
6. — Perturbation of spectra in Hilbert space. Am. Math. Soc. Lectures in Appl. Math. vol. 3 (1965).
7. GELFAND, I. M., and G. E. SHILOV: Generalized functions, vol. 3, English translation. Academic Press 1967.
7a. —, and N. YA. VILENKIN: Generalized functions, vol. 4, English translation. Academic Press 1964.
8. HØEGH-KROHN, J. R.: Partly gentle perturbation with application to perturbation by annihilation-creation operators. Proc. Nat. Acad. Sci. US **58**, 2189—2192 (1967).
9. HOWLAND, J. S.: Banach space techniques in the perturbation theory of self-adjoint operators with continuous spectra, J. Math. Anal. Appl. **20**, 22—47 (1967).
10. — A perturbation-theoretic approach to eigenfunction expansions. J. Functional Analysis **2** ,1—23 (1968).
11. IKEBE, T.: Eigenfunction expansions associated with the Schroedinger operators and their applications to scattering theory. Arch. Rational Mech. Anal. **5**, 1—34 (1960).
12. JAUCH, J. M.: Theory of the scattering operator. Helv. Phys. Acta **31**, 127—158 (1958).
13. KATO, T.: On finite-dimensional perturbation of selfadjoint operators. J. Math. Soc. Japan **9**, 239—249 (1957).
14. — Perturbation of continuous spectra by trace class operators, Proc. Japan Acad. **33**, 260—264 (1957).
15. — Wave operators and unitary equivalence. Pacific J. Math. **15**, 171—180 (1965).
16. — Wave operators and similarity for non-selfadjoint operators. Math. Ann. **162**, 258—279 (1966).
17. — Perturbation theory for linear operators. Berlin-Heidelberg-New York: Springer 1966.
18. KURODA, S. T.: On the existence and the unitary property of the scattering operator. Nuovo Cimento **12**, 431—454 (1959).

19. Kuroda, S. T.: Finite-dimensional perturbation and a representation of scattering operator. Pacific J. Math. **13**, 1305—1318 (1963).
20. — On a stationary approach to scattering problem. Bull. Am. Math. Soc. **70**, 556—560 (1964).
21. — Stationary methods in the theory of scattering, Proceedings of the seminar on: Perturbation theory and its applications in quantum mechanics, pp. 185—214. Wiley 1966.
22. — An abstract stationary approach to perturbation of continuous spectra and scattering theory. J. d'Analyse Math. **20**, 57—117 (1967).
23. — Perturbation of eigenfunction expansions, Proc. Nat. Acad. Sci. US **57**, 1213—1217 (1967).
24. Rejto, P. A.: On gentle perturbations, I. Comm. Pure Appl. Math. **16**, 279—303 (1963).
25. — On gentle perturbations, II. Comm. Pure Appl. Math. **17**, 257—292 (1964).
26. — On partly gentle perturbations, I. J. Math. Anal. Appl. **17**, 453—462 (1967).
27. — On partly gentle perturbations, II. J. Math. Anal. Appl. **20**, 145—187 (1967).

Induced Representations of Locally Compact Groups and Applications

By George W. Mackey

Harvard University, Cambridge

1. Introduction

Roughly the first half of this paper is an introductory exposition of some of the main ideas of the theory of induced representations with special emphasis on the influence of the work of M. H. Stone. The second half deals with applications and with certain extensions and refinements of known results demanded by these applications.

Specifically it is shown in Section 9 that it is possible to obtain very concrete and explicit information about the structure of an induced representation when the group G is a semi direct product $N \circledS H$ with N commutative and the subgroup G_0 is of the form $N_0 \circledS H_0$ with $N_0 \subseteq N$ and $H_0 \subseteq H$. As explained in Section 10 this result has applications to the study of the „energy bands" of solid state physics. In Section 11 similar results are obtained; (a) about the restriction of an irreducible representation of $N \circledS H$ to $N_0 \circledS H_0$ and (b) about the tensor product of two irreducible representations of $N \circledS H$. These differ from the analogous results in [19] in that our decomposition is now into *irreducible* representations. These new results are corollaries of the theorem of Section 9 and the results of [19]. Some physical applicat ons of the tensor product result are briefly indicated.

Sections 12 and 13 are again expository and explain briefly how the theory of the first eight sections can be generalized so as to apply to projective representations. Section 14 contains a theorem about the structure of induced projective representations of commutative groups which specializes to one needed in a variant of Cartier's recent approach to the theory of theta functions [6]. The section begins with an account of this variant of Cartier's theory. The final section describes certain aspects of the relationship between induced projective representations and the theory of automorphic forms with emphasis on the close parallel between this theory and the theory of theta functions.

2. Some basic definitions

Let S be a space in which a σ field of „Borel sets" has been singled out and let μ be a (countably additive) measure defined on all Borel sets. We assume the Borel structure of S to be such that the Hilbert space $\mathscr{L}^2(S, \mu)$ is separable. Now suppose that a separable locally compact group G „acts" on S in the sense that for each $x \in G$, $s \to s x$ is a one to one map of S onto S and $(s x_1) x_2 = s x_1 x_2$ for all $s \in S$ and all x_1 and x_2 in G. We suppose that $s, x \to s x$ is a Borel function and that μ is G invariant in the sense that $\mu(E x) = \mu(E)$ for all x in G and all Borel subsets E of S. Then for each $x \in G$ we obtain a unitary operator U_x acting in the Hilbert space $\mathscr{L}^2(S, \mu)$ by setting $U_x(f)(s) = f(s x)$. It can be shown that $x \to U_x$ is a homomorphism of G into the group of all unitary operators in $\mathscr{L}^2(S, \mu)$ and that $x \to U_x(f)$ is a continuous function from G to $\mathscr{L}^2(S, \mu)$ for all f in $\mathscr{L}^2(S, \mu)$. As such U is what is called a (strongly continuous) *unitary representation* of G.

We may also assign an operator in $\mathscr{L}^2(S, \mu)$ to each Borel subset E of S. We denote by P_E the projection operator which takes each f in $\mathscr{L}^2(S, \mu)$ into the function f' which is zero in $S - E$ and coincides with f in E. The mapping $E \to P_E$ from Borel sets into projections is a countably additive homomorphism of the Boolean algebra of all Borel subsets of S into a Boolean algebra of projection operators. Specifically, it has the properties:

(i) $P_{E \cap F} = P_E P_F$

(ii) $P_S = I$, $P_0 = 0$

(iii) $P_{E_1 \cup E_2 \cup \cdots} = P_{E_1} + P_{E_2} + \cdots$

wherever E, F and the E_j are Borel subsets of S and $E_i \cap E_j = 0$ for $i \neq j$. As such it is what we shall call a *projection valued measure.*

The unitary representation U of G and the projection valued measure P on S satisfy a simple and obvious identity which will be fundamental in what follows. It is

$$U_x^{-1} P_E U_x = P_{[E]x} \tag{1}$$

for all x in G and all Borel subsets E of S. By a *system of imprimitivity* for an arbitrary unitary representation U of G we shall mean a projection valued measure P defined on a G space S which satisfies (1). Neither U nor P need be of the concrete form described above.

3. Unitary group representations and Stone's theorem

The first person to have made a serious study of the infinite dimensional unitary representations of a non compact locally compact group seems to have been M. H. Stone. In a note [34] published in

1930 he announced a theorem — now known as STONE's theorem — which sets up a natural one-to-one correspondence between the unitary representations of the additive group of the real line and the projection valued measures on the real line. If P is a projection valued measure and f is a vector in the underlying Hilbert space then $E \to (P_E(f), f)$ is a real valued measure α_f and one may form $\int e^{ixt} d\alpha_f(t)$ for all x. The correspondence in question is uniquely characterized by the fact that $(U_x(f), f) = \int e^{ixt} d\alpha_f(t)$ is an identity in x for all f. Here U is the unitary representation corresponding to the projection valued measure P.

Since the celebrated spectral theorem* sets up a similar correspondence between self-adjoint operators and projection valued measures on the line, one has also a natural one-to-one correspondence between unitary representations of the real line U and self-adjoint operators H. This correspondence is such that $U_t = e^{-itH}$ and reduces most problems about unitary representations of the additive group of the real line to problems about self-adjoint operators. In particular, the problem of classifying all possible unitary representations of the additive group of the real line is reduced to the corresponding problem about self-adjoint operators. Hence, it is completely solved by the Hahn-Hellinger spectral multiplicity theory as abstracted and generalized** to unbounded operators in Chapter VII of STONE's now classic treatise [35].

This early work of Stone had several important applications almost immediately and in addition was influential in more than one way in the later development of the theory of unitary group representations. Two of these applications are to quantum mechanics and are mentioned in the paper [34]. First of all it follows from general considerations that the time evolution of a quantum mechanical system is given by a unitary representation of the line and STONE's theorem provides a rigorous proof that the corresponding differential equation (SCHRÖDINGER's equation) must take the form $\frac{\partial \psi}{\partial t} = -iH\psi$ where H is a suitable self-adjoint operator. Secondly, STONE's theorem makes it possible to replace the HEISENBERG commutation relations for the self-adjoint operators defining the position and momentum observables by analogous commutation relations for the corresponding unitary representations of the real line. In this form it is possible to prove rigorously that these commutation relations have a unique irreducible solution. That this is so was announced in [34] and proved later by VON NEUMANN in [26]. It is often called the STONE-VON NEUMANN uniqueness theorem. We shall have more to say about this result below.

* Stone himself gave one of the earlier proofs of this theorem. Cf. Chapter V of [35].

** See HALMOS [10] for a recent treatment.

A third important application of STONE's theorem was made by KOOPMAN in 1931 [12]. He made the (then novel) observation that a constant energy hyper-surface in the phase space of a classical dynamical system may be taken as the S in the general construction of Section 2. The additive group of the real line acts on the constant energy hypersurface through the time development of the system and the Liouville measure is invariant under the action. Thus one obtains a unitary representation of the additive group of the real line and hence via STONE's theorem a classification of dynamical systems. A few months later this application of STONE's theorem by KOOPMAN led directly to the ergodic theorems of VON NEUMANN [27] and BIRKHOFF [4] and thus to the creation of the subject of ergodic theory.

In 1943 and 1944, NEUMARK [29], AMBROSE [1] and GODEMENT [9] all independently discovered that STONE's theorem could be generalized to arbitrary separable* locally compact commutative groups G. Introducing the so called dual group $\hat{G}$, that is, the locally compact commutative group of all continuous homomorphisms of G into the group K of all complex numbers of modulus one, they showed that there is a natural one-to-one correspondence between the unitary representations U of G and the projection valued measures P on $\hat{G}$ such that $(U_x(f), f) = \int \chi(x)\, d\,\alpha_f(\chi)$ where α_f is the measure $E \to (P_E(f), f)$, $x \in G$, E is an arbitrary Borel subset of $\hat{G}$ and f is an arbitrary element in the appropriate Hilbert space. Since the continuous homomorphisms of the additive group of the line into K are just the functions $x \to e^{itx}$ for real t it is clear that this theorem reduces to STONE's when G is the additive group of the line. Not long afterwards, it was recognized that the HAHN-HELLINGER theory, as generalized by STONE, could be applied to give a complete classification of the unitary representations of an arbitrary separable locally compact commutative group G. Indeed, by the theorem of NEUMARK, AMBROSE, and GODEMENT cited above, it suffices to classify the projection valued measures on $\hat{G}$ and in this latter problem, the group structure of $\hat{G}$ is irrelevant. The analysis for the case in which $\hat{G}$ is the additive group of the real line extends word for word to the general case. Actually, whenever G is not compact, there exists a one-to-one Borel preserving map of $\hat{G}$ on the additive group R^+ of the real line — so that one can in fact deduce a classification of projection valued measures in $\hat{G}$ from that of projection valued measures in R^+. Of course, the classification problem is trivial when G is compact.

* They actually treated the non separable case but we shall be interested only in the separable result.

The elements of $\hat{G}$ of course correspond one-to-one to the one-dimensional unitary representations of G and hence to those unitary representations of G which are *irreducible* in the sense of there being no proper closed invariant subspaces of the underlying Hilbert space. One-dimensional representations are obviously irreducible and for commutative groups it is not difficult to show that every irreducible unitary representation is one-dimensional. In stating that the generalized Hahn-Hellinger theory classifies the unitary representations of G we are of course tacitly assuming that $\hat{G}$ is known; i.e., we classify all unitary representations assuming that the irreducible ones are known.

While the problem of finding $\hat{G}$ is easily solved for most of the interesting commutative groups G the analogous problem of finding all (equivalence classes of) irreducible unitary representations is one of the main problems of the theory when G is non-commutative. On the other hand, more or less independently of solving this difficult problem, one can ask to what extent one can find an analogue of the Hahn-Hellinger theory which classifies all unitary representations once the irreducible ones have been found. It is a useful, interesting, and rather surprising fact that an almost perfect analogue exists — provided that G belongs to the important but by no means exhaustive class of separable locally compact groups known as the type I groups. We shall not give details, but content ourselves with the remark that it is based upon the von Neumann direct integral theory [28] as applied to group representations by MAUTNER [25] and upon a natural Borel structure in the space $\hat{G}$ of all equivalence classes of irreducible unitary representations of G. A description will be found in [7], [20], [21], and [23].

4. The Stone-von Neumann uniqueness theorem

Let us look more closely at the STONE-VON NEUMANN uniqueness theorem. In the one-dimensional case, it says that (up to unitary equivalence) there is just one pair U, V of unitary representations of the real line which satisfies the identity

$$U_t V_s = V_s U_t e^{its} \tag{2'}$$

for all real t and s, and is irreducible in the sense that the underlying Hilbert space admits no proper closed subspaces which are invariant under all U_s and V_t. Actually, as shown by VON NEUMANN, every solution of (2′) is a discrete direct sum of irreducible solutions. Now, for any fixed s, $t \to e^{ist}$ is a character of the additive group of the real line and as s varies, we get every character. This suggests that (2′) may be generalized to

$$U_t V_\chi = V_\chi U_t \chi(t) \tag{2}$$

for all $t \in G$ and all $\chi \in \hat{G}$ where G is a locally compact commutative group, $\hat{G}$ is its dual group, and U and V are unitary representations of G and $\hat{G}$ respectively. Indeed, when G is the additive group of an n-dimensional vector space, the most general member of $\hat{G}$ is

$$t_1, t_2 \cdots t_n \to e^{i(t_1 s_1 + \cdots + t_n s_n)}$$

where $s_1 \ldots s_n$ is an n-tuple of real numbers and the n-dimensional version of the Stone-von Neumann theorem says that (2) has a unique irreducible solution for this choice of G.

It is of course natural to wonder whether (2) has a unique irreducible solution for any locally compact commutative group G. This question occured to me in 1948 when I read [34] for the first time and I was pleased to discover that certain techniques I had learned from generalizing the results of [8] and studying the typescript of [28] were directly applicable and yielded a proof of uniqueness [16] for any separable G. These techniques used the Hahn-Hellinger theory in an essential way as well as the Neumark-Ambrose-Godement generalization of STONE's theorem. The latter theorem of course makes it possible to replace the representation V of $\hat{G}$ by a projection valued measure on $\hat{\hat{G}}$. On the other hand, the Pontryagin-von Kampen duality theorem allows us to identify $\hat{\hat{G}}$ with G and thus to replace V by a projection valued measure P on G. Moreover, an easy calculation shows that U and V satisfy the identity (2) if and only if U and P satisfy

$$U_x P_E U_x^{-1} = P_{[E]x^{-1}} \tag{3}$$

for all x in G and all Borel subsets E of G.

But (3) is just (1) of Section 2 in the special case in which $S = G$, G is commutative and the action of G on $S (= G)$ is right multiplication. In other words, to say that U and V satisfy (2) is the same as to say that P is a system of imprimitivity for U based on G and the (generalized) Stone-von Neumann uniquess theorem may be reformulated to read as follows: With S and G as above, there is to within unitary equivalence a unique irreducible pair U, P where U is a unitary representation of G and P is a system of imprimitivity for U based on G. So stated, the theorem is meaningful for any separable* locally compact group G whether commutative or not, and this much more general theorem turns out to be true [16].

Thus, the Stone-von Neumann uniqueness theorem which seems quite special in its original formulation is in fact just the specialization to the real line of a general theorem about arbitrary locally compact groups.

* See LOOMIS [15] for an extension to the non separable case.

5. The imprimitivity theorem and induced representations

When formulated in terms of systems of imprimitivity, the Stone-von Neumann uniqueness theorem continues to be meaningful not only when we cease to assume commutativity for G but also when we drop the assumption that $S=G$. Given a separable locally compact group G and a Borel space S on which G acts as in Section 2, one can ask about the possible pairs U, P where U is a unitary representation of G and P is a system of imprimitivity for U based on S and whether any two such pairs are necessarily unitarily equivalent. In the special case in which the action of G on S is transitive, these questions can be answered rather completely. Let s_0 be a fixed origin in S and let H be the subgroup consisting of all x in G with $s_0 x = s_0$. Then $x \to s_0 x$ sets up a one-to-one correspondence between S and the set G/H of all right H cosets in G. Moreover, the action of G on S is carried by this correspondence into the action of G on G/H defined by the equation

$$(H y)\, x = H(y x)\,.$$

Finally, whenever S satisfies mild regularity conditions, it can be shown that H is closed and that $x \to s_0 x$ is a Borel isomorphism. In other words, we may assume that S is a coset space G/H. For each fixed G and $S=G/H$, the *imprimitivity theorem* [17], [20], [22] sets up a natural one-to-one correspondence between the equivalence classes of pairs U, P which satisfy (1) and the equivalence classes of unitary representations of the subgroup H. This correspondence commutes with the taking of direct sums and hence preserves irreducibility. Thus, when $S=G/H$, there are as many inequivalent pairs satisfying (1) as there are inequivalent irreducible unitary representations of H. The case in which $S=G$ is of course that in which $H=\{e\}$ so that there is a unique irreducible unitary representation of H. The imprimitivity theorem thus implies the non-commutative generalized Stone-von Neumann uniqueness theorem, and explains why it usually fails when $S \neq G$.

The construction of the pair U, P associated with a given unitary representation L of H is a more or less natural generalization of the construction given in Section 2. If G acts transitively on S in Section 2, then we may identify S with G/H for some closed subgroup H of G and functions on S may be identified with functions on G which satisfy the identity

$$f(\xi x) = f(x) \tag{4}$$

for all $\xi \in H$ and all $x \in G$. Given a unitary representation L of H let us replace complex valued functions on G by functions on G with values in the Hilbert space $\mathcal{H}(L)$ in which the operators L_x act. Then it is natural to replace the identity (4) by

$$f(\xi x) = L_\xi(f(x)) \tag{5}$$

Given a Borel function satisfying (5) let us consider the real valued function

$$x \rightarrow (f(x), f(x))$$

obtained by taking its scalar product with itself. We have

$$(f(\xi x), f(\xi x)) = (L_\xi(f(x)), L_\xi(f(x))) = (f(x), f(x))$$

by the unitarity of the L_ξ. Hence $x \rightarrow (f(x), f(x))$ is a constant on the right H cosets and so may be identified with a function on $S = G/H$. Hence we may consider its integral over S with respect to μ and ask whether or not it is finite. We define $\mathcal{F}^L$ to be the space of all Borel functions from G to $\mathcal{H}(L)$ which satisfy (5) and have

$$\int_{G/H} (f(x), f(x))\, d\mu < \infty .$$

Identifying functions equal almost everywhere we obtain a Hilbert space with

$$\|f\|^2 = \int_{G/H} (f(x), f(x))\, d\mu .$$

For each fixed y let $U_y^L(f)(x) = f(xy)$. Then U_y^L is a unitary operator in this Hilbert space and $y \rightarrow U_y^L$ is a unitary representation of G. It clearly reduces to the U of Section 2 when L is the one-dimensional identity of H. To obtain a system of imprimitivity for U^L which reduces to the P of Section 2, we need only define P_E^L to be the projection operator $f \rightarrow \varphi_{E'} f$ where E' is the inverse image in G of the Borel subset E of S and $\varphi_{E'}$ is the characteristic function of E'. Our construction seems to depend upon the measure μ but in fact does not. If G/H has an invariant measure at all this measure is unique up to a multiplicative constant, and this constant does not affect the equivalence class of U^L, P^L.

Actually the pair U^L, P^L may be constructed whether or not G/H admits an invariant measure. Let ν be any finite measure in G having the same null sets as Haar measure and let $\tilde{\nu}(E) = \nu(E')$ for each Borel subset E of $S = G/H$. Here, as above, E' is the inverse image of E in G. Then $\tilde{\nu}$ is a measure in S which is quasi-invariant in the sense that $\tilde{\nu}$ and $E \rightarrow \tilde{\nu}(Ey)$, have the same null sets for all y. Let $x \rightarrow \varrho_y(x)$ be the density or Radon-Nikodym derivative of the y translate of $\tilde{\nu}$ with respect to $\tilde{\nu}$. Then we may use $\tilde{\nu}$ as a substitute for μ in defining the pair U^L, P^L provided only that we restore the unitariness of U_y^L by changing the definition to

$$(U_y^L f)(x) = \sqrt{\varrho_y(x)}\, f(xy).$$

It is not difficult to show that changing ν changes U^L, P^L to a pair which is unitarily equivalent to the original. Thus the unitary equivalent class of U^L, P^L is uniquely determined by L — in fact by its unitary equivalence class. We call U^L the *unitary representation of G induced by the unitary representation* L of H and we call P^L the system of imprimitivity *canonically associated with* U^L.

In these terms, the imprimitivity theorem may be formulated as follows: given the closed subgroup H of the separable locally compact group G and given a pair consisting of a unitary representation U of G and a system of imprimitivity P for it based on $S = G/H$ there exists a unitary representation L of H and a unitary operator W such that $W U_x W^{-1} = U_x^L$ and $W P_E W^{-1} = P_E^L$ for all $x \in G$ and all Borel subsets E of S. L is uniquely determined up to unitary equivalence. A useful subsidiary result states that the ring of all bounded linear operators which commutes with all U_x^L and all P_E^L is isomorphic to the ring of all bounded linear operators which commutes with all L_ξ. Thus the pair U^L, P^L is irreducible if and only if L is irreducible.

6. Multiplier representations as induced representations

Let G, S and μ be as in Section 2 and let a be a Borel function from $S \times G$ to the complex numbers of modulus one which satisfies the cocycle identity

$$a(s, x_1 x_2) = a(s, x_1)\, a(s x_1, x_2). \tag{6}$$

Define $U_x^a(f)(s) = a(s, x) f(s x)$. Then each U_x^a is a unitary operator and $x \to U_x^a$ is a unitary representation of G. Indeed, apart from almost everywhere considerations (6) is a necessary and sufficient condition that

$$U_{x_1 x_2}^a = U_{x_1}^a U_{x_2}^a.$$

Following BARGMANN [2] we shall call U^a a *multiplier representation*. When $a(s, x) \equiv 1$ we recover the U of Section 2. On the other hand, defining P_E just as in Section 2, we find that for any a, P is a system of imprimitivity for U^a based on S.

Now consider the special case in which $S = G/H$ for some closed subgroup H of G. By the imprimitivity theorem, the pair U^a, P must be equivalent to the pair U^L, P^L for some unitary representation L of H and L can depend only on the cocycle a. To see what L is, notice that if $s_0 = He$ then

$$a(s_0, \xi_1 \xi_2) = a(s_0, \xi_1)\, a(s_0 \xi_1, \xi_2) = a(s_0, \xi_1)\, a(s_0, \xi_2)$$

whenever ξ_1 and $\xi_2 \in H$. Thus $\xi \to a(s_0, \xi)$ is a character χ_a of H. Moreover

$$a(s_0, \xi x) = a(s_0, \xi)\, a(s_0, x) = \chi_a(\xi)\, a(s_0, x)$$

for all $\xi \in H$ and all $x \in G$. Consider now the representation U^{χ_a} induced by χ_a. If f is in $\mathscr{H}(U^{\chi_a})$, the space of U^{χ_a}, then $f(\xi x) = \chi_a(\xi) f(x)$ for all $\xi \in H$, $\xi \in G$ so that $f(\xi x)/a(s_0, \xi x)$ is constant on the right H cosets and may be identified with a Borel function in S. It is easy to see that $f \to f'$ where $f'(x) = f(x)/(a(s_0, x))$ defines a unitary map of $\mathscr{L}^2(S, \mu)$ on $\mathscr{H}(U^{\chi_a})$ and that this unitary map sets up an equivalence between the pair U^a, P and U^{χ_a}, P^{χ_a}.

Since one can show that every χ arises from some a one sees that the systems U^a, P where U^a is a multiplier representation, coincide (modulo unitary equivalence) with the systems U^L, P^L where L varies over the *one-dimensional* unitary representations of H. When $S = G/H$ does not have an invariant measure, one can still identify representations induced by one-dimensional representations with multiplier representations but now a will take on values which are not on $|z| = 1$. Indeed, the Radon-Nikodym derivative needed to compensate for the non-invariance of μ will satisfy (6) and what we must use is the product of its square root with a cocycle taking values on $|z| = 1$.

It is of course not difficult to generalize the notion of multiplier representation so that the above considerations apply to arbitrary induced representations U^L. When L is not one-dimensional $\mathscr{L}^2(S, \mu)$ must be replaced by $\mathscr{L}^2(S, \mu, \mathscr{H}_0)$ where $\mathscr{H}_0$ is a Hilbert space whose dimension is equal to that of the space of L and a must be replaced by a Borel function from $S \times G$ to the unitary operators in $\mathscr{H}_0$. We leave further details to the reader.

7. The imprimitivity theorem and quantum mechanics

We have been led to the imprimitivity theorem by starting with the Stone-von Neumann uniqueness theorem and generalizing three times. The end result seems far removed from the very special case which inspired it. However, the connection is actually rather close. If one seeks a deeper meaning for the Heisenberg commutation relations one finds that they are a consequence of the assumption that the laws of physics are invariant under the group $\mathscr{E}$ of all isometries of physical space. Indeed the formulation of this invariance leads one in a natural way to postulate that the position observables of a particle in quantum mechanics are defined by a projection valued measure P in physical space S and that this projection valued measure is a system of imprimitivity for the unitary representation* U of $\mathscr{E}$ which implements

* Strictly speaking one must allow "projective" unitary representations. However we shall postpone our discussion of this refinement to Section 12.

the invariance. Here P_E is the self-adjoint operator corresponding to the "observable" which is one when the particle is in the set E and zero when it is not. Given any "coordinate" in the sense of a real valued continuous function f on space the self-adjoint operator associated with the corresponding observable is that whose projection valued measure is $A \to P_{f^{-1}(A)}$. Thus P determines the operator assigned to the position observables. To say that P is a system of imprimitivity for U is just to say that the transform by U_x of P_E is the P assigned to the x translate of E. Surely this is just what one means by Euclidean invariance.

Use of the imprimitivity theorem now leads to a classification of all possibilities for the position and momentum observables and (in the case of Euclidean space) to the conclusion that the Heisenberg commutation relations necessarily hold for the operators describing the position and momentum observables. However, one gets much more. One finds what the analogue of the Heisenberg commutation relations should be when space is not Euclidean and one is led in a natural way to the concept of spin. For further details, see [23], [34], and [37].

8. The irreducible unitary representations of semi-direct products

As indicated toward the end of Section 2, there is an extension of the Hahn-Hellinger theory which allows one to reduce the problem of classifying the unitary representations of a type I separable locally compact group G to the problem of classifying the *irreducible* unitary representations of G. This latter problem is more or less trivial for the interesting commutative groups but can be very difficult when G is non-commutative — even when G is finite. In this section we shall indicate briefly how the imprimitivity theorem can be applied to yield a solution to the problem for a special but important class of examples.

Let the group G admit a commutative closed normal subgroup N and let G also contain a closed subgroup H such that $N \cap H = \{e\}$ and $NH = G$. Then H is isomorphic to G/N and every element in G is uniquely a product nh where $n \in N$ and $h \in H$. One says that G is a *semi-direct product* of N and H. Clearly

$$(n_1 h_1)(n_2 h_2) = n_1 h_1 n_2 h_1^{-1} h_1 h_2 = n_1 \alpha_{h_1}(n_2) h_1 h_2$$

where α_h is the automorphism $n \to h n h^{-1}$ of N. Thus G is completely known when we know N, H and the homomorphism $h \to \alpha_h$ of H into the group of automorphisms of N. As examples, we mention the group of all permutations on three objects, the group of all transformations of the real line of the form $x \to ax + b$ where $a \neq 0$, the group $\mathscr{E}$ of all isometries of physical space, and the inhomogeneous Lorentz group. In the first two examples, N and H are both commutative. $\mathscr{E}$ is a semi-direct product of a three-dimensional vector group and a compact group and the last example is a semi-direct product of a four-dimensional vector group and the homogeneous Lorentz group.

Let V be any unitary representation of G where G is a semi-direct product of N and H. Let $A_n = V_{ne'}$, $B_h = V_{eh}$. Then $V_{nh} = A_n B_h$ so that V is completely determined by the representations A and B of N and H respectively. A simple computation shows that $nh \to A_n B_h$ defines a representation of G if and only if $B_h^{-1} A_n B_h = A_{\alpha_h(n)}$ for all h and n. This identity is very like the identity defining a system of imprimitivity. Moreover, it reduces to the latter if we use the Neumark-Ambrose-Godement generalization of STONE's theorem to describe A by the corresponding projection valued measure P^A on $\hat{N}$. In this connection we note of course that each automorphism α_h of N has a dual α_h^* which is an automorphism of $\hat{N}$. Specifically, $[\chi]\,\alpha_h^*$ is the character $n \to \chi(\alpha_h(n))$. Clearly $\hat{N}$ becomes an H space if we define $[\chi]\,h = [\chi]\,\alpha_h^*$ and the identity is easily seen to be equivalent to the statement that P^A is a system of imprimitivity for B based in $\hat{N}$.

In order to apply the imprimitivity theorem we must have a transitive system of imprimitivity and H does not usually act transitively on $\hat{N}$. On the other hand, H restricted to any "orbit" of H in $\hat{N}$ does act transitively and under appropriate circumstances we may concentrate on the restriction of P^A to an orbit. We define the *orbit* $\pi(\chi)$ of $\chi \in \hat{N}$ to be the set of all $[\chi]\,h$ with $h \in H$ and let $\hat{\hat{N}}$ denote the space of all orbits. Let us define a subset F of $\hat{\hat{N}}$ to be a Borel set if $\pi^{-1}(F)$ is a Borel subset of $\hat{N}$ and let us say that $\hat{\hat{N}}$ has a countably separated Borel structure if there exist countably many Borel sets which separate points. This condition holds in particular whenever there exists a Borel subset of $\hat{N}$ which meets each orbit just once. Whenever it does hold we shall say that G is a *regular* semi-direct product of N and H. The importance of this condition is that it implies that $P^A_{\hat{N}-\mathcal{O}} = 0$ for some unique orbit $\mathcal{O}$ whenever V is irreducible. Thus every irreducible unitary representation of a regular semi-direct product is described by a pair A, B where P^A is a transitive system of imprimitivity for B based on an orbit of $\hat{N}$ under H.

We refer the reader to the published literature [17], [19], [20] for further details and content ourselves with a statement of the theorem that results when one applies the imprimitivity theorem as indicated.

Theorem. *Let G be a semi-direct product of N and H where N is normal and commutative and N and H are separable and locally compact.*

For each $\chi \in \hat{N}$ *let* H_χ *denote the subgroup of all* $h \in H$ *for which* $[\chi]\, h = \chi$. *Then* H_χ *is closed and for each irreducible unitary representation* L *of* H_χ, $n, h \to \chi(n)\, L_h$ *is a unitary representation* χL *of the subgroup* $N H_\chi$. *Form the induced representation* $U^{\chi L}$ *of* G. *Let* C *be a set which meets each* H *orbit in* $\hat{N}$ *once and only once. Then*

(a) $U^{\chi L}$ *is irreducible for all* χ *and* L.

(b) As χ *varies over* C *and* L *varies over the inequivalent irreducible representations of* H_χ *we get inequivalent irreducible representations of* G *and we get one equivalent to every* $U^{\chi L}$ *whether or not* χ *lies in* C.

(c) If G *is a regular semi-direct product then every irreducible representation of* G *is equivalent to some* $U^{\chi L}$.

When G is the group of all transformations of the additive group of the real line of the form $x \to a x + b$ where $a > 0$ then $\hat{N}$ is the additive group of the real line and the action of H on $\hat{N}$ is such that we may take C to be the set consisting of -1, 0 and 1. Correspondingly H_χ is $\{e\}$, H and $\{e\}$. The representations associated with 0 and H are just the one-dimensional representations of H "lifted" to G. There is just one irreducible unitary representation associated with -1, $\{e\}$ and just one associated with 1, $\{e\}$ because $\{e\}$ has just one irreducible unitary representation. They are infinite dimensional representations induced by one-dimensional representations of N. To within equivalence, there are no other irreducible unitary representations of G. These results about G are a reformulation of the main results of the paper [8] of GELFAND and NEUMARK referred to in Section 4.

When G is the inhomogeneous Lorentz group the orbit structure of $\hat{N}$ under H is more complicated and there are four distinct possibilities for H_χ. It is isomorphic either to the homogenous Lorentz group, to the homogeneous Lorentz group in three-dimensional space time, to the Euclidean group in the plane or to the rotation group in three space. The Euclidean group in the plane is a semi-direct product of two commutative groups and its irreducible unitary representations are determinable by a second application of our theorem. Since the representations of the rotation group in three space are easily deduced from the classical theory of spherical harmonics our theorem reduces to the problem of classifying the irreducible unitary representations of the inhomogeneous Lorentz group to the corresponding problem for the homogeneous Lorentz group in three and four dimensional space time. Actually these representations for which H_χ is a homogeneous Lorentz group are in a certain sense "physically uninteresting". Thus our theorem provides a complete classification of the "physically interesting" irreducible unitary representations of the inhomogeneous Lorentz group. These consequences of our theorem for the inhomogeneous Lorentz group are a reformulation of the main results of a celebrated paper of WIGNER [38]. WIGNER was actually the first person to analyze the irreducible unitary representations of a group having *infinite-dimensional* irreducible unitary representations*.

9. On the structure of certain induced representations

We saw in the last section that the irreducible unitary representations of many groups can be put into the form U^L where L is an irreducible representation of a suitable subgroup. On the other hand U^L certainly need not be irreducible when L is, and in fact there are a number of problems in mathematics and mathematical physics whose resolution depends upon analyzing a representation of the form U^L into its irreducible constituents. In this section we shall show that a very complete solution to the problem may be given in the special case in which the group is a semi-direct product $N \circledS H$, the subgroup from which L is induced is of the form $N_0 \circledS H_0$ where N_0 and H_0 are closed subgroups of N and H respectively and certain auxiliary conditions are satisfied.

Let G, N, H, N_0, H_0 be as just indicated and let $G_0 = N_0 \circledS H_0$. Let χ be any member of $\hat{N}_0$ and let $H_{0,\chi}$ be the closed subgroup of all $h \in H_0$ for which $[\chi]h = \chi$. Let L be any irreducible unitary representation of H_χ. Then $n, h \to \chi(n) L_h$ is an irreducible unitary representation of $N_0 H_{0,\chi}$ and so is the representation $U^{\chi L}$ of G_0 which it induces. Moreover, whenever G_0 is a regular semi-direct product of N_0 and H_0 we know from the theory of Section 8 that every irreducible unitary representation of G_0 is of the form $U^{\chi L}$. Let $W = U^{U^{\chi L}}$ denote the unitary representation of G induced by the irreducible unitary representation $U^{\chi L}$ of G_0.

Theorem. *Let $N_0^{\perp}$ denote the annihilator of N_0 in $\hat{N}$ and let $N_0^{\perp}\chi$ denote the $N_0^{\perp}$ coset of all $f \in \hat{N}$ which agree on N_0 with χ. For each $f \in N_0^{\perp}\chi$ let H_f denote the closed subgroup of all $h \in H$ with $[f]h = f$ and consider the representation L'' of H_f induced by the restriction of L to $H_{0,\chi} \cap H_f$. Let W^f denote the representation of G induced by $n, h \to f(n) L''_h$. Then $N_0^{\perp}\chi$ is invariant under $H_{0,\chi}$ and $W^{[f]h} \sim W^{[f]}$ for all $h \in H_{0,\chi}$. If there are only countably many $H_{0,\chi}$ orbits in $N_0^{\perp}\chi$ then W is a direct sum of the W^{f_j} where we choose one f_j from each $H_{0,\chi}$ orbit which is not of measure zero with respect to Haar measure in $N_0^{\perp}\chi$. More generally suppose that the space O_χ of all $H_{0,\chi}$ orbits in $N_0^{\perp}\chi$ is „countably separated" in the sense that there are countably many subsets of O_χ whose inverse images*

* WIGNER actually considered projective representations as well. However as explained in Section 12 our theory extends to the projective case.

in $N_0^{\perp}\chi$ are Borel sets and which separate the points of O_χ. Then W is a direct integral over O_χ of representations $\tilde{W}^f$ where $\tilde{W}^f$ is equivalent to W^f for all f in the orbit.

Finally for each f let $L'' = \int M^\lambda dv(\lambda)$ be a decomposition into irreducibles of L''. Then $W^f = \int U^{fM^\lambda} dv(\lambda)$ and each U^{fM^λ} is irreducible.

Remark. This theorem allows us to decompose $W = U^{v\chi L}$ into irreducibles whenever we know how to decompose the representations L'' of H_f into irreducibles. As with the theory of Section 8 we do not actually solve our problem but instead reduce it to one about closed subgroups of H.

Since this theorem does not appear in the literature we present a proof. Our representation W is the representation of G induced by a representation of G_0 which is in turn induced by the representation χL of the subgroup $N_0 H_{0,\chi}$ of G_0. Thus by the theorem on inducing in stages ([19] Theorem 4.1) W is equivalent to the representation of G induced by the representation χL of $N_0 H_{0,\chi}$.

As our next step we apply the stages theorem again using $NH_{0,\chi}$ as intermediate subgroup instead of G_0. Let V be the representation of $NH_{0,\chi}$ induced by χL. Then W is the representation of G induced by V and any decomposition of V will be reflected in a corresponding decomposition of W. We shall find a decomposition of V and then study the representations of G which the components induce. To find this decomposition we first restrict V back to N. The spectral theorem (Section 3) assigns a projection valued measure P on $\hat{N}$ to this restriction which is a system of imprimitivity for V with respect to a certain action of $N Ⓢ H_{0,\chi}$ on $\hat{N}$. This action is obtained from the semi-direct product action of $H_{0,\chi}$ on N by transfering to $\hat{N}$ as described in Section 8 and then "lifting" to $N Ⓢ H_{0,\chi}$. On the other hand by the "restriction theorem" (Theorem 12.1 of [19]) V restricted back to N is a direct integral of characters in a certain $N_0^{\perp}$ coset in $\hat{N}$; namely the coset $N_0^{\perp}\chi$ described in the statement of the theorem. It follows that our projection valued measure P must be concentrated in $N_0^{\perp}\chi$. Now, by hypothesis, the space of all $H_{0,\chi}$ orbits in $N_0^{\perp}\chi$ is countably separated. Hence this is true of the space of all $N Ⓢ H_{0,\chi}$ orbits since N acts trivially. Thus, just as in the proof of Theorem 12.1 of [19], V is a direct integral over the space O_χ of $H_{0,\chi}$ orbits in $N_0^{\perp}\chi$. The contribution of the orbit containing f has the restriction of P to this orbit as a system of imprimitivity and this system of imprimitivity is transitive. This contribution is induced by a representation of the subgroup of $NH_{0,\chi}$ leaving f fixed. This subgroup is clearly $NH_{0,f}$ where $H_{0,f}$ is the closed subgroup of H_0 consisting of all h with $[f]h = f$.

Of course $H_{0,f}=H_f\cap H_0=H_f\cap H_{0,\chi}$ and it is easy to see that the inducing representation is just fL^1 where L^1 is the restriction of L to $H_{0,f}$.

We have now decomposed V as a direct integral of representations of the form U^{fL^1} where f is a character of N, L^1 is the restriction of L from $H_{0,\chi}$ to $H_{0,f}$ and the inducing is from $NH_{0,f}$ to $NH_{0,\chi}$. Because of the fact that inducing "commutes" with the taking of direct integrals we have correspondingly a decomposition of W as a direct integral of representations of the form U^{fL^1} where now the inducing is from $NH_{0,f}$ to NH.

Our next task is to decompose the representation $U^{fL'}$. To do this we apply the stages theorem once again, this time using NH_f as the intermediate subgroup. It follow at once from the definitions concerned that the representation of NH_f induced by fL' is fL'' where L'' is the representation of H_f induced by the representation L' of $H_{0,f}=H_f\cap H_{0,\chi}$. Suppose that $L''=\int M^\lambda d\nu(\lambda)$ where the M^λ are irreducible. Then the representation of G induced by fL' is $\int U^{fM^\lambda d\nu(\lambda)}$ and by the theory of 8. each U^{fM^λ} is irreducible since M is an irreducible representation of H_f. This completes the proof.

10. An application to solid state physics

In studying the quantum mechanics of a solid body it seems to be a useful first approximation to assume that the nuclei are fixed point charges and that the electrons move independently of one another in the potential field produced by these fixed charges. The fixed positions form a pattern capable of unlimited replication and in the limiting case of an infinite number of particles this pattern is invariant under the operations of a certain discrete subgroup Γ of the Euclidean group $\mathscr{E}$. This discrete subgroup is such that the coset space $\mathscr{E}/\Gamma$ is compact and is what is known as a *space group*.

Under the above simplifying assumptions the main problem to be solved is that of determining the eigenvalues of the Hamiltonian operator H for a single electron moving in the indicated potential field. Moreover it turns out to be convenient to concentrate attention on the limiting case in which there are an infinite number of particles and the pattern is completely filled. In this limit the operator H has a continuous spectrum but its determination in suitable terms tells us all we need to know about the eigenvalues of H in the physically more realistic finite case.

Ignoring any influence of electron spin our problem then is thus: Let the Euclidean group $\mathscr{E}$ act on physical space E^3 in the usual fashion and let v be a real valued function defined on E^3 which is invariant under the action of the space group $\Gamma \subset \mathscr{E}$. Determine the spectrum of the operator H:

$$\psi \to \frac{-\hbar^2}{2m}\left(\frac{\partial^2 \psi}{\partial X^2} + \frac{\partial^2 \psi}{\partial y^2} + \frac{\partial^2 \psi}{\partial z^2}\right) + v\psi$$

where m is the mass of the electron and $2\pi\hbar$ is PLANCK'S constant. The domain of H must of course be properly specified as a subspace of $\mathscr{L}^2(E^3)$ in such a manner that it is self-adjoint.

Let U be the unitary representation of $\mathscr{E}$ defined by the natural action of $\mathscr{E}$ on E^3 and let Γ_T denote the commutative group of all translations in Γ. Then $U_\alpha H = HU_\alpha$ for all α in Γ and hence in particular for all α in Γ_T. Let $(U)^{\Gamma_T}$ denote the restriction of U to Γ_T and let P denote the projection valued measure in $\hat{\Gamma}_T$ associated with $(U)^{\Gamma_T}$ by the spectral theorem. Then H commutes with all P_E so the direct integral decomposition of $H(U) = \mathscr{L}^2(E^3)$ defined by P is invariant under H. In other words we have a direct integral decomposition of $H = \int H^\chi \, d\nu(\chi)$ which is parametrized by the characters χ of Γ_T. It turns out (for reasons which will become clearer below) that each H^χ has a pure point spectrum and is described by a set of eigenvalues. Clearly then the (continuous) spectrum of H will be completely described by giving the eigenvalues of H^χ for each χ. Strictly speaking H^χ is only defined for almost all χ — at least according to the analysis we have sketched. However, a closer look, exploiting the fact that H is a differential operator, makes it possible to define H^χ for every χ as a definite differential operator. Now $\hat{\Gamma}_T$ is a compact, connected topological group isomorphic to a three dimensional torus. Thus we may speak of continuous dependence on χ and it turns out that the eigenvalues of the H^χ vary continuously with χ. Letting $E_1(\chi) \leqq E_2(\chi) \leqq \ldots$ denote the eigenvalues of H^χ, we see that we may describe the spectrum of H by a countable family of continuous real valued functions defined on the compact connected group $\hat{\Gamma}_T$. The spectrum as a point set is just the union of the ranges of the E_j and hence the union of a countable number of compact connected sets. In most interesting cases this union is not connected and in fact has infinitely many connected components. Thus there exists a sequence $x_1 < x_2 < x_3 < \ldots$ of real numbers such that x is in the spectrum if and only if x is in one of the closed intervals $x_{2j+1} \leqq x \leqq x_{2j}$. In other words the spectrum has a "band" structure. The physicists refer to the problem of finding the functions $\chi \to E_1(\chi), E_2(\chi), \ldots$ as "energy band theory".

If we replace the commutative group Γ_T by the full group Γ (which usually will not be commutative) we may repeat much of the above argument obtaining operators H^L parametrized by the irreducible representatives L of Γ. It is not difficult to see that each H^L is a direct sum of a finite number of the H^χ and that each eigenvalue in H^L occurs with a multiplicity which is divisible by the dimension of L. A detailed study of the H^L and their decomposition into H^χ thus yields information about relationships among the $E_j(\chi)$ attributable to the action of the symmetry group Γ/Γ_T. Knowledge of such relationships is of course of considerable help in actually finding the E_j and approximation thereto for a given v.

Now the decomposition of the H^L into H^χ's depends only on the restriction of L to Γ_T and not on v. Moreover as v changes continuously to zero through the values $\varepsilon v\,(0 \leqq \varepsilon \leqq 1)$, it will usually happen that the spectrum of the H^L and H^χ nove continuously into that for $v=0$. Thus we may parameterize the eigenvalues of the H^L and H^χ by those of H_0^L and H_0^χ. Moreover if we know explicitly the eigenvalues of the H_0^χ and H_0^L we will know the eigenvalue structure of the H^χ and H^L as far as the parameters of these eigenvalues are concerned.

The point of interest to us can now be stated as follows: The problem of determining the eigenvalues of the H_0^L (and H_0^χ) can be completely reduced to (a part of) the problem of finding the structure of the induced representations U^L and U^χ of $\mathscr{E}$. To describe this reduction we observe first that $H_0 U_\alpha = U_\alpha H_0$ for all $\alpha \in \mathscr{E}$. Now it is well known and easily proved that U is multiplicity free and hence uniquely a direct integral of inequivalent irreducible representations U^λ. It follows that H_0 is correspondingly a direct integral of constant operators $C(\lambda) I$ where $C(\lambda) I$ acts in the "space" of U^λ. Actually the U^λ may be parametrized in a natural way by the positive real numbers and if λ is the parameter then $C(\lambda)$ is a constant times λ^2 [in physical terms $C(\lambda)$ is the energy of a free particle whose linear moment vector has length λ]. Thus the spectral values of H_0 correspond one to one to certain irreducible unitary representations of $\mathscr{E}$ and it is possible to prove the following theorem.

Theorem. *For each irreducible unitary representation L of Γ the induced representation U^L is a discrete direct sum of irreducibles and the number of times it contains U^λ is the multiplicity of $C(\lambda)$ in H_0^L divided by* $\dim(L)$. *Of course a parallel statement relates H_0^χ and U^χ.*

The theorem of Section 9 allows us to compute U^χ in all cases and to compute U^L whenever Γ is a so-called "symmorphic" space group. Roughly one third of the possible space groups are symmorphic. It would probably not be difficult to extend the theorem of Section 9 so that it applies to the general case.

For the physical literature on group theoretical aspects of the energy band problem the reader is referred to the reprints collected in [11] especially the classic paper of Bouckaert, Smoluchowski and Wigner.

11. Restrictions and tensor products

There is a certain duality between forming the representation U^L of a separable locally compact group G induced by a unitary representation L of a closed subgroup H and forming a unitary representation of H from a unitary representation M of G by considering only the M_x for $x \in H$, i.e. by "restricting M to H". Indeed for finite groups a classical theorem of FROBENIUS says that when L and M are irreducible the multiplicity with which U^L contains M is equal to the multiplicity with which L is contained in the restriction of M to H. This suggests that we seek a "dual" of the theorem of Section 9 which tells us how to decompose the restrictions to $G_0 = N_0 \circledS H_0$ of the irreducible unitary representations of $G = N \circledS H$. A first step in the proof of such a theorem is provided by the author's "restriction theorem" mentioned in Section 9. This theorem tells us how to decompose the restriction of an induced representation to a subgroup into a direct sum or direct integral of other induced representations. Of course these summands in general need not be irreducible. In the case at hand however the subgroups are such that an easy special case of the theorem of Section 9 applies. [We have to induce from subgroups of $N_0 \circledS H_0$ of the form $N_0 \circledS H'_0$ where $H'_0 \leqq H_0$ and thus the orbit space has only one element.] In other words we may obtain our desired dual of the theorem of Section 9 as a corollary of the restriction theorem of [19] and the theorem of Section 9 itself. We will content ourselves with stating the result. The reader should have no difficulty in verifying that this result is a consequence of the argument just indicated.

Let G be a semi-direct product $N \circledS H$ where N and H are separable and locally compact and N is commutative. Let N_0 and H_0 be closed subgroups of N and H respectively such that N_0 is invariant under $n \to h(n)$ for all $h \in H_0$. Let f be any member of $\hat{N}$ and let H_f be the subgroup of all h for which $[f]h = f$. Let M be any irreducible unitary representation of H_f and let V denote the (irreducible) representation of G induced by the representation $n, h \to f(n) M_h$ of $N \circledS H_f$. The theorem we are about to state gives the reduction into irreducibles of the restriction V' of V to $G_0 = N_0 \circledS H_0$.

Theorem. *Suppose that H_0 and H_f are regularly related subgroups of H in the sense that the space of all $H_0 : H_f$ double cosets is countably separated. Then V' is a direct integral over these double cosets (with respect to the image of the Haar measure class) and the contribution of the double coset containing h may be computed as follows: Let M^h denote the representation of $H_{[f]h} = h^{-1} H_f h$ which takes $h^{-1} \xi h$ into M_ξ. Restrict M^h to $H_{[f]h} \cap H_0$ and then induce to $H_{0,\chi}$ where χ is the restriction of $[f]h$ to N_0 and $H_{0,\chi}$ is the subgroup of all $h \in H_0$ with $[\chi]h = \chi$. Let*

$\int L^\lambda dv(\lambda)$ denote a reduction of this induced representation into irreducibles. Then for all λ the representations $nh \to \chi(n) L_h^\lambda$ of $N_0 H_0$ induce irreducible representations $U^{\chi L}$ of $G_0 = N_0 \circledS H_0$ and $\int U^{\chi L^\lambda} dv(\lambda)$ is *the desired contribution of the double coset containing h.*

This theorem like its dual does not actually solve the reduction problem. It reduces it to questions about the subgroup H which often have easy answers. We note also that wherever $G = N \circledS H$ is a regular semi-direct product the representations V include all irreducible unitary representations of G.

Let L and M be irreducible unitary representations of the separable locally compact group G. Then their (outer) tensor product $L \times M$ is an irreducible unitary representation of $G \times G$. If we restrict $L \times M$ to the diagonal $\tilde{G}$ of $G \times G$ consisting of all x, y with $x = y$ and note that $\tilde{G}$ is naturally isomorphic to G we obtain a unitary representation $L \otimes M$ of G which is usually reducible. It is called the (inner) tensor product of L and M.

One of the basic problems about any G is that of determining the reductions of the tensor products $L \otimes M$ of its irreducible unitary representations. Now suppose that $G = N \circledS H$ where H and N are as above. If the semi-direct product is regular then the theory of Section 8 gives us all irreducible unitary representations of G in terms of the irreducible unitary representations of the subgroups H_χ of H. In view of what has gone before it is natural to seek to reduce the problem of finding the (inner) tensor products of the irreducible unitary representations of G to problems about the subgroup H. A theorem doing just this can be obtained as a corollary of the theorem stated above. Indeed we have already seen that reducing an inner tensor product can be looked upon as a special case of reducing a restriction to a subgroup (from $G \times G$ to $\tilde{G}$). Moreover in the special case in which $G = N \circledS H$ the restriction problem which arises is one to which our earlier theorem applies. We leave the details to the reader and, as above, content ourselves with stating the result.

Let $G = N \circledS H$ where N is commutative and N and H are separable and locally compact. Let χ_1 and χ_2 be characters of N and for $j = 1, 2$ let H_{χ_j} be the closed subgroup of H consisting of all $h \in H$ with $[\chi_j] h = \chi_j$. Let L^1 and L^2 be irreducible unitary representations of H_{χ_1} and H_{χ_2} respectively and let $\chi_j L^j$ be the representation $nh \to \chi_j(n) L_h^j$ of $N \circledS H^{\chi_j}$. By the theory of Section 8 the induced representations $U^{\chi_1 L^1}$ and $U^{\chi_2 L^2}$ are irreducible and if G is a regular semi-direct product every pair of irreducible unitary representations of G may be so obtained.

Theorem. *Suppose that H_{χ_1} and H_{χ_2} are regularly related in the sense explained above. Then the (inner) tensor product $U^{\chi_1 L^1} \otimes U^{\chi_2 L^2}$ is* *a direct integral over the* $H_{\chi_1}:H_{\chi_2}$ *double cosets (with respect to the Haar measure class) and the contribution of the double coset containing h may be computed as follows: Let* $\chi_3=[\chi_1]h$ *and let* $\chi_4=\chi_2\chi_3$. *Let* M^j *be the restriction of* L^j *to* $H_{\chi_3}\cap H_{\chi_2}\leqq H_{\chi_4}$. *Form the (inner) tensor product* $M^1\otimes M^2$ *and then the induced representation* $U^{M^1\otimes M^2}$ *of* H_{χ_4}. *Let* $\int W^\lambda d\omega(\lambda)$ *be a decomposition of* $U^{M^1\otimes M^2}$ *as a direct integral of irreducible representations. Then the induced representations* $U^{\chi_4 W^\lambda}$ *will be irreducible unitary representations of G and* $\int U^{\chi_4 W^\lambda}d\omega(\lambda)$ *will be the contribution to* $U^{\chi_1 L^1}\otimes U^{\chi_2 L^2}$ *of the double coset containing h.*

This theorem on tensor products has applications to physics. A free relativistic particle is intrinsically associated with an irreducible unitary representation of the inhomogeneous Lorentz group. Accordingly certain questions about the kinematics of relativistic two particle interactions require knowledge of the decomposition of the (inner) tensor products of the corresponding representations. Our theorem provides these decompostitions which however have already been determined by other methods [14], [31]. There are also questions in the theory of the solid state whose resolution can be made to depend upon decomposing the (inner) tensor product of two irreducible unitary representations of the appropriate space group.

When the space group is symmorphic our theorem applies and it would probably not be difficult to adopt it to the most general space group. Again though, the problem has been more or less solved in the physical literature. See [5] and the papers cited therein.

12. Induced projective representations

We have already mentioned that physical applications require that the notion of unitary representation be generalized somewhat. This is because many unitary operators in physics actually appear as operators implementing automorphisms of the lattice of all closed subspaces of the underlying Hilbert space. Clearly a unitary operator is determined by the lattice automorphism which it defines only up to a multiplicitive constant of modulus one. Thus if we have a homomorphism of a group G into the group of automorphisms and attempt to describe it by a mapping of G into unitary operators we will have an arbitrary choice to make at each group element x. Once we have made it and have a mapping $x\rightarrow W_x$ we cannot conclude that $W_{xy}=W_x W_y$ since the choices "may not match". We can conclude only that W_{xy} and $W_x W_y$ define the same automorphism i.e. that $W_{xy}=\sigma(x, y) W_x W_y$ where for each x and y, $\sigma(x, y)$ is a complex number of modulus one. Of course we can try to get rid of the multiplier σ by making our arbitrary choices in a sufficiently clever way. However it turns out that this is not always possible and we must learn to deal with unitary operator valued functions on a group which satisfy identities of the form $W_{xy} = \sigma(x, y) W_x W_y$. They are called *projective representations* or *ray representations*.

Projective representations force themselves on our attention in another way also. It is natural to attempt to extend the theory of Section 8 to the case in which the normal subgroup N is not necessarily commutative and there need exist no subgroup H. Such an extension is in fact possible and is worked out in detail in [22]. By analogy one would expect a theory allowing one to describe the irreducible unitary representations of G in terms of those of the normal subgroup N and certain closed subgroups of the quotient group G/N. In fact this is almost what one gets but there is an important change. Given a relevant subgroup of G/N the corresponding irreducible unitary representations of G may be parameterized not by the ordinary irreducible unitary representations of G/N but by those irreducible projective representations having a fixed non trivial multiplier σ.

It is easy to show that any function σ on $G \times G$ which occurs as a projective multiplier must satisfy (A) $\sigma(x, y)\sigma(xy, z) = \sigma(x, yz)\sigma(y, z)$ and hence must be what the homological algebraists call a two-cocycle. Conversely if σ is any Borel function from $G \times G$ to the complex numbers of modulus one which satisfies this identity it can be shown that there exists a unitary operator valued function $x \to W_x$ such that: (B) $W_{xy} = \sigma(x, y) W_x W_y$ for all x and y in G and such that (C) $x \to (W_x(\varphi), \psi)$ is a Borel function for all φ and ψ in the underlying Hilbert space. It is customary to normalize by always choosing W_e to be the identity and then we have the additional condition (D) $\sigma(x, 1) = \sigma(1, x) = 1$.

Given any function σ from $G \times G$ to the complex numbers of modulus 1 which satisfies (A) and (D) we define a σ representation of G to be a unitary operator valued function on G which satisfies (B) and (C). For each fixed σ there is a theory of the σ representations of G which is almost completely parallel to the theory of ordinary representations. We refer to [22] for details but remark that one even has an analogue of inducing and that the obvious analogue of the imprimitivity theorem is true.

The definition of inducing for σ representations is easy to give and we do so as we shall need the notion in what follows. Note first that a multiplier σ for a group G is also a multiplier for every closed subgroup H. Let σ be such a multiplier and let L be any σ representation of H. To define the σ representation of G induced by L we repeat the definition given in Section 5 with two changes. We replace the identity $f(\xi x) = L_\xi f(x)$ by $f(\xi x) = L_\xi f(x) \sigma(\xi, x)$ and the identity $(U_x^L f)(y) = f(yx)$ by $(U_x^L f)(y) = \frac{f(yx)}{\sigma(y, x)}$. Notice that σ may be identically one in certain subgroups H. When this happens ordinary representations of H can induce projective representations of G. Of course σ is always one at the identity so we may always define the σ representation of G induced by the identity representation of $\{e\}$. For obvious reasons it is called the σ regular representation of G.

Let W by a σ representation and let g be an arbitrary Borel function from G to the complex numbers of modulus one such that $g(e)=1$. Then if $W_x^1 = g(x) W_x$, W^1 will be a projective representation with multiplier σ^1 where $\sigma^1(x, y) = [g(xy)/g(x)g(y)]\,\sigma(x, y)$. Clearly, multiplication by g sets up a one-to-one correspondence between σ representations and σ^1 representations which preserve all essential properties. Thus there is little need to distinguish between two multipliers when one can be obtained from the other by multiplication by a function of the form $x, y \to g(xy)/g(x)g(y)$. Such pairs of multipliers are said to be *similar* or *equivalent*. From the standpoint of homological algebra similar multipliers are of course just cohomologous cocycles.

In two of the groups of greatest interest in physics, namely the Euclidean group and the inhomogeneous Lorentz group, there are to within equivalence just two multipliers one of which is the identity. Thus the theory of ordinary representations takes care of half the cases. The other half may be taken care of by making the small modification necessary to include σ representations. When there are just two multipliers it is easy to show that the one not equivalent to the identity is equivalent to a multiplier σ which takes on only the values one and minus one. Thus if W is a σ representation we have $W_{xy} = \pm W_x W_y$. Assigning both W_x and $-W_x$ to x we obtain a so called *two valued representation.*

13. Projective representations and the Stone-von Neumann theorem

Let G be a separable locally compact commutative group and let $\hat{G}$ be its dual. Let U and V be unitary representations of G and $\hat{G}$ respectively and let U and V satisfy the identity:

$$U_t V_\chi = \chi(t)\, V_\chi\, U_t \quad \text{of Section 4}$$

Let $W_{t,\chi} = U_t V_\chi$ for all t, χ in the product group $A = G \times \hat{G}$. Then

$$\begin{aligned} W_{(t_1\chi_1)(t_2\chi_2)} &= U_{t_1} V_{\chi_1} U_{t_2} V_{\chi_2} = U_{t_1} U_{t_2} V_{\chi_1} V_{\chi_2} \overline{\chi_1(t_2)} \\ &= U_{t_1 t_2} V_{\chi_1\chi_2} \overline{\chi_1(t_2)} = W_{(t_1\chi_1)(t_2\chi_2)} \overline{\chi_1(t_2)}\,. \end{aligned}$$

Thus W is a projective representation of A whose multiplier σ is defined by the equation $\sigma(t_1, \chi_1; t_2, \chi_2) = \chi_1(t_2)$. Conversely given any σ

representation W of A (with the σ just defined) we verify at once that $W_{t,\chi} = U_t V_\chi$ where U and V are the restrictions of W to $G \times e$ and $e \times \hat{G}$ respectively, and U and V satisfy the identity in question. Thus the first generalization of the Stone-von Neumann uniqueness theorem may be reinterpreted as stating that for the particular σ defined above the commutative group $G \times \hat{G}$ has to within equivalence just one irreducible σ representation. It follows in particular that changing from one σ to another can have quite profound effects on the representation theory of a group. The ordinary representations of $G \times \hat{G}$ are all one dimensional and there are as many inequivalent ones as there are elements of $G \times \hat{G}$.

The theory of Section 8 carries over to σ representations without essential change whenever $\sigma \equiv 1$ on the normal subgroup N. Applying it with $N = G \times e$ we arrive once more at the uniqueness theorem as well as the additional information that our unique irreducible σ representation is equivalent to the σ representation of $G \times \hat{G}$ induced by the identity representation of $G \times e$. More generally let Γ be a closed subgroup of G and let $\Gamma^{\perp}$ be the group of all $\chi \in \hat{G}$ which reduce to 1 on Γ. Then $\Gamma \times \Gamma^{\perp}$ is a closed subgroup of $G \times \hat{G}$ on which $\sigma \equiv 1$ and we may speak of the σ representation of $G \times \hat{G}$ induced by the identity representation of $\Gamma \times \Gamma^{\perp}$. It follows from the theory of projective representations developed in [22] that this σ representation is also irreducible and hence equivalent to W. Though Weil does not use this language or point of view the facts just outlined about the uniqueness of W and the different ways of realizing it play a significant role in his recent paper [36]. (Cf. the author's review MR 29, No. 2324.)

14. Projective representations and Cartier's treatment of theta functions

Let V be a finite dimensional complex vector space and let Γ be a closed subgroup of the additive group of V which is isomorphic to a direct product of a finite number of replicas of the integers and is such that V/Γ is compact. Let H be a positive definite Hermitean inner product in V and, assuming it exists,* let χ be a function from Γ to the complex numbers of modulus one such that

$$\chi(\gamma_1 + \gamma_2) \equiv \chi(\gamma_1)\, \chi(\gamma_2)\, e^{-\pi i \mathscr{I}(H(\gamma_1, \gamma_2))} \tag{7}$$

for all $\gamma_1, \gamma_2 \in \Gamma$. Here $\mathscr{I}$ denotes "imaginary part of". By definition a theta function is a holomorphic function f on V which satisfies the

* It is not difficult to show that a χ satisfying (7) exists if and only if $\mathscr{I}(H)$ is integer valued on $\Gamma \times \Gamma$.

identity

$$f(v+\gamma)=f(v)\,\chi(\gamma)\,e^{\pi\left[\frac{H(\gamma,\gamma)}{2}+H(\gamma,v)\right]} \tag{8}$$

for all v in V and all $\gamma\in\Gamma$. The quotient of two such (with the same H and χ) is a meromorphic function on V invariant under Γ and hence a meromorphic function on the compact quotient space V/Γ. This is how one constructs meromorphic functions on V/Γ when they exist and a central question that arises is as to the dimension of the vector space of all theta functions for a given H, Γ and χ.

Inspired by WEIL's paper [36], CARTIER [6] has recently shown how the (known) answer to this question may be derived from the Stone-von Neumann theorem and certain auxiliary considerations involving distributions. We outline here a closely related alternative treatment which exploits the theory of projective representations as such and avoids the use of distributions.

In this paragraph we describe the central idea. A more detailed account follows. Let $\sigma(v_1,v_2)=e^{-\pi i\mathscr{I}(H(v_1,v_2))}$. Then σ is multiplicative in each variable and so is multiplier for V. Moreover (7), says that χ is a one dimensional σ representation of Γ. Hence we may construct the σ representation U^χ of V induced by χ as well as the σ regular representation of V. We may realize the latter concretely in such a fashion that it has an irreducible subspace consisting entirely of holomorphic functions and conclude from this that every bounded linear operator from $\mathscr{H}(U^\chi)$ to this irreducible subspace is defined by a "kernel". This kernel is of the form $v_1, v_2 \to A'(v_1-v_2)\,e^{\pi H(v_2,v_1)}\,e^{-\pi\frac{H(v_1,v_2)}{2}}$ for a holomorphic complex valued function A' on V and the condition that the operator defined by this kernel should intertwine U^χ with the σ regular representation is precisely that it should satisfy the identity (8) defining the theta functions. In this way one shows that the dimension of the space of theta functions is equal to the dimension of the indicated space of intertwining operators and hence to the multiplicity with which U^χ contains the irreducible in question. Actually it is easy to see that V has (to within equivalence) a unique irreducible σ representation so that the determination of the desired dimension is completely equivalent to the problem of determining the structure of U^χ. This structure can be determined by applying the general theory of induced σ representations. Note that the mere existence of a one dimensional σ representation of Γ implies that σ restricted to Γ is equivalent to the multiplier which is identically one. As indicated above this need not be true for every pair H, Γ and when it is not there are no χ's and no theta functions.

We begin with some general facts about intertwining operators for induced projective representations. Let G be a finite group and let σ be a projective multiplier for G. Let L and M be σ representations of the subgroups H_1 and H_2 of G respectively. Let U^L and U^M be the σ representations of G induced by L and M. Finally let $R(U^M, U^L)$ denote the vector space of all linear operators from the space $\mathcal{H}(U^M)$ of U^M to the space $\mathcal{H}(U^L)$ of U^L for which $TU_x^M = U_x^L T$ for all x. $R(U^M, U^L)$ is the space of all intertwining operators. We have then the:

Theorem. *There is a vector space isomorphism of $R(U^M, U^L)$ with the space of all functions $x \to A(x)$ from G to the linear operators from $\mathcal{H}(M)$ to $\mathcal{H}(L)$ which satisfy the identity*

$$A(\xi x \eta) = \sigma(\xi, x) L_\xi A(x) M_\eta \sigma(x, \eta) \qquad \text{for all } x \in G, \quad \xi \in H_1, \eta \in H_2. \tag{9}$$

The member T_A of $R(U^M, U^L)$ maps $f \in \mathcal{H}(U^M)$ into f_A where

$$f_A(x) = \frac{0(H_2)}{0(G)} \sum_{y \in G} \frac{A(xy^{-1}) f(y)}{\sigma(x, y^{-1})}$$

Actually a short calculation shows that:

$$\frac{A(x(\eta y)^{-1}) f(\eta y)}{\sigma(x_1 (\eta y)^{-1})} = \frac{A(xy^{-1}) f(y)}{\sigma(x, y^{-1})}$$

for all $\eta \in H_2$ so that we may sum over the right cosets and remove the factor $\frac{0(H_2)}{0(G)}$.

For the special case in which $\sigma \equiv 1$ this theorem is proved in [18]. The adaptation of the argument given there is straightforward and will be left to the reader.

When G is not finite one can still form intertwining operators from solutions of (9), at least when G/H_2 admits an invariant measure. One simply integrates with respect to this invariant measure instead of summing over the cosets. However one must restrict to those A's for which the integrals converge and lead to functions in G with finite $\mathcal{H}(U^L)$ norm. Moreover, this is no guarantee that every intertwining operator may be constructed from a suitable A — essentially because not every bounded linear operator from $L^2(G/H_2)$ to $L^2(G/H_1)$ need be an integral operator. On the other hand when auxiliary circumstances allow us to confine attention to integral operators it is not hard to show that our kernel will define an intertwining operator if and only if (9) is satisfied.

Returning to our main problem let us see what (9) reduces to if we let L be the one dimensional identity representation of $\{0\} \subseteq V$ and let

$M = \chi I$ where χ is a σ character of Γ. Our function A on V is then complex valued and (9) becomes (10) $A(v+\gamma) \equiv \chi(\gamma) A(v) e^{-\pi i \mathscr{I}(H(v,\gamma))}$. This is similar to but not identical with the identity (8) defining theta functions. However let $g(v) = e^{-\frac{\pi H(v,v)}{2}}$. A simple calculation shows that:

$$g(v+\gamma) = g(v) e^{-\frac{\pi H(\gamma,\gamma)}{2}} e^{-\pi R H(v,\gamma)}$$

where R denotes "real part of". Hence if $A^1 = A/g$ then A satisfies (10) if and only if A^1 satisfies (8).

Of course we can write the formula for the intertwining operator defined by T_A in terms of the function A^1. If $T_A(f) = f_A$ then

$$f_A(v) = \int_{V/\Gamma} \frac{A'(v-v_1)\, g(v-v_1)\, f(v_1)\, dv_1}{\sigma(v, -v_1)}$$

Moreover, it turns out to be advantageous to do so. Indeed writing $g(v-v_1)/\sigma(v, -v_1)$ in terms of H we find that

$$g(v-v_1)/\sigma(v, -v_1) \equiv g(v)\, g(v_1)\, e^{\pi H(v_1, v)}$$

so that

$$f_A(v) = g(v) \int_{V/\Gamma} e^{\pi H(v_1, v)} A'(v-v_1)\, g(v_1)\, f(v_1)\, dv_1 .$$

Since $v \to e^{-\pi H(v_1, v)}$ is holomorphic this suggests that f_A/g will be holomorphic wherever A^1 is holomorphic and that one should look at the operators $\frac{f \to U_x^L(fg)}{g}$. A simple computation shows that

$$\frac{U_x^L(fg)}{g} = f^1 \text{ where } f^1(v) = f(v+x)\, e^{-\frac{\pi}{2} H(x,x)}\, e^{-\pi H(x,v)}$$

Since $v \to e^{-\pi H(x,v)}$ is holomorphic we see that f^1 is holomorphic wherever f is and hence that the functions f in $H(U^L)$ for which f/g is holomorphic form an invariant subspace. That such exist is clear since for any "polynomial" P on V, Pg is in $\mathscr{L}^2(V)$. Actually it is not hard to show that the f for which f/g is holomorphic form a closed invariant subspace equal to the closure of the space of all Pg where P is a polynomial.

Let W denote the subrepresentation of U^L defined by this closed invariant subspace. The mapping $f \to f/g$ allows us to realize it in the space of all holomorphic functions which are square summable with respect to the measure $g^2\, dv$. Now it is well known and easily verified that in such a space of holomorphic functions the "evaluation functionals" $h \to h(v_0)$ are all continuous in the Hilbert space topology. Hence if T is a bounded linear operator from another Hilbert space to this one, $\varphi \to T(\varphi)(v_0)$ is a continuous linear functional on this other

Hilbert space. Thus there exists ψ_{v_0} in this Hilbert space such that $T(\varphi)(v_0) \equiv (\varphi, \psi_{v_0})$. Applying these considerations to the case at hand one shows without difficulty that any intertwining operator from $H(U^\chi)$ to U^L whose range lies in $\mathscr{H}(W)$ must be defined by a kernel A^1 satisfying (8) as indicated above. It is now routine to show that A^1 must be holomorphic and hence a theta function. On the other hand it is straightforward to verify that for every theta function A^1 the mapping $f \to f_{A^1}$ where $f_{A^1}(v) = \int\limits_{V/\Gamma} \frac{A^1(v-v_1)\, g(v-v_1)\, dv_1}{\sigma(v, -v_1)}$ is indeed an intertwining operator for U^χ and W. Thus we prove the

Theorem. *The dimension of the space of all intertwining operators for U^χ and W is equal to the dimension of the space of all theta functions belonging to H, Γ and χ.*

To complete the program we must show that W is irreducible, that V has a unique equivalence class of irreducible σ representations and we must determine the multiplicity with which the unique irreducible occurs in U^χ. Note first that $\mathscr{I} H(v_1, v_2)$ is an anti symmetric non degenerate real bilinear form on V and by elementary linear algebra is equivalent to any other such. It follows easily that σ is equivalent to a σ' arising from the set up of the Stone-von Neumann theorem. The desired uniqueness is thus a consequence of the Stone-von Neumann theorem. That W is irreducible can be proved in several ways. We shall content ourselves with the remark that it is so related to the "holomorphic function representation" of the Heisenberg commutation relations, recently studied by SEGAL and BARGMANN [3], [32] that its irreducibility is an immediate consequence of their work.

We deduce the required dimension from the following more general

Theorem. *Let G be a separable locally compact commutative group such that $x \to x^2$ is onto and let α be an anti-symmetric isomorphism of G with its dual $\hat{G}$. Form the projective multiplier σ defined by the equation $\sigma(x, y) = f_x(y)$ where $f_x = \alpha(x)$ and suppose α such that G has a unique equivalence class of irreducible σ representations. Let Γ be any closed subgroup of G such that σ restricted to Γ is equivalent to the multiplier which is identically one. Let $\Gamma^1 \supseteq \Gamma$ be the closed subgroup of all $x \in G$ for which $\alpha(x)(\gamma) = \alpha(\gamma)(x)$ for all $\gamma \in \Gamma$. Let χ be any σ character of Γ and let n be the cardinal number of Γ^1/Γ. Then if n is finite it is a square and U^χ, the σ representation of G induced by χ is the direct sum of $\sqrt{n}$ replicas of the unique irreducible σ representation of G. On the other hand if n is infinite then U^χ is the direct sum of countably many replicas of the unique irreducible σ representation of G.*

Proof. We apply the general theory of [22] to analyze the irreducible σ representations of G using Γ as normal subgroup. By hypothesis $\hat{\Gamma}^\sigma$ is just the set of all σ characters and the action of $x \in G$ on $\chi \in \hat{\Gamma}^\sigma$ takes it into the product with χ of the character $\gamma \to \alpha(\gamma)(x)/\alpha(x)(\gamma) = \alpha(x^{-2})(\gamma)$. Thus there is just one orbit in $\hat{\Gamma}^\sigma$ and for all χ the subgroup of G leaving χ fixed is Γ^1. Moreover, it follows from Theorem 8.2 of [22] that for each χ there exists a Borel function ϱ from Γ^1 to the complex numbers of modulus one such that if $\nu(x, y) = \frac{\sigma(x, y)\,\varrho(x)\,\varrho(y)}{\varrho(x y)}$ then $\nu(\gamma, x_1, \gamma_2 y) = \nu(x, y)$ for all $\gamma_1 \in \Gamma, \gamma_2 \in \Gamma$ and such that $\varrho(\gamma) = \chi(\gamma)$ for all $\gamma \in \Gamma$. Let ν^0 be the multiplier on Γ^1/Γ such that $\nu(x, y) = \nu^0(\bar{x}, \bar{y})$ where $\bar{x}$ denotes the canonical image of x in Γ^1/Γ and similarly for $\bar{y}$. It then follows from Theorem 8.3 of [22] that for each fixed χ we obtain a one-to-one map of the equivalence class of irreducible $1/\nu^0$ representations of Γ^1/Γ onto the equivalence classes of irreducible σ representations of G as follows. Given the irreducible $1/\nu^0$ representation L of Γ^1/Γ "lift" it to an irreducible $1/\nu$ representation L^1 of Γ^1 and let $L''_x = \varrho(x) L^1_x$. Then L'' will be an irreducible σ representation of Γ^1 and we may form $U^{L''}$ the σ representation of G induced by L''. Theorem 8.3 of [22] asserts that $U^{L''}$ is irreducible and that the map $L \to U^{L''}$ is one-to-one and onto between equivalence classes. Since G has a unique equivalence class of σ representations it follows that Γ^1/Γ has a unique equivalence class of $1/\nu^0$ representations. Now it follows readily from the definition concerned that the σ representation of Γ^1 induced by χ is $x \to \varrho(x) M_x$ where M_x is the regular $1/\nu_0$ representation of Γ^1/Γ lifted to a $1/\nu$ representation of Γ^1. Hence by the theorem on inducing in stages U^χ contains the unique irreducible σ representation of G just as many times as the regular $1/\nu_0$ representation of Γ^1/Γ contains the unique irreducible $1/\nu_0$ representation of Γ^1/Γ. When Γ^1/Γ is finite we have only to apply the Frobenius reciprocity theorem for projective representations of finite groups to conclude that the $1/\nu_0$ regular representation contains every $1/\nu_0$ irreducible a number of times equal to its dimension and hence that $0(\Gamma^1/\Gamma) = d^2$ where d is the dimension in question. More generally let A be any separable locally compact commutative group and let ω be a projective multiplier for A such that there is a unique equivalence class of irreducible ω representations. In $A \times A$ consider the multiplier ω^1 defined by the equation $\omega'(x_1, y_1, x_2, y_2) = \frac{\omega(x_2\, y_2)}{\omega(x_1\, y_1)}$. Let $\tilde{A}$ be the diagonal of $A \times A$; that is the subgroup of all x, y with $x = y$. Restricted to this subgroup ω' is identically one. Hence we may consider U^I the ω' representation of $A \times A$ induced by the one dimensional identity representation of $\tilde{A}$. Now an immediate computation shows that $U^I_{x,e}$ and $U^I_{e,y}$ commute for all x and y, that $y \to U^I_{e,y}$ is the ω regular representation and that $x \to U^I_{x,e}$ is the $1/\omega$ regular representation. Thus if the ω regular representation contains its unique irreducible constituent only a finite number of times then the $U^I_{x,e}$ generate a finite dimensional algebra of operators and the $1/\omega$ regular representation must have finite dimensional irreducible constituents. But $x \to V^*_{x^{-1}}$ is a $1/\omega$ representation whenever $x \to V_x$ is an ω representation. Thus there can be no finite dimensional $1/\omega$ irreducible representation unless A is finite and we conclude that whenever A is infinite the ω regular representation must contain its unique irreducible constituent infinitely many times. This completes the proof of the theorem.

15. Induced projective representations and automorphic forms

A meromorphic function on V periodic with respect to Γ is of course just a meromorphic function on the compact complex manifold V/Γ. In the special case of one complex dimension we may describe the most general compact complex manifold (compact Riemann surface) in a closely related fashion and construct the most general meromorphic function on such a surface as a quotient of holomorphic functions which have properties analogous to those of theta functions. These functions are called automorphic forms and we wish to conclude by giving some brief indications concerning the extent to which one can develop a theory of these analogous to that of Section 14.

Every Riemann surface (compact or not) has a simply connected covering surface which is conformally equivalent to either the whole complex plane, the upper half plane or the complex plane compactified by a point at ∞. We shall consider only the second case here. (The first leads back to the theory of theta functions — and the third is less interesting.) Let G then denote the group of all automorphisms of V as a Riemann surface. These automorphisms are just the transformation $z \to \frac{az+b}{cz+d}$ where a, b, c and d are real and $ad-bc>0$ and it is clear that we may restrict a, b, c, d so that $ad-bc=1$. Indeed G may be identified with the quotient of $SL(2, R)$ by its two element center. Then the fundamental group of our surface may be identified with a closed subgroup Γ of G in such a way that the surface is obtained from V by identifying two points whenever one is carried into the other by an element of G. Thus the meromorphic functions on the surface may be identified with the meromorphic functions on V which are invariant under the action of G — the so called automorphic functions on V.

In defining automorphic forms on V as analogues of theta functions we note that in the theta function situation the vector space V plays the role of both G and the upper half plane — the additive group of V acting on itself via the group operation. We note also that the positive definite inner product H is completely determined by the function $v_1, v \to e^{\pi\left(H(v_1, v) + \frac{H(v_1, v_1)}{2}\right)} = B(v_1, v)$ and that B satisfies the identity

$$B(v_1 + v_2, v) = B(v_2, v + v_1)\, B(v_1, v)\, \sigma(v_1, v_2)$$

and is holomorphic in v for each fixed v_1. Its analogue for automorphic forms is a complex valued function J on $G \times V$ which is holomorphic as a function of v for each fixed α in G and satisfies an identity of the form

$$J(\alpha_1 \alpha_2, v) = J(\alpha_1, v)\, J(\alpha_2, (v)\alpha_1)\, \sigma(\alpha_1, \alpha_2)$$

where σ is a projective multiplier for G. We define such a function J^{-r} for each real r as follows. For each $\alpha \in G$ choose a representative $\begin{pmatrix} a & b \\ c & d \end{pmatrix} \in SL(2, R)$ such that $c \geqq 0$ and let $L(\alpha, v)$ be that determination of $\log(cv + d)$ for which $L(\alpha, i) = \log\sqrt{c^2 + d^2} + i\theta$ where $0 \leqq \theta < \pi$. Then $(cv + d)^n = e^{nL(\alpha, v)}$ for every integer n and we define $(cv + d)^r = e^{rL(\alpha, v)}$ for every real number r. Finally we set $J^{-r}(\alpha, v) = (cv + d)^r$. When r is an even integer it is straightforward to verify that $J^{-r}(\alpha_1 \alpha_2, v) \equiv J^{-r}(\alpha_1, v)\, J^{-r}(\alpha_2, (v)\alpha_1)$. More generally the left side is equal to the right side multiplied by a factor which depends only on α_1 and α_2 and is of absolute value one. We denote it by $\sigma_r(\alpha_1, \alpha_2)$. An easy calculation shows that σ_r is a projective multiplier for all real r and that $\sigma_{r+2n} = \sigma_r$ for all integers n. Actually it can be shown that σ_{r_1} and σ_{r_2} are inequivalent whenever $r_1 - r_2$ is not an even integer and that every multiplier for G is equivalent to some σ_r.

Suppose now that Γ admits σ_r characters and let χ be any one of these. Then an *automorphic form* of dimension r and "multiplier system" χ is by definition a meromorphic function f on the upper half plane V such that

$$f((v)\gamma) \equiv J^r(\gamma, v)\, \chi(\gamma)\, f(v)$$

for all $v \in V$ and all $\gamma \in \Gamma$.

In the theory of theta functions our group $G = V$ had a unique equivalence class of irreducible σ representations and the dimension of the space of all theta functions turned out to be equal to the multiplicity of occurence of this unique irreducible in the σ representation U^χ induced by the σ character χ. In the present instance we may define U^χ exactly as before but for every r the group G has infinitely many inequivalent irreducible σ_r representations. On the other hand for each σ_r there are infinitely many values of s for which $\sigma_r = \sigma_s$ and these values are in one-to-one correspondence in a natural way with those equivalence classes of irreducible σ_r representations of G which appear as *discrete* direct summands of the σ_r regular representation. Thus giving a real number s amounts to giving a multiplier σ_s *and* a particular equivalence class of irreducible σ_s representations. It turns out that there is a natural isomorphism between the space of all (entire) automorphic forms of dimension s and multiplier system χ and the space of all intertwining operators for U^χ and the irreducible σ_s representation with index s.

Specifically for each real r let $\mathcal{H}^r$ denote the Hilbert space of all complex valued functions on the upper half plane which are square summable with respect to the measure $y^{-(2+r)}\,dx\,dy$. For each $\alpha \in SL(2, R)$ and each $f \in \mathcal{H}^r$ let $W_\alpha^r(f)(v) = f((v)\alpha)\,J^{-r}(\alpha, v)$. Then a routine computation shows that each W_α^r is unitary and that $\alpha \to W_\alpha^r$ is a σ_r representation of G. When r is an even integer W^r is an ordinary representation and is clearly a "multiplier representation" in the sense discussed in Section 6. (Note that we have used the word "multiplier" for two rather different objects. One is a one cocycle and the other is a two cocycle.) Since the action of G on V is transitive these multiplier representations must be the representations of G induced by characters of the subgroup K of G leaving a point of V fixed. If we choose this point to be the complex unit i we find that K is the image in G of the commutative subgroup of all members of $SL(2, R)$ of the form $\begin{pmatrix} \cos\theta & \sin\theta \\ -\sin\theta & \cos\theta \end{pmatrix}$. It is now easy to verify that W^r is the representation of G induced by the character $\theta \to e^{-ir\theta}$. Moreover it is clear that as r varies over the even integers these characters vary over all of the characters of K. Now for $r = -2, -4, -6, \ldots$ the space $\mathcal{H}^r$ contains non zero holomorphic functions and these functions clearly form an invariant subspace. Actually it is known that this subspace is closed and irreducible and hence defines an irreducible subrepresentation $W^{r,0}$ of W^r. The irreducible representations $W^{-2,0}, W^{-4,0}, W^{-6,0} \ldots$ are of course members of the well known discrete series of Bargmann [2]. As far as automorphic forms of negative even dimension are concerned the analogue of the theorem relating U^χ to the space of theta functions states the following. For any character χ of Γ (in particular $\chi \equiv 1$) and any negative even integer r there is a natural isomorphism between the space of all entire automorphic forms of character χ and dimension r and the space $R(U^\chi, W^{r,0})$ of all intertwining operators for U^χ and the irreducible representation $W^{r,0}$ of G.

When r is a real number other than an even integer the multiplier $\sigma_r \not\equiv 1$ and Γ need not admit any σ_r characters. When it does we may proceed much as above. The map $\theta \to e^{i(r+2k)\theta}$ is a σ_r character of K for all integers k and we get all σ_r characters of K in this way. Moreover W^{r+2k} is the σ_r representation of G induced by the character $\theta \to e^{i(r+2k)\theta}$ and whenever $r+2k<-1$; that is, whenever k is an integer less than $-(r+1)/2$ we may define a holomorphic subrepresentation $W^{r+2k,0}$ of W^{r+2k}. Finally for every real $r<-1$ one finds that there exists an isomorphism between the space of all entire automorphic forms of dimension r and multiplier system χ and the space $R(U^\chi, W^{r,0})$ of all intertwining operators for the induced σ_r representation U^χ and the irreducible σ_r representation $W^{r,0}$.

Once again we find ourselves confronted with the problem of finding the structure of an induced representation. However whether or not we have a trivial multiplier this problem is much more difficult than the analogue which is solved by the final theorem of Section 14. This is because we have replaced a commutative group by a semi simple Lie group. We refer the reader to [13] and Section 4 of [33] for further details.

Looking at automorphic forms as intertwining operators for induced representations is useful in interpreting certain concepts. For example if T_1 and T_2 are both intertwining operators for U^χ and $W^{r,0}$ then $T_2^* T_1$ is in the commuting algebra of the irreducible representation $W^{r,0}$ and hence is a constant c times the identity. Clearly $T_1, T_2 \to c(T_1, T_2)$ is a positive definite Hermitean inner product in the space of intertwining operators and we thus define such an inner product in the corresponding space of automorphic forms. This is of course just the celebrated inner product introduced in 1940 by PETERSSON [30]. Also if T_1 is a member of the commuting algebra of U^χ and T_2 is an intertwining operator for U^χ and $W^{r,0}$ then $T_2 T_1$ is an intertwining operator for U^χ and $W^{r,0}$. In this way the space of automorphic forms becomes an R module where R is the commuting algebra of U^χ. In the special case in which $\chi=1$ it follows from the general theory of induced representations that there is a member of the commuting algebra of U^χ canonically associated with every $\Gamma:\Gamma$ double coset which contains only a finite number of right and left Γ cosets. Thus each such $\Gamma:\Gamma$ double coset defines a linear operator in the space of automorphic forms. These are the well known Hecke operators.

The considerations of this section of course barely scratch the surface of the rich interrelationship that exists between the theory of automorphic forms and the theory of induced representations. One can consider non compact Riemann surfaces, higher dimensional σ_r representations of Γ, non holomorphic automorphic forms, complex manifolds of higher dimension etc. etc. The whole subject is in a state of rapid development and we shall not attempt to describe it further here.

References

1. Ambrose, W.: Spectral resolution of groups of unitary operators. Duke Math. J. **11**, 589–595 (1944).
2. Bargmann, V.: Irreducible unitary representations of the Lorentz group. Ann. of Math. (2) **48**, 568–640 (1947).
3. — On a Hilbert space of analytic functions and an associated integral transform. Comm. Pure Appl. Math. **14**, 187–214 (1961).
4. Birkhoff, G. D.: Proof of a recurrence theorem for strongly transitive systems. Proof of the ergodic theorem. Proc. Nat. Acad. Sci. US **17**, 650–660 (1931).
5. Bradley, C. J.: Space groups and selection rules. J. Math. Phys. **7**, 1145–1152 (1966).
6. Cartier, P.: Quantum mechanical commutation relations and theta functions. Algebraic groups and discontinuous subgroups. Proc. Symp. Pure Math. Boulder Colorado 1965, p. 361–383. Providence, R. I.: Amer. Math. Soc. 1966.
7. Dixmier, J.: Les C^* algebres et leurs représentations. Cahiers Scientifiques, Fasc. XXIX. Paris: Gauthier-Villars 1964.
8. Gelfand, I., and M. Neumark: Unitary representations of the group of linear transformations of the straight line. Dokl. Akad. Nauk SSSR **11**, 411–504 (1947) [Russian].
9. Godement, R.: Sur une généralization d'un théorème de Stone. Compt. Rend. **218**, 901–903 (1944).
10. Halmos, P. R.: Introduction to Hilbert space and the theory of spectral multiplicity. New York: Chelsea Publ. Co. 1951.
11. Knox, R. S., and A. Gold: Symmetry in the solid state. New York: W. A. Benjamin Inc. 1964.
12. Koopman, B. O.: Hamiltonian systems and Hilbert space. Proc. Nat. Acad. Sci. US **17**, 315–318 (1931).
13. Langlands, R. P.: The dimension of spaces of automorphic forms. Am. J. Math. **85**, 99–125 (1963).
14. Lomont, J. S.: Decomposition of direct products of representations of the inhomogeneous Lorentz group. J. Math. Phys. **1**, 237–243 (1960).
15. Loomis, L. H.: Note on a theorem of Mackey. Duke Math. J. **19**, 641–645 (1952).
16. Mackey, G. W.: A theorem of Stone and von Neumann. Duke Math. J. **16**, 313–326 (1949).
17. — Imprimitivity for representations of locally compact groups I. Proc. Nat. Acad. Sci. US **35**, 537–545 (1949).
18. — On induced representations of groups. Am. J. Math. **73**, 576–592 (1951).
19. — Induced representations of locally compact groups. I. Ann. of Math. (2) **55**, 101–139 (1952).
20. — The theory of group representations (mimeographed notes by J. M. G. Fell and D. Lowdenslager). Chicago: Chicago University Press 1955.
21. — Borel structure in groups and their duals. Trans. Am. Math. Soc. **85**, 134–165 (1957).
22. — Unitary representations of group extensions I. Acta Math. **99**, 265–311 (1958).

23. MACKEY, G.W.: Infinite dimensional group representations. Bull. Am. Math. Soc. **69**, 628–686 (1963).
24. — Induced representations of groups and quantum mechanics. New York: W. A. Benjamin Inc. 1968.
25. MAUTNER, F. I.: Unitary representations of locally compact groups. I. Ann. of Math. (2) **51**, 1–25 (1950).
26. NEUMANN, J. v.: Die Eindeutigkeit der Schrodingerschen Operatoren. Math. Ann. **104**, 570–578 (1931).
27. — Proof of the quasi ergodic hypothesis. Proc. Nat. Acad. Sci. US **18**, 70–82 (1932).
28. — On rings of operators. Reduction theory. Ann. of Math. (2) **50**, 401–485 (1949).
29. NEUMARK, M. A.: Positive definite operator functions on a commutative group. Izv. Akad. Nauk SSSR **7**, 237–244 (1943). [Russian].
30. PETERSSON, H.: Über eine Metrisierung der Automorphen Formen und die Theorie der Poincareschen Reihen. Math. Ann. **117**, 453–537 (1940).
31. PUKANSZKY, L.: On the Kronecker product of irreducible unitary representations of the inhomogeneous Lorentz group. J. Math. Mech. **10**, 475–491 (1961).
32. SEGAL, I. E.: Mathematical problems of relativistic physics. Lectures in applied mathematics (Proceedings of the Summer Seminar, Boulder Colorado, 1960), vol. II. Providence, R. I.: Amer. Math. Soc. 1963.
33. SELBERG, A.: Harmonic analysis and discontinuous groups in weakly symmetric Riemannian spaces with applications to Dirichlet series. J. Indian Math. Soc. **20**, 47–87 (1956).
34. STONE, M. H.: Linear transformations in Hilbert space. III. Operational methods and group theory. Proc. Nat. Acad. Sci. US **16**, 172–175 (1930).
35. — Linear transformations in Hilbert space and their applications to analysis. Am. Math. Soc. Coll. Publ. XV. New York 1932.
36. WEIL, A.: Sur certains groupes d'opérateurs unitaires. Acta Math. **111**, 143–211 (1964).
37. WIGHTMAN, A. S.: On the localizability of quantum mechanical systems. Rev. Mod. Phys. **34**, 845–872 (1962).
38. WIGNER, E. P.: On unitary representations of the inhomogeneous Lorentz group. Ann. of Math. (2) **40**, 149–204 (1939).

Convolution Operators in Spaces of Nuclearly Entire Functions on a Banach Space*

By LEOPOLDO NACHBIN **

University of Rochester, Rochester, New York

We shall be concerned with the theorem stated below. Its proof depends on the two propositions indicated afterwards.

Let us start by explaining some of the pertinent notation and terminology. For additional information on the background material, we refer the reader to the bibliography quoted at the end of this article.

We shall be dealing with a complex Banach space E. For each positive integer $m = 0, 1, \ldots$, $\mathscr{P}(^m E)$ will denote the Banach space of all continuous m-homogeneous complex-valued polynomials on E. Moreover, $\mathscr{H}(E)$ will represent the vector space of all complex-valued functions on E which are holomorphic on the entire E. For each $f \in \mathscr{H}(E)$, we have its Taylor series at the origin

$$f(x) = \sum_{m=0}^{\infty} \frac{1}{m!} \hat{d}^m f(0)(x)$$

for every $x \in E$, and the corresponding differentials of order $m = 0, 1, \ldots$

$$\hat{d}^m f(0) \in \mathscr{P}(^m E).$$

If E' indicates the dual Banach space to E, we shall have that $\varphi^m \in \mathscr{P}(^m E)$ for every $\varphi \in E'$. We shall denote by $\mathscr{P}_f(^m E)$ the vector subspace of $\mathscr{P}(^m E)$ generated by all φ^m when φ runs over E'. It consists of those elements of $\mathscr{P}(^m E)$ each of which may be represented as a finite sum

$$\varphi_1^m + \cdots + \varphi_r^m,$$

where the φ_j belong to E' for each $j = 1, \ldots, r$. An element of $\mathscr{P}(^m E)$ is said to be of finite type in case it lies in $\mathscr{P}_f(^m E)$.

* The preparation of this work was sponsored in part by the U.S.A. National Science Foundation through a grant to the University of Rochester, Rochester, New York, U.S.A.

** On leave from the Centro Brasileiro de Pesquisas Físicas and the Instituto de Matemática Pura e Aplicada, Universidade do Brasil, Rio de Janeiro, Guanabara, Brasil.

The Banach space $\mathscr{P}_N(^mE)$ of all nuclear m-homogeneous complex-valued polynomials on E is characterized by the following requirements:

(1) $\mathscr{P}_N(^mE)$ is a vector subspace of $\mathscr{P}(^mE)$;

(2) $\mathscr{P}_N(^mE)$ is a Banach space with respect to a norm denoted by $\|\cdot\|_N$ and called the nuclear norm; it is to be distinguished from the current norm on $\mathscr{P}(^mE)$ which is denoted simply by $\|\cdot\|$;

(3) $\mathscr{P}_f(^mE)$ is contained and dense in $\mathscr{P}_N(^mE)$ with respect to the nuclear norm and the inclusion mapping of $\mathscr{P}_N(^mE)$ into $\mathscr{P}(^mE)$ is continuous from the nuclear norm to the current one;

(4) For each $P \in \mathscr{P}_f(^mE)$, its nuclear norm $\|P\|_N$ is equal to the infimum of the sums

$$\|\varphi_1\|^m + \cdots + \|\varphi_r\|^m$$

for all possible representations

$$P = \varphi_1^m + \cdots + \varphi_r^m,$$

where the φ_j belong to E' for each $j = 1, \ldots, r$.

A nuclear complex-valued polynomial on E is by definition a complex-valued polynomial on E all of whose homogeneous components are nuclear in the above sense.

A nuclear complex-valued exponential-polynomial on E is defined to be a function on E of the form Pe^φ, where P is a nuclear complex-valued polynomial on E and $\varphi \in E'$.

In order to introduce the locally convex space $\mathscr{H}_N(E)$ of all nuclearly entire complex-valued functions on E, let us make the following preliminary considerations.

For every norm α on E which is equivalent to the one originally given on the same vector space, and every subset X of E, we shall say for short that α is X-centered if X is contained in the open ball with respect to α of center at 0 and radius equal to 1.

We shall define $f \in \mathscr{H}(E)$ to be nuclearly entire in case we have

$$\hat{d}^m f(0) \in \mathscr{P}_N(^mE)$$

for each $m = 0, 1, \ldots$, and, corresponding to every compact subset K of E, there exists an equivalent norm α on E which is K-centered and is such that

$$\sum_{m=0}^{\infty} \left\| \frac{1}{m!} \hat{d}^m f(0) \right\|_{N\alpha} < +\infty;$$

here and in the sequel $\|\cdot\|_{N\alpha}$ stands for the nuclear norm of a nuclear homogeneous complex-valued polynomial on E when this vector space is endowed with α rather than its originally given norm. We shall denote by $\mathscr{H}_N(E)$ the vector subspace of $\mathscr{H}(E)$ of all nuclearly entire complex-valued functions on E.

For the purpose of describing the natural locally convex topology on $\mathscr{H}_N(E)$ that we shall use, let us introduce the following concepts.

A seminorm p on the vector space $\mathscr{H}_N(E)$ is said to be nuclearly ported by a compact subset K of E provided that, corresponding to every equivalent norm α on E which is. K-centered, there exists some real number $c(\alpha) > 0$ for which the following estimate

$$p(f) \leqq c(\alpha) \cdot \sum_{m=0}^{\infty} \left\| \frac{1}{m!} \hat{d}^m f(0) \right\|_{N\alpha}$$

holds true for an arbitrary $f \in \mathscr{H}_N(E)$. We notice, as it is standard in similar situations of seminorms ported by compact subsets, that the right-hand side of the above estimate is not necessarily always finite. However, once $f \in \mathscr{H}_N(E)$ and the compact subset K of E are given, there exists some equivalent norm α on E for which the mentioned right-hand side turns out to be finite; for such a choice of α the indicated estimate will give us an information on $p(f)$, hence on the seminorm p.

The locally convex topology $\mathscr{I}_{\omega N}$ on $\mathscr{H}_N(E)$ that we shall use is the one defined by all seminorms on $\mathscr{H}_N(E)$ each of which is nuclearly ported by some compact subset of E.

A convolution operator $\mathcal{O}$ in $\mathscr{H}_N(E)$ is defined to be a continuous linear mapping of $\mathscr{H}_N(E)$ into itself commuting with all translations in E, that is $\mathcal{O}\,\tau_t = \tau_t\,\mathcal{O}$ for all $t \in E$; here the translation mapping τ_t of $\mathscr{H}_N(E)$ onto itself is defined by $(\tau_t f)(x) = f(x-t)$ for all $x \in E$ and an arbitrary $f \in \mathscr{H}_N(E)$, where $t \in E$. Such a convolution operator is actually a constant coefficient linear differential operator of finite or infinite order acting in $\mathscr{H}_N(E)$.

Theorem. *The vector subspace $\mathcal{O}^{-1}(0)$ on which a convolution operator $\mathcal{O}$ in $\mathscr{H}_N(E)$ does vanish is the closure of its vector subspace generated by the nuclear exponential-polynomials that it contains.*

As it is standard in this type of an approximation result, the proof of the theorem is reduced via the Hahn-Banach theorem to two propositions: one of them concerns a characterization of Borel transforms of all the elements in the dual space $\mathscr{H}'_N(E)$ to $\mathscr{H}_N(E)$; and the other one refers to a division property between such Borel transforms. We pass now to a description of these propositions.

If $T \in \mathscr{H}'_N(E)$, that is T is a continuous linear form on $\mathscr{H}_N(E)$, its Borel transform $\hat{T}$ is the complex-valued function on E' defined by

$$\hat{T}(\varphi) = T(e^{\varphi})$$

for every $\varphi \in E'$. The mapping $T \mapsto \hat{T}$ is linear and one-to-one. In order to characterize its image set, we shall introduce the following concepts.

Let α be a seminorm on the vector space E. Denote by E_α the completion of the normed space associated to E when this vector space is seminormed by α. We shall say that α is compact in case the natural linear mapping $E \to E_\alpha$ is compact. This means that the closed ball in E with respect to the norm originally given in E of center at 0 and radius equal to 1 is totally bounded with respect to α; that is, given any $\varepsilon > 0$, there are $x_1, \ldots, x_r \in E$ such that, for every $x \in E$ for which $\|x\| \leqq 1$, we can find some $j = 1, \ldots, r$ satisfying $\alpha(x_j - x) < \varepsilon$.

An entire function $f \in \mathscr{H}(E)$ is said to be of compact exponential type in case there exists a compact seminorm α on E such that, for every $\varepsilon > 0$ we can find a real number $c(\varepsilon) \geqq 0$ for which the following estimate

$$|f(x)| \leqq c(\varepsilon) \cdot \exp[\alpha(x) + \varepsilon \cdot \|x\|]$$

holds true for every $x \in E$.

In case, however, instead of having an entire function on E, we are dealing with an entire function on E', as it will be precisely our case, we will have the following more stringent notion besides the already acquired one of an entire function of compact exponential type on E'. Letting α denote now a seminorm on E', denote by E'_α the completion of the normed space associated to E' when this vector space is seminormed by α. In case the natural linear mapping $E' \to E'_\alpha$ is continuous, that is α is continuous, we may consider the continuous linear transpose mapping $(E'_\alpha)' \to E''$, which is actually one-to-one in our case, where $(E'_\alpha)'$ stands for the dual Banach space to E'_α and E'' represents the double dual Banach space to E. We shall say that α is E-compact if not only α is compact, that is $E' \to E'_\alpha$ is compact, hence continuous, but in addition the transpose mapping $(E'_\alpha)' \to E''$ which is necessarily compact does map $(E'_\alpha)'$ into the natural image of E in E''; that is, Φ being any α-continuous linear form on E', there is a necessarily unique $x \in E$ such that $\Phi(\varphi) = \varphi(x)$ for every $\varphi \in E'$.

An entire function $F \in \mathscr{H}(E')$ is said to be of E-compact exponential type in case there exists an E-compact seminorm α on E' such that, for every $\varepsilon > 0$ we can find a real number $c(\varepsilon) \geqq 0$ for which the following estimate

$$|F(\varphi)| \leqq c(\varepsilon) \cdot \exp[\alpha(\varphi) + \cdot \varepsilon \|\varphi\|]$$

holds true for every $\varphi \in E'$.

Proposition 1. *A complex-valued function F on E' is the Borel transform $\hat{T}$ of some continuous linear form T on $\mathscr{H}_N(E)$ if and only if F is an entire function of E-compact exponential type on E'.*

Proposition 2. *F_1, F_2 and F_3 being entire complex-valued functions on E' such that $F_1 = F_2 F_3$ and F_2 is not identically zero, then F_3 will be of E-compact exponential type along with F_1 and F_2.*

GUPTA's previous work was concerned with results analogous to the preceding theorem and propositions for convolution operators in the Frechet space of all nuclearly entire complex-valued functions of bounded type $\mathscr{H}_{Nb}(E)$. The results indicated in the present note were obtained from the bounded nuclear case by a sort of an inductive limit process.

There are other natural possible candidates for the concepts of nuclearly entire functions and of nuclearly entire functions of bounded type. It is not yet known whether they are equivalent to the definitions given here or in GUPTA's work, and whether results similar to those we proved for the considered versions of $\mathscr{H}_{Nb}(E)$ and of $\mathscr{H}_N(E)$ can also be established for such alternative concepts.

The same kind of a theory should be developped for the spaces $\mathscr{H}(E)$ of all entire complex-valued functions on E, and $\mathscr{H}_b(E)$ of all entire complex-valued functions of bounded type on E, that is for the current holomorphy type.

Once these various cases are settled, there will be hope of establishing a similar theory by means of the concept of a holomorphy type Θ to collect $\mathscr{H}_N(E)$ and $\mathscr{H}(E)$ into $\mathscr{H}_\Theta(E)$, and to collect $\mathscr{H}_{Nb}(E)$ and $\mathscr{H}_b(E)$ into $\mathscr{H}_{\Theta b}(E)$.

Finally, through the use of weights on E, it should become possible to collect $\mathscr{H}_\Theta(E)$ and $\mathscr{H}_{\Theta b}(E)$ into a single type of a locally convex space of entire complex-valued functions on E leading to a synthesis of these various aspects of the theory.

Detailed proofs of the above mentioned results will appear elsewhere in a joint paper with GUPTA.

References

1. GROTHENDIECK, A.: Produits tensoriels topologiques et espaces nucléaires. Mem. Am. Math. Soc. **16**, 1—140 (1955).
2. GUPTA, C. P.: Malgrange theorem for nuclearly entire functions of bounded type on a Banach space. Notas de Matemática. Inst. Mat. Pura e Apl. (Rio de Janeiro) **37**, 1—50 (1968).
3. HÖRMANDER, L.: Linear partial differential operators. Berlin-Göttingen-Heidelberg: Springer 1963.
4. MALGRANGE, B.: Existence et approximation des solutions des équations aux dérivées partielles et des équations de convolution. Ann. Inst. Fourier **6**, 271—355 (1955—1956).
5. NACHBIN, L.: Topology on spaces of holomorphic mappings. Berlin-Heidelberg-New York: Springer 1969.
6. SCHWARTZ, L.: Théorie des distributions, tomes 1 & 2. Paris: Hermann 1950—1951.
7. TRÈVES, F.: Linear partial differential equations with constant coefficients. New York-London-Paris: Gordon & Breach 1966.

Operants: A Functional Calculus for Non-Commuting Operators*

By Edward Nelson

Princeton University

Dedicated to Professor Marshall Harvey Stone

Abstract

Quantum theory has stimulated the study of functions of non-commuting operators, giving rise in particular to the Weyl calculus for functions of the momentum and position operators and the Feynman calculus of time-ordered integrals. Here we describe a general functional calculus for non-commuting operators on a Banach space and examine the calculus in the simplest cases. For finite matrices the theory has an interesting connection with symmetric hyperbolic systems with constant coefficients.

1. Introduction

Operants are new mathematical objects which we introduce in order to study the functional calculus of non-commuting operators. The word "operant" is used to suggest the potentiality of operating. The advantage of operants is that the ordinary commutative functional calculus may be applied to them quite freely. Operators are obtained from operants by applying a certain linear mapping $\mathscr{T}$, which is a kind of integration process.

There is essentially only one functional calculus for non-commuting operators. Let $A_1, \ldots, A_n$ be operators and suppose we wish to define a mapping

$$f \mapsto f(A_1, \ldots, A_n) \tag{1}$$

from polynomials f in n variables to operators, which is linear in f, and such that if f is a polynomial p in one variable composed with a linear function:

$$f(z_1, \ldots, z_n) = p(a_1 z_1 + \ldots + a_n z_n), \tag{2}$$

then

$$f(A_1, \ldots, A_n) = p(a_1 A_1 + \ldots + a_n A_n).$$

* Research supported by NSF Grant GP-7309.

These two simple requirements determine the mapping, since over a field of characteristic zero every polynomial is a sum of polynomials of the form (2). For example,

$$z_1 z_2 = \tfrac{1}{2}(z_1+z_2)^2 - \tfrac{1}{2} z_1^2 - \tfrac{1}{2} z_2^2,$$

so that if $f(z_1, z_2) = z_1 z_2$ then

$$f(A_1, A_2) = \tfrac{1}{2}(A_1+A_2)^2 - \tfrac{1}{2} A_1^2 - \tfrac{1}{2} A_2^2 = \tfrac{1}{2}(A_1 A_2 + A_2 A_1).$$

In general, (1) may be computed by writing f in symmetric form and then substituting the operators for the variables. For functions f more general than polynomials, (1) may be defined by a limiting process, and this is one of the main topics of this paper.

The operational calculus which we have outlined suffers from the unavoidable defect that it is not multiplicative in f. For example, if $g(z_1, z_2) = z_1$, $h(z_1, z_2) = z_2$ and $f = gh$ then $f(A_1, A_2) = g(A_1, A_2)\, h(A_1, A_2)$ — that is, $\tfrac{1}{2}(A_1 A_2 + A_2 A_1) = A_1 A_2$ — if and only if A_1 and A_2 commute. To circumvent this difficulty we associate to each operator A an operant $\tilde{A}$, multiply operants by means of a formal commutative multiplication, and recover operators from operants by means of a linear mapping $\mathcal{T}$ such that

$$f(A_1, \ldots, A_n) = \mathcal{T} f(\tilde{A}_1, \ldots, \tilde{A}_n).$$

Any operant α such that $\mathcal{T}\alpha\beta = 0$ for all operants β is identified with 0. This rather trivial construction leads even in the simplest cases to complicated and interesting commutative algebras of operants.

We also consider operators $A_1, \ldots, A_n$ depending on a real time parameter t and construct an operant algebra and a linear mapping $\mathcal{T}$ which is time-ordered integration. In this way we make rigorous the functional calculus of FEYNMAN [2].

2. The complete symmetric tensor algebra

Let $\mathcal{V}$ be a normed linear space and let $\mathcal{S}_0 = \mathcal{S}_0(\mathcal{V})$ be the symmetric tensor algebra over $\mathcal{V}$. We shall find it convenient not to identify $\mathcal{V}$ with a subset of $\mathcal{S}_0$, and we represent the canonical injection of $\mathcal{V}$ into $\mathcal{S}_0$ by

$$A \mapsto \tilde{A}.$$

(The reason for this is that in the applications $\mathcal{V}$ will be a space of operators and the notation AB would be ambiguous.) Thus every element α of $\mathcal{S}_0$ is a finite sum

$$\alpha = a + \sum_j \tilde{A}_{i_1}^{(j)} \ldots \tilde{A}_{i_{n_j}}^{(j)} \tag{3}$$

where a is in $\boldsymbol{C}$ and each $A_i^{(j)}$ is in $\mathscr{V}$. We give $\mathscr{S}_0$ the norm

$$\|\alpha\| = \inf\left\{|a| + \sum_j \|A_{i_1}^{(j)}\| \cdots \|A_{i_{n_j}}^{(j)}\|\right\} \tag{4}$$

where the infimum is over all representations of α of the form (3). It is easy to see that $\|\alpha\beta\| \leqq \|\alpha\| \, \|\beta\|$ for all α, β in $\mathscr{S}_0$, so that $\mathscr{S}_0$ is a commutative normed algebra. Consequently its completion $\mathscr{S} = \mathscr{S}(\mathscr{V})$ is a commutative Banach algebra. We call $\mathscr{S}$ the *complete symmetric tensor algebra over* $\mathscr{V}$.

By the *spectrum* of a commutative Banach algebra we mean the Gelfand maximal ideal space, which may be identified with the set of multiplicative linear functionals not identically zero, in the weak* topology. This basic construction was first conceived by STONE for Boolean algebras and commutative self-adjoint operator algebras on Hilbert space.

We let $\mathscr{V}^*$ be the Banach space of all continuous linear functionals on $\mathscr{V}$ and recall that its unit ball is compact in the weak* topology (the topology induced by the completion of $\mathscr{V}$).

Theorem 1. *Let $\mathscr{S}$ be the complete symmetric tensor algebra over the normed linear space $\mathscr{V}$. There is a homeomorphism $\phi \mapsto \phi_0$ of the spectrum of $\mathscr{S}$ onto the unit ball of $\mathscr{V}^*$ such that*

$$\phi(\tilde{A}) = \phi_0(A) \tag{5}$$

for all A in $\mathscr{V}$.

Proof. If ϕ_0 is in the unit ball of $\mathscr{V}^*$ we define ϕ on the image of $\mathscr{V}$ in $\mathscr{S}$ by (5). Then ϕ_0 has a unique multiplicative linear extension ϕ to $\mathscr{S}_0$. By (4), $\|\phi\| \leqq 1$, so that ϕ has a unique continuous multiplicative linear extension (again denoted by ϕ) to $\mathscr{S}$. Conversely, any element ϕ of the spectrum of $\mathscr{S}$ is of norm $\leqq 1$ (this is true for any commutative Banach algebra), so its restriction to the image of $\mathscr{V}$ in $\mathscr{S}$ is of norm $\leqq 1$. But it is easy to see that $\|A\| = \|\tilde{A}\|$ for all A in $\mathscr{V}$, so that ϕ_0 defined by (5) is of norm $\leqq 1$. If $\phi_\nu \to \phi$ in the spectrum of $\mathscr{S}$ then $\phi_{\nu 0}(A) \to \phi_0(A)$ for all A in $\mathscr{V}$, and consequently for all A in the completion of $\mathscr{V}$ since the $\phi_{\nu 0}$ are uniformly bounded in norm. Therefore the mapping $\phi \to \phi_0$ is continuous, and since it is a bijective mapping of one compact Hausdorff space onto another it is a homeomorphism. This concludes the proof.

Theorem 1 will be used in the sequel. The next theorem is included only to show that the functional calculus for $\mathscr{S}$ gives nothing of interest (in contrast to the functional calculus to be developed for the algebra of operants).

Theorem 2. *Let $\mathcal{S}$ be the complete symmetric tensor algebra over the normed linear space $\mathcal{V}$, let A be in $\mathcal{V}$, and let*

$$f(z)=\sum_{n=0}^{\infty} a_n z^n$$

be a power series which is absolutely convergent for $|z| \leqq \|A\|$. Define $f(\tilde{A})$ by

$$f(\tilde{A})=\sum_{n=0}^{\infty} a_n \tilde{A}^n .$$

Then

$$\|f(\tilde{A})\|=\sum_{n=0}^{\infty} |a_n|\, \|A\|^n . \tag{6}$$

Proof. By (4), $\|\tilde{A}^n\| \leqq \|A\|^n$. Let ϕ_0 be in the unit ball of $\mathcal{V}^*$ with $\phi_0(A)=\|A\|$. (Such a ϕ_0 exists by the Hahn-Banach theorem.) Then, if ϕ is the corresponding element of the spectrum of $\mathcal{S}$,

$$\phi(\tilde{A}^n)=\phi(\tilde{A})^n=\|A\|^n ,$$

and since $\|\phi\| \leqq 1$ this implies that $\|\tilde{A}^n\|=\|A\|^n$. The space $\mathcal{S}_0$ is a graded vector space, and by (4) it is easy to see that the norm of an element of $\mathcal{S}_0$ is the sum of the norms of its homogeneous components. Hence for each N,

$$\left\|\sum_{n=0}^{N} a_n \tilde{A}^n\right\| = \sum_{n=0}^{N} |a_n|\, \|A\|^n ,$$

from which (6) follows.

3. Definition of operants

We denote the unit of any Banach algebra with unit by 1.

Theorem 3. *Let $\mathcal{C}$ be a Banach algebra with unit, $\mathcal{V}$ a linear subspace of $\mathcal{C}$ containing 1, $\mathcal{S}$ the complete symmetric tensor algebra over $\mathcal{V}$. There is a unique continuous linear mapping*

$$\mathcal{T}: \mathcal{S} \to \mathcal{C}$$

such that for all A in $\mathcal{V}$ and polynomials p of one variable,

$$\mathcal{T} p(\tilde{A})=p(A) . \tag{7}$$

For all α in $\mathcal{S}$, $\|\mathcal{T}\alpha\| \leqq \|\alpha\|$. For all $A_1, \ldots, A_n$ in $\mathcal{V}$,

$$\mathcal{T} \tilde{A}_1 \ldots \tilde{A}_n=\frac{1}{n!} \sum_{\pi} A_{\pi(1)} \ldots A_{\pi(n)} \tag{8}$$

where the summation extends over all permutations π of $1, \ldots, n$.

Proof. The right hand side of (8) is symmetric and multilinear in $A_1, \ldots, A_n$. Hence there is a unique linear mapping $\mathcal{T}: \mathcal{S}_0 \to \mathcal{C}$ such that (8) holds and $\mathcal{T} a = a$ for all a in $\boldsymbol{C}$. By (4), $\|\mathcal{T}\alpha\| \leqq \|\alpha\|$ for all α in $\mathcal{S}_0$, so that $\mathcal{T}$ extends by continuity to $\mathcal{S}$. For this mapping $\mathcal{T}$, (7) clearly holds. This proves the existence of $\mathcal{T}$.

The uniqueness of $\mathcal{T}$ follows from the well-known fact that every polynomial in n variables over a field of characteristic zero is a sum of polynomials in one variable composed with linear functions. In fact, one has the easily verified identity

$$\prod_{i=1}^{n} z_i = \frac{1}{n!} \sum_{r=1}^{n} (-1)^{n-r} \sum_{i_1 < \cdots < i_r} (z_{i_1} + \cdots + z_{i_r})^n,$$

in which the variables z_i need not be distinct. This completes the proof.

Given $\mathcal{V}$ as in Theorem 3, we define $\mathcal{N} = \mathcal{N}(\mathcal{V})$ by

$$\mathcal{N} = \{\alpha \in \mathcal{S} : \mathcal{T}\alpha\beta = 0 \text{ for all } \beta \text{ in } \mathcal{S}\}.$$

It is immediate that $\mathcal{N}$ is a closed ideal in $\mathcal{S}$ and that $\mathcal{T}$ is 0 on $\mathcal{N}$. We define $\mathcal{A} = \mathcal{A}(\mathcal{V})$ by

$$\mathcal{A} = \mathcal{S} / \mathcal{N}$$

Then $\mathcal{A}$ is a commutative Banach algebra and $\mathcal{T}$ induces a linear mapping (again denoted by $\mathcal{T}$)

$$\mathcal{T} : A \to \mathcal{C}$$

such that $\|\mathcal{T}\alpha\| \leqq \|\alpha\|$ for all α in $\mathcal{A}$. Elements of $\mathcal{A}$ are called *operants over* $\mathcal{V}$.

If A is in $\mathcal{V}$ then we shall use the symbol $\tilde{A}$ to denote either the image of A in $\mathcal{S}$ or in $\mathcal{A}$, it being made clear from the context which is meant. The mapping $A \mapsto \tilde{A}$ of $\mathcal{V}$ into $\mathcal{A}$ is injective; in fact, it is isometric since by (4), $\|\tilde{A}\| \leqq \|A\|$ and also $\|A\| = \|T\tilde{A}\| \leqq \|\tilde{A}\|$.

By assumption the unit 1 of $\mathcal{C}$ is in $\mathcal{V}$, so $\tilde{1}$ is in $\mathcal{S}$. The element $1 - \tilde{1}$ of $\mathcal{C}$ is not 0 (since 1 is homogeneous of degree 0 in $\mathcal{S}_0$ and $\tilde{1}$ is homogeneous of degree 1 in $\mathcal{S}_0$). However, it is clear from (8) that $1 - \tilde{1}$ is in $\mathcal{N}$. Consequently, $1 = \tilde{1}$ in $\mathcal{A}$.

4. Time-ordered integration and the Feynman calculus

FEYNMAN [2] discovered a beautiful operator calculus in which the basic idea is that the order of operation of operators is indicated by a time parameter rather than by the position of the operators in an expression. This idea has been quite useful in formal developments related to quantum electrodynamics. For a discussion of the calculus we refer to [2]; here we show how the notions involved can be made rigorous.

Let $\mathscr{C}$ be a Banach algebra with unit. By $L^1_{\mathscr{C}}(\boldsymbol{R})$ we mean the Banach space of all Bochner integrable functions from $\boldsymbol{R}$ to $\mathscr{C}$ (modulo functions equal a.e. to 0) with the norm

$$\|A\| = \int_{-\infty}^{\infty} \|A(t)\| \, dt.$$

Theorem 4. *Let $\mathscr{C}$ be a Banach algebra with unit, let $\mathscr{V}$ be a linear subspace of $L^1_{\mathscr{C}}(\boldsymbol{R})$, and let $\mathscr{S}$ be the complete symmetric tensor algebra over $\mathscr{V}$. There is a unique continuous linear mapping.*

$$\mathscr{T}: \mathscr{S} \to \mathscr{C}$$

such that $\mathscr{T}1 = 1$ and for all $A_1, \ldots, A_n$ in $\mathscr{V}$,

$$\mathscr{T}\tilde{A}_1 \ldots \tilde{A}_n = \int \cdots \int A_{\pi(1)}(t_{\pi(1)}) \cdots A_{\pi(n)}(t_{\pi(n)}) \, dt_1 \ldots dt_n \tag{9}$$

where for each n-tuple of distinct real numbers $t_1, \ldots, t_n$, π is the permutation of $1, \ldots, n$ such that

$$t_{\pi(1)} > \cdots > t_{\pi(n)}.$$

For all α in $\mathscr{S}$, $\|\mathscr{T}\alpha\| \leqq \|\alpha\|$.

Proof. Almost every n-tuple consists of distinct $t_1, \ldots, t_n$, so the integral in (9) is well-defined. It is symmetric and multilinear in $A_1, \ldots, A_n$, so there is a unique linear mapping $\mathscr{T}: \mathscr{S}_0 \to \mathscr{C}$ such that (9) holds and $\mathscr{T}1 = 1$. By (4), $\|\mathscr{T}\alpha\| \leqq \|\alpha\|$ for all α in $\mathscr{S}_0$, so $\mathscr{T}$ extends by continuity to $\mathscr{S}$. This concludes the proof.

Again, we let

$$\mathscr{N} = \{\alpha \in \mathscr{S} : \mathscr{T}\alpha\beta = 0 \text{ for all } \beta \text{ in } \mathscr{S}\}.$$

This is a closed ideal in $\mathscr{S}$ on which $\mathscr{T}$ is 0, so that $\mathscr{T}$ induces a linear mapping of norm $\leqq 1$ on the quotient Banach algebra $\mathscr{A} = \mathscr{S}/\mathscr{N}$. Elements of $\mathscr{A}$ are called *operants over* $\mathscr{V}$ and $\mathscr{T}$ is called *time-ordered integration.*

5. Spectral properties of operants

Let $\mathscr{V}$ be a linear subspace, containing 1, of the Banach algebra $\mathscr{C}$. The operant algebra $\mathscr{A}$ is by definition a quotient of $\mathscr{S}$, so the spectrum of $\mathscr{A}$, which we denote by $\sigma(\mathscr{A})$, is the subset of the spectrum of $\mathscr{S}$ consisting of those elements which vanish on $\mathscr{N}$. By Theorem 1 we may identify $\sigma(\mathscr{A})$ with a subset of the unit ball of $\mathscr{V}^*$. Since $1 = \tilde{1}$ in $\mathscr{A}$, any ϕ in $\sigma(\mathscr{A})$ satisifies $\phi(1) = 1$. We call a linear functional ϕ of norm $\leqq 1$ on $\mathscr{V}$ such that $\phi(1) = 1$ a *state* of $\mathscr{V}$. By the Hahn-Banach theorem a state of $\mathscr{V}$ has an extension which is a state of $\mathscr{C}$ (and in case $\mathscr{C}$ is a $\mathscr{C}^*$ algebra a state is automatically real on self-adjoint elements and positive on positive elements). Thus $\sigma(\mathscr{A})$ is a subset of the space of states on $\mathscr{V}$ in the weak* topology.

If $\mathscr{V}=\mathscr{C}=L(\mathscr{X})$ is the algebra of all bounded operators on the Banach space $\mathscr{X}$ we refer to operants over $\mathscr{V}$ as operants over $\mathscr{X}$.

Theorem 5. *Let $\mathscr{A}$ be the algebra of operants over a Banach space $\mathscr{X}$. For each u in $\mathscr{X}$ and u^* in $\mathscr{X}^*$ such that $\|u^*\|=\|u\|=(u^*, u)=1$ there is a unique element ϕ_u in $\sigma(\mathscr{A})$ such that*

$$\phi_u(\tilde{A})=(u^*, Au) \tag{10}$$

for all A in $L(\mathscr{X})$.

Proof. For each such u and u^* the functional $\phi_u: A\mapsto(u^*, Au)$ is a state on $L(\mathscr{X})$, so we need only show that ϕ_u is 0 on $\mathscr{N}$.

Define E in $L(\mathscr{X})$ by

$$Ev=(u^*, v)\,u.$$

Then E is an idempotent of norm 1. We clain that for all α in $\mathscr{S}$,

$$\lim_{m\to\infty}\mathscr{T}\tilde{E}^m\alpha \tag{11}$$

exists in $L(\mathscr{X})$. The mapping $\alpha\mapsto\mathscr{T}E^m\alpha$ is norm decreasing, so we need only prove this for α of the form

$$\alpha=\tilde{A}_1\ldots\tilde{A}_n, \tag{12}$$

since these span a dense set in $\mathscr{S}$. But as $m\to\infty$, all but an arbitrarily small proportion of the $(m+n)!$ terms in the expansion by (8) of $\mathscr{T}\tilde{E}^m\tilde{A}_1\ldots\tilde{A}_n$ have at least one factor of E before and after each A_i. Therefore

$$\begin{aligned}\lim_{m\to\infty}\mathscr{T}\tilde{E}^m\tilde{A}_1\ldots\tilde{A}_n&=\mathscr{T}(EA_1E)^\sim\cdots(EA_nE)^\sim\\&=(u^*, A_1u)\cdots(u^*, A_nu)E.\end{aligned}$$

Thus for all α of the form (12), and consequently for all α in $\mathscr{S}$,

$$\phi_u(\alpha)=\lim_{m\to\infty}\big(u^*, (\mathscr{T}\tilde{E}^m\alpha)\,u\big).$$

But if α is in $\mathscr{N}$, $\mathscr{T}\tilde{E}^m\alpha=0$ so that ϕ_u is 0 on $\mathscr{N}$. This concludes the proof.

If A is in $L(\mathscr{X})$ we define its *numerical range* $W(A)$ to be the set of all complex numbers of the form (u^*, Au) where $\|u^*\|=\|u\|=(u^*, u)=1$. This agrees with the usual notion [5] in case $\mathscr{X}$ is a Hilbert space. (We choose the inner product in a Hilbert space to be linear in the second variable.)

Theorem 6. *Let $\mathscr{A}$ be the algebra of operants over a Banach space $\mathscr{X}$. For all A in $L(\mathscr{X})$,*

$$\|e^{\tilde{A}}\|=\sup_{w\in W(A)}|e^w|. \tag{13}$$

Proof. By subtracting a real number from A we may assume without loss of generality that $W(A)$ is contained in the left half-plane, so that the right hand side of (13) is $\leqq 1$. Semigroup theory implies then that $\|e^{tA}\| \leqq 1$ for $t \geqq 0$. (Since A is bounded a short direct proof of this fact may also be given.)

As in any Banach algebra with unit,

$$e^{\tilde{A}} = \lim_{n\to\infty}\left(1+\frac{\tilde{A}}{n}\right)^n.$$

Since $1=\tilde{1}$ in $\mathscr{A}$ and the mapping $A \mapsto \tilde{A}$ is isometric,

$$\left\|1+\frac{\tilde{A}}{n}\right\| = \left\|1+\frac{A}{n}\right\|.$$

But $1+\frac{A}{n} = e^{\frac{A}{n}} + o\left(\frac{1}{n}\right)$ and $\|e^{\frac{A}{n}}\| \leqq 1$. Therefore $\left\|1+\frac{\tilde{A}}{n}\right\| \leqq 1 + o\left(\frac{1}{n}\right)$

and so $\|e^{\tilde{A}}\| \leqq 1$. This proves the inequality $\leqq$ in (13).

On the other hand, if w is in $W(A)$ then by Theorem 5 there is a ϕ in $\sigma(\mathscr{A})$ such that $\phi(\tilde{A}) = w$, and consequently

$$\|e^{\tilde{A}}\| \geqq |\phi(e^{\tilde{A}})| = |e^{\phi(\tilde{A})}| = |e^w|.$$

This concludes the proof.

Notice that we have shown that for any element of $\mathscr{A}$ of the form $e^{\tilde{A}}$ for A in $L(\mathfrak{X})$ the norm and spectral radius coincide.

By Theorem 2, in the Banach algebra $\mathscr{S}$ we would have $\|e^{\tilde{A}}\| = e^{\|A\|}$. Theorem 6 gives a much more useful result. For example, if H is any bounded self-adjoint operator on a Hilbert space and t is real, $\|e^{it\tilde{H}}\| = 1$. This estimate does not depend in any way on the norm of H. By STONE's theorem, any strongly continuous one-parameter group of unitary operators is of the form e^{itH} for H a possibly unbounded self-adjoint operator, and this suggests the possibility of defining operants $e^{it\tilde{H}}$ for such H. Also, the analogue of Theorem 6 holds for the time-dependent operants of § 4. These facts enable one to construct a theory of operants corresponding to unbounded operators in such a way as to make rigorous most of the manipulations of the Feynman calculus [2]. We shall not do this here, but proceed instead with the discussion of the spectral properties of operants in the bounded time-independent case.

Theorem 7. *Let $\mathscr{A}$ be the algebra of operants over a Banach space $\mathfrak{X}$. For all A in $L(\mathfrak{X})$, the spectrum of $\tilde{A}$ contains the closure $\overline{W}(A)$ of the numerical range of A and is contained in the convex hull of $\overline{W}(A)$. If $\mathfrak{X}$ is a Hilbert space the spectrum of $\tilde{A}$ is $\overline{W}(A)$.*

Proof. The spectrum $\sigma(\tilde{A})$ of $\tilde{A}$ is the set of all numbers of the form $\phi(\tilde{A})$ where ϕ is in $\sigma(\mathscr{A})$. By Theorem 5, $\sigma(\tilde{A})$ contains $W(A)$ and since it is closed it contains $\overline{W}(A)$. Now let λ be outside the convex hull of $\overline{W}(A)$. Without loss of generality we may assume that $\overline{W}(A)$ is contained in the left half-plane and $\lambda > 0$. By Theorem 6,

$$\int_0^\infty e^{t\tilde{A}} e^{-\lambda t}\, dt$$

converges, and so it converges to $(\lambda - \tilde{A})^{-1}$. Thus $\sigma(\tilde{A})$ is contained in the convex hull of $\overline{W}(A)$.

The last statement of the theorem follows from the theorem of STONE [5, p. 131] [6] that $W(A)$ is convex if A is an operator on Hilbert space.

6. The Weyl calculus

Let $A = (A_1, \ldots, A_n)$ be an n-tuple of bounded self-adjoint operators on the Hilbert space $\mathscr{H}$, let $\mathscr{V}$ be the linear subspace of $L(\mathscr{H})$ spanned by them and 1, and let $\mathscr{A}(\mathscr{V})$ be the operant algebra over $\mathscr{V}$. By Theorem 6, if $\lambda = (\lambda_1, \ldots, \lambda_n)$ is in $\boldsymbol{R}^n$ then $\|e^{i\lambda\cdot\tilde{A}}\| = 1$, where $\lambda \cdot \tilde{A} = \lambda_1 \tilde{A}_1 + \cdots + \lambda_n \tilde{A}_n$. Consequently, if f is any function on $\boldsymbol{R}^n$ with integrable Fourier transform $\hat{f}$, the inverse Fourier integral

$$f(\tilde{A}) = \frac{1}{(2\pi)^n} \int e^{i\lambda\cdot\tilde{A}} \hat{f}(\lambda)\, d\lambda$$

converges in $\mathscr{A}(\mathscr{V})$. The mapping defined for such f by

$$f \mapsto f(A) = \mathscr{T} f(\tilde{A}) = \frac{1}{(2\pi)^n} \int e^{i\lambda\cdot A} \hat{f}(\lambda)\, d\lambda = (R, f),$$

where R is the inverse Fourier transform of $e^{i\lambda\cdot A}$, is called the *Weyl calculus*. It was introduced by HERMANN WEYL [8; § 45] in connection with the quantization problem of defining $f(P, Q)$ where P and Q are the (unbounded) momentum and position operators. It has been studied in this context by IRVING SEGAL [4] and others, and in the general case of self-adjoint operators by MICHAEL TAYLOR [7] and ROBERT F. V. ANDERSON [1]. The theory of operants is a convenient framework for studying the Weyl calculus.

Theorem 8. *With A as above, let R be the $L(\mathscr{H})$-valued distribution which is the inverse Fourier transform of $e^{i\lambda\cdot A}$ on $\boldsymbol{R}^n$. The spectrum of $\tilde{A}_1, \cdots, \tilde{A}_n$ in $\mathscr{A}(\mathscr{V})$ is the support of R. If W is any linear subspace of $L(\mathscr{H})$ containing $\mathscr{V}$ then the spectrum of $\tilde{A}_1, \ldots, \tilde{A}_n$ in $\mathscr{A}(\mathscr{W})$ contains the support of R.*

Proof. We denote the support of R by supp R and the spectrum of $\tilde{A}_1, \ldots, \tilde{A}_n$ in $\mathscr{A}(\mathscr{W})$ by $\sigma_{\mathscr{W}}(\tilde{A})$. As we remarked at the beginning of

§ 5, the state ϕ in $\sigma(\mathcal{A}(\mathcal{W}))$ is real on self-adjoint operators, so that $\sigma_{\mathcal{W}}(\tilde{A})$ is contained in $\boldsymbol{R}^n$.

Now suppose that x in $\boldsymbol{R}^n$ is in supp R but is not in $\sigma_{\mathcal{W}}(\tilde{A})$. Then there is an f in $\mathcal{D}$ (the space of C^∞ functions on $\boldsymbol{R}^n$ with compact support) vanishing on $\sigma_{\mathcal{W}}(\tilde{A})$ such that $f(A)=(R,f)$ is not 0, so that $f(\tilde{A})\neq 0$. We may then find a g in $\mathcal{D}$ vanishing on $\sigma_{\mathcal{W}}(\tilde{A})$ such that g is 1 on the support of f. Then $g^m f=f$ so that

$$\|f(\tilde{A})\|=\|(g^m f)(\tilde{A})\|\leqq\|g(\tilde{A})^m\|\,\|f(\tilde{A})\|. \tag{14}$$

Since the Fourier transform $\hat{g}$ of g is integrable, we have that for any ϕ in $\sigma(\mathcal{A}(\mathcal{W}))$,

$$\begin{aligned}\phi(g(\tilde{A})) &= \frac{1}{(2\pi)^n}\int \phi(e^{i\lambda\cdot\tilde{A}})\,\hat{g}(\lambda)\,d\lambda\\ &= \frac{1}{(2\pi)^n}\int e^{i\lambda\cdot\phi(\tilde{A})}\,\hat{g}(\lambda)\,d\lambda = g(\phi(\tilde{A}))=0.\end{aligned}$$

Thus $g(\tilde{A})$ is in the radical of $\mathcal{A}(\mathcal{W})$, and by the spectral radius formula

$$\|g(\tilde{A})^m\|^{\frac{1}{m}}\to 0.$$

This implies that $\|g(\tilde{A})^m\|<1$ for some m, and by (14) this means that $f(\tilde{A})=0$, a contradiction. Therefore supp $R\subset\sigma_{\mathcal{W}}(\tilde{A})$.

Since $\mathcal{V}$ is the span of $A_1,\ldots,A_n$ and 1, it follows that $\tilde{A}_1,\ldots,\tilde{A}_n$ and 1 generate $\mathcal{A}(\mathcal{V})$. Suppose that x in $\boldsymbol{R}^n$ is in $\sigma_{\mathcal{V}}(\tilde{A})$ but not in supp R. Then there is an f in $\mathcal{D}$ such that $f(x)\neq 0$ but $(R, fp)=0$ for all polynomials p. Therefore $\mathcal{T}f(\tilde{A})\,p(\tilde{A})=0$ for all polynomials p; that is, $\mathcal{T}f(\tilde{A})\,\alpha=0$ for a dense set of α in $\mathcal{A}(\mathcal{V})$ and hence for all α in $\mathcal{A}(\mathcal{V})$. Consequently, $f(\tilde{A})=0$. But x is in $\sigma_{\mathcal{V}}(\tilde{A})$, so there is a ϕ in $\sigma(\mathcal{A}(\mathcal{V}))$ such that $\phi(\tilde{A})=x$, and as we have already seen, $\phi(f(\tilde{A}))=f(\phi(\tilde{A}))=f(x)\neq 0$, so that $f(\tilde{A})\neq 0$, a contradiction. Therefore supp $R=\sigma_{\mathcal{V}}(\tilde{A})$. This completes the proof.

Consider the hyperbolic system

$$\frac{\partial u}{\partial t}=A\cdot\nabla u=\left(A_1\frac{\partial}{\partial x_1}+\cdots+A_n\frac{\partial}{\partial x_n}\right)u \tag{15}$$

where A is an n-tuple of self-adjoint matrices. By performing a Fourier transformation in the space variables we see that the fundamental solution of (15) is the inverse Fourier transform of $e^{i\lambda\cdot At}$. Thus the support of the fundamental solution of (15) is the cone in $\boldsymbol{R}^{n+1}$ through the spectrum of $\mathcal{A}(\mathcal{V})$. This connection with symmetric hyperbolic systems was pointed out by L. Hörmander (see [1]).

For results on the Weyl calculus we refer to Taylor [7] and Anderson [1]. We shall describe briefly one result of Anderson as it gives the structure of the operant algebra for the simplest of all non-commutative Banach algebras: $L(\mathcal{H})$ where $\mathcal{H}$ is a two-dimensional Hilbert space.

Let $\sigma_1, \sigma_2, \sigma_3$ be the Pauli matrices:

$$\sigma_1 = \begin{pmatrix} 0 & 1 \\ 1 & 0 \end{pmatrix}, \quad \sigma_2 = \begin{pmatrix} 0 & i \\ -i & 0 \end{pmatrix}, \quad \sigma_3 = \begin{pmatrix} 1 & 0 \\ 0 & -1 \end{pmatrix}.$$

Together with 1, these form a basis for $L(\mathcal{H})$ (where $\mathcal{H}$ is a two-dimensional Hilbert space with a given orthonormal basis). Anderson shows that the $L(\mathcal{H})$-valued distribution R, the inverse Fourier transform of $e^{i\lambda\cdot\sigma}$, is given by

$$(R, f) = \left(\mu, \left(1 + r\frac{\partial}{\partial r} + \sigma \cdot \nabla\right) f\right)$$

where $r^2 = x_1^2 + x_2^2 + x_3^2$ on $\boldsymbol{R}^3$ and μ is normalized surface measure on the unit sphere Σ of $\boldsymbol{R}^3$. If u is any unit vector in $\mathcal{H}$,

$$(u, \sigma_1 u), \quad (u, \sigma_2 u), \quad (u, \sigma_3 u)$$

are the components of a point on Σ. Thus the spectrum of the operant algebra $\mathcal{A}$ over $\mathcal{H}$ is the set of all pure states of $L(\mathcal{H})$ (those of the form $A \mapsto (u, Au)$) and not their convex combinations. This example also shows that the ideal $\mathcal{N}$ is highly non-trivial: for any smooth f which vanishes together with its gradient on Σ, $f(\tilde{\sigma}) = 0$ in $\mathcal{A}$. The ideal $\mathcal{N}$ is generated by $1 - \tilde{1}$ and by α^2 where

$$\alpha = 1 - \tilde{\sigma}_1^2 - \tilde{\sigma}_2^2 - \tilde{\sigma}_3^2.$$

Notice that $\alpha \neq 0$ since $\mathcal{T}\alpha = -2$, but for any polynomial β in $\tilde{\sigma}_1, \tilde{\sigma}_2, \tilde{\sigma}_3$ if we multiply α^2 by β, write the result in symmetric form, and substitute the matrices for the variables we get the zero matrix.

7. Operants over a finite-dimensional Hilbert space

Let $\mathcal{H}$ be an n-dimensional Hilbert space, $\mathcal{A}$ the algebra of operants over $\mathcal{H}$. We shall find the spectrum of $\mathcal{A}$ and the mapping $\mathcal{T}$ explicitly, generalizing the result of ANDERSON for the case $n = 2$.

We let $L(\mathcal{H})^*$ be the dual space to $L(\mathcal{H})$, Σ in $L(\mathcal{H})^*$ the set of all linear functionals on $L(\mathcal{H})$ of the form $A \mapsto (u, Au)$ where u is a unit vector in $\mathcal{H}$, μ the unitarily invariant positive measure on Σ normalized so that $\mu(\Sigma) = 1$, $\mathcal{V}_0$ the real vector space of all self-adjoint elements of $L(\mathcal{H})$ and $\mathcal{V}_0'$ its real dual, so that $\mathcal{V}_0'$ is all elements of $L(\mathcal{H})^*$ which are real on $\mathcal{V}_0$.

The symmetric tensor algebra $\mathcal{S}_0(L(\mathcal{H}))$ may be identified with the algebra of polynomial functions on $L(\mathcal{H})^*$, and elements of $\mathcal{S} = \mathcal{S}(L(\mathcal{H}))$ are functions on $L(\mathcal{H})^*$. Let R be the inverse Fourier transform of the $L(\mathcal{H})$-valued function on $\mathcal{V}_0$ given by $A \mapsto e^{iA}$. Then R is an $L(\mathcal{H})$-valued distribution on $\mathscr{V}_0'$, for all α in $\mathscr{S}$ we have $\mathscr{T}\alpha = (R, \alpha)$, and the support of R is the spectrum of $\mathscr{A}$, by Theorem 8. We shall show that this support is Σ and shall exhibit R explicitly. The method we use is due to ANDERSON [1].

The gradient ∇ on $\mathscr{V}_0'$ takes complex-valued functions on $\mathscr{V}_0'$ into $L(\mathscr{H})$-valued functions on $\mathscr{V}_0'$. If $e_1, \ldots, e_n$ is a basis for $\mathscr{H}$, so that elements A of $L(\mathscr{H})$ are represented by matrices (a_{ij}) with respect to this basis, then x_{ij} defined by $x_{ij}(A) = a_{ij}$ are a basis for $L(\mathscr{H})^*$, and ∇ is the matrix with entries $\partial/\partial x_{ij}$. We let ∇^k be the k-th matrix power of ∇. It is the inverse Fourier transform of multiplication by $(iA)^k$ on $\mathscr{V}_0$.

Theorem 9. *Let R be the inverse Fourier transform of the function e^{iA} on $\mathscr{V}_0$. The support of R, and the spectrum of $\mathscr{A}$, is Σ. Explicitly,*

$$R = \sum_{k=0}^{n-1} \sum_{j=0}^{n-k-1} \sum_{m=0}^{j} (-1)^{n+j+m+1} \binom{j}{m} \frac{1}{(n-1-j+m)!} \cdot \nabla^k \phi_{n-k-j-1}(\nabla)(\nabla \cdot x)^m \mu \tag{16}$$

where $\phi_0(\nabla) = 1$ and $\phi_j(\nabla)$ is the sum of the principal minors of order j of ∇, for $j = 1, \ldots, n$.

Proof. The relations

$$\begin{aligned} e^{r\lambda_1} &= g_0 + g_1\lambda_1 + \cdots + g_{n-1}\lambda_1^{n-1} \\ &\vdots \\ e^{r\lambda_n} &= g_0 + g_1\lambda_n + \cdots + g_{n-1}\lambda_1^{n-1} \end{aligned} \tag{17}$$

determine unique functions $g_0, \ldots, g_{n-1}$ of $r, \lambda_1, \ldots, \lambda_n$ provided the λ's are distinct. We may solve for the g_k by Cramer's rule; in particular,

$$g_0 = \frac{\begin{vmatrix} e^{r\lambda_1} \lambda_1 \ldots \lambda_1^{n-1} \\ \vdots \\ e^{r\lambda_n} \lambda_n \ldots \lambda_n^{n-1} \end{vmatrix}}{D} = (-1)^{n-1} \frac{\begin{vmatrix} \lambda_1 \lambda_1^2 \ldots e^{r\lambda_1} \\ \vdots \\ \lambda_n \lambda_n^2 \ldots e^{r\lambda_k} \end{vmatrix}}{D}, \tag{18}$$

$$g_{n-1} = \frac{\begin{vmatrix} 1\ \lambda_1 \ldots e^{r\lambda_1} \\ \vdots \\ 1\ \lambda_n \ldots e^{r\lambda}\, n \end{vmatrix}}{D}, \tag{19}$$

where D is the Vandermonde determinant

$$D = \prod_{i<j} (\lambda_j - \lambda_i) = \begin{vmatrix} 1\ \lambda_1 \ldots \lambda_1^{n-1} \\ \vdots \\ 1\ \lambda_n \ldots \lambda_n^{n-1} \end{vmatrix}.$$

If we differentiate the numerator on the right hand side of (18) we obtain a determinant which is the same as the numerator of (19) except that the i-th row is multiplied by λ_i. Let ϕ_j by the j-th elementary symmetric function of $\lambda_1, \ldots, \lambda_n$:

$$\phi_j = \sum_{i_1 < \cdots < i_j} \lambda_{i_1} \ldots \lambda_{i_j}.$$

Then we have by the above argument that

$$\frac{\partial g_0}{\partial r} = (-1)^{n-1} \phi_n g_{n-1}. \tag{20}$$

Now let

$$X = \frac{\partial}{\partial \lambda_1} + \cdots + \frac{\partial}{\partial \lambda_n}.$$

If we apply X to both sides of (17) we find that

$$\begin{aligned} r e^{r\lambda_k} = & X g_0 + X g_1 \cdot \lambda_k + \cdots + X g_{n-1} \cdot \lambda_k^{n-1} \\ & + g_1 + 2 g_2 \cdot \lambda_k + \cdots + 0 \end{aligned}$$

for $k = 1, \ldots, n$. Therefore

$$\begin{aligned} r g_0 &= X g_0 + g_1 \\ r g_1 &= X g_1 + 2 g_2 \\ &\vdots \\ r g_{n-2} &= X g_{n-2} + (n-1) g_{n-1} \\ r g_{n-1} &= X g_{n-1}, \end{aligned}$$

so that

$$g_k = \frac{1}{k!} (r - X)^k g_0. \tag{21}$$

Easy inductions show that $\phi_{n-k} = k!\, X^k \phi_n$ and

$$(-1)^k \frac{(r-X)^k}{k!} \phi_n g_{n-1} = \phi_{n-k} g_{n-k}. \tag{22}$$

By (21) for $k = n-1$,

$$\begin{aligned} \frac{\partial g_{n-1}}{\partial r} &= \frac{\partial}{\partial r} \frac{1}{(n-1)!} (r-X)^{n-1} g_0 \\ &= \frac{1}{(n-2)!} (r-X)^{n-2} g_0 + \frac{1}{(n-1)!} (r-X)^{n-1} \frac{\partial g_0}{\partial r}, \end{aligned}$$

and by (21), (20), and (22) this is equal to

$$g_{n-2} + \frac{(-1)^{n-1}}{(n-1)!} (r-X)^{n-1} \phi_n g_{n-1} = g_{n-2} + \phi_1 g_{n-1}.$$

Therefore

$$g_{n-2} = \frac{\partial g_{n-1}}{\partial r} - \phi_1 g_{n-1}.$$

If we apply $\partial/\partial r$ again we find in the same way that

$$g_{n-3} = \frac{\partial^2 g_{n-1}}{\partial r^2} - \phi_1 \frac{\partial g_{n-1}}{\partial r} + \phi_2 g_{n-1},$$

and by induction on $n-k$,

$$g_k = \sum_{j=0}^{n-k-1} (-1)^{n+k+j+1} \phi_{n-k-j-1} \frac{\partial^j}{\partial r^j} g_{n-1}, \tag{23}$$

where $\phi_0 = 1$.

Let $\lambda = (\lambda_1, \ldots, \lambda_n)$. By (17), the functions g_k have the homogeneity property that

$$g_k(r, \lambda) = r^k g(1, r\lambda).$$

Let

$$\gamma_k(\lambda) = g_k(1, i\lambda)$$

so that

$$e^{i\lambda_j} = \gamma_0(\lambda) + \gamma_1(\lambda)(i\lambda_j) + \cdots + \gamma_{n-1}(\lambda)(i\lambda_j)^{n-1}$$

for $j = 1, \ldots, n$. Then $g_{n-1}(r, i\lambda) = r^{n-1}\gamma_{n-1}(r\lambda)$, so that by (23), the Leibnitz formula for the derivative of a product, and the fact that

$$\frac{\partial}{\partial r}\gamma_{n-1}(r\lambda) = \lambda \cdot \nabla\gamma_{n-1}(r\lambda),$$

we have

$$\gamma_k(\lambda) = \sum_{j=0}^{n-k-1} (-1)^{n+k+j+1} \phi_{n-k-j-1}(i\lambda) \sum_{m=0}^{j} \binom{j}{m} \frac{(n-1)!}{(n-1-j+m)!} \cdot (\lambda\cdot\nabla)\, \gamma_{n-1}(\lambda). \tag{24}$$

If A is any matrix we let $\phi_j(A)$ be the sum of the principal minors of order j of A (so that $\phi_1(A) = \operatorname{tr} A$ and $\phi_n(A) = \det A$) and $\phi_0(A) = 1$. We let $\gamma_k(A)$ be the functions such that

$$e^{iA} = \gamma_0(A) + \gamma_1(A)(iA) + \cdots + \gamma_{n-1}(A)(iA)^{n-1}.$$

Then we may substitute A for λ in (24). Now we claim that

$$\frac{1}{(n-1)!}\hat{\mu}(A) = \gamma_{n-1}(A). \tag{25}$$

If we show this, and use the fact that the inverse Fourier transform of $A\cdot\nabla$ is $-\nabla\cdot x$, then (16) follows at once from (24) with A substituted for λ.

The Fourier transform of μ is

$$\hat{\mu}(A) = \int e^{-ix\cdot A}\, d\mu(x).$$

Let us find $\hat{\mu}(A)$ under the assumption that A is i times the diagonal matrix with distinct entries $\lambda_1, \ldots, \lambda_n$. By definition of μ we then have that

$$\hat{\mu}(A) = \int e^{-i(u, Au)}\, d\nu(u) = \int e^{\Sigma\lambda_j |u_j|^2}\, d\nu(u)$$

where ν is normalized surface measure on the unit sphere S of $\mathscr{H}$.

The mapping

$$q: (u_1, \ldots, u_n) \mapsto (|u_1|^2, \ldots, |u_n|^2)$$

of $\boldsymbol{C}^n$ into $\boldsymbol{R}^n$ maps S onto the simplex T in $\boldsymbol{R}^n$ of all positive $(w_1, \ldots, w_n)$ with sum 1. If we use polar coordinates $|u_k|$, θ_k for each complex variable u_k then Lebesgue measure on $\boldsymbol{C}^n$ is

$$|u_1|\, d|u_1|\, d\theta_1 \ldots |u_n|\, d|u_n|\, d\theta_n,$$

and so the measure on $\boldsymbol{R}_n$ induced by the mapping q is a constant times

$$d|u_1|^2 \ldots d|u_n|^2.$$

Under the mapping q a shell of constant thickness about S goes onto a slab of constant thickness about T. Consequently the measure on the simplex T induced by the mapping q applied to ν is normalized Lebesgue measure on T; that is, the measure

$$(n-1)!\, dw_1 \ldots dw_{n-1}.$$

Therefore

$$\frac{1}{(n-1)!} \hat{\mu}(A) = \int\limits_T e^{\Sigma \lambda_j w_j}\, dw_1 \ldots dw_{n-1}. \tag{26}$$

Denote the right hand side of (26) by $\Phi_n(\lambda_1, \ldots, \lambda_n)$. Explicit integration of the last variable w_{n-1} shows that

$$\Phi_n(\lambda_1, \ldots, \lambda_n) = \frac{\Phi_{n-1}(\lambda_1, \ldots, \lambda_{n-2}, \lambda_{n-1}) - \Phi_n(\lambda_1, \ldots, \lambda_{n-2}, \lambda_n)}{\lambda_n - \lambda_{n-1}}. \tag{27}$$

Now A has entries $-i\lambda_1, \ldots, -i\lambda_n$ so $\gamma_{n-1}(A) = g_{n-1}(1, \lambda)$, which is given by (19) with $r=1$. If we expand the numerator of (19) along the last column we find that $g_{n-1}(1, \lambda)$ satisfies the same recursion relation (27) as Φ_n. Hence to prove that $\Phi_n(\lambda) = g_{n-1}(1, \lambda)$ it suffices to prove this for $n=1$, and it is immediate that $\Phi_1(\lambda_1) = g_0(1, \lambda_1) = e^{\lambda_1}$. Thus (25) holds for diagonal A with distinct entries. Since both sides of (25) are continuous and unitarily invariant, (25) holds for all A in $\mathscr{V}_0$. This proves (16).

Obviously supp $R \subset \Sigma$. By Theorem 8, supp $R = \sigma(\mathscr{A})$ and by Theorem 5, $\Sigma \subset \sigma(\mathscr{A})$. Therefore supp $R = \sigma(\mathscr{A}) = \Sigma$. This concludes the proof.

As a corollary, the support of the fundamental solution of the hyperbolic system

$$\frac{\partial u}{\partial t} = A \cdot \nabla u \tag{28}$$

is the cone in $\boldsymbol{R}^{n^2+1}$ through Σ. This result is due to GÅRDING [3], who used Riesz integrals rather than Fourier transforms to solve (28). If A

is an n-tuple of self-adjoint matrices, define the numerical range $W(A)$ to be the subset of $\boldsymbol{R}^n$ of all points with coordinates of the form

$$(u, A_1 u), \ldots, (u, A_n u)$$

where u is a unit vector. Then the support of the fundamental solution of (15) is contained in the cone in $\boldsymbol{R}^{n+1}$ through the numerical range of A. The numerical range of A is the spectrum of $\tilde{A}$ in the operant algebra over the underlying finite-dimensional Hilbert space but the spectrum of $\tilde{A}$ in the operant algebra over the span $\mathscr{V}$ of A and 1 may be smaller.

References

1. Anderson, R. F. V.: The Weyl functional calculus. J. Functional Analysis (to appear).
2. Feynman, R. P.: An operator calculus having applications in quantum electrodynamics. Phys. Rev. **84**, 108—128 (1951).
3. Gårding, L.: The solution of Cauchy's problem for two totally hyperbolic linear differential equations by means of Riesz integrals. Ann. of Math. **48**, 785—826 (1947), Correction **52**, 506—507 (1950).
4. Segal, I. E.: Transforms for operators and symplectic automorphisms over a locally compact Abelian group. Math. Scand. **13**, 31—43 (1963).
5. Stone, M. H.: Linear transformations in Hilbert space and their applications to analysis. Am. Math. Soc. Coll. Publ. **15** (1932).
6. — Hausdorff's theorem concerning Hermitian forms. Bull. Am. Math. Soc. **36**, 259—261 (1930).
7. Taylor, M. E.: Functions of several self-adjoint operators. Proc. Am. Math. Soc. **19**, 91—98 (1968).
8. Weyl, H.: Gruppentheorie und Quantenmechanik. Leipzig 1928.

Local Non-linear Functions of Quantum Fields

By Irving Segal*

Brandeis University and Massachusetts Institute of Technology

1. Introduction

I think it is good that one of us is treating a topic that impinges explicitly on relations with theoretical physics, because the complementarity between mathematics as a pure discipline and mathematics as a universal distillation from the experiential universe is one of the striking features in some of Marshall Stone's line of work. In addition to this primary consideration, there is the general one that the Sovereign Nation of Mathematics — so to speak — is now surely sufficiently secure and vital that it befits it to explore relations with other Nations, such as Physics. The subject of quantum fields has seemed particularly appropriate in these connections. Purely mathematically, it provides a proving ground for the development and testing of methods for dealing with some of the most novel and exciting problems of contemporary analysis, — to name only some of these, the problems of highly singular perturbations, falling outside the scope of conventional theories of operators in Hilbert space and generalized functions, of which some of the work of Feldman [1] and Nelson [8] is representative; the development of real and complex analysis for functions on infinite-dimensional spaces, of which some of the work of Gross [5], Shale [16], and myself [10] is representative; and the problem, not only of analysis but of meaning, of the application of non-linear operations to weak functions (e.g., the raising of a Schwartzian distribution to a power) (cf. e.g. [5], [11] and [14]). On the other hand, in relation to potential commerce with the Nation of Physics, one has in quantum fields probably the most sophisticated and far-reaching idea of contemporary theoretical physics.

2. What is a quantum field?

The term "quantum field" means many things, not all of them tangible, to many people; as a consequence it seems desirable, in the mathematical treatment, to use more neutral terminology. What we

* Research conducted in part during the tenure of a Guggenheim Fellowship and in part with the support of the Office of Scientific Research (U.S.A.F.).

shall call a "quantum process" can be regarded as a mathematical abstraction from a modal, often heuristic, usage of the term "quantum field".

Let M be a given set, and $\boldsymbol{P}$ a given real linear vector space consisting of functions, or linear equivalence classes of functions, from M to a given real linear vector space $\boldsymbol{L}$. (For brevity, and/or a suggestion of the role of these concepts in concrete instances, M may be called the *underlying geometrical manifold*; $\boldsymbol{P}$, the *probe space* (after the discussion of the role of probes in field measurements in a well-known paper of BOHR and ROSENFELD); and $\boldsymbol{L}$, the *spin space*). An *operational process* in M with probe space $\boldsymbol{P}$ is defined as an equivalence class of linear mappings Φ from $\boldsymbol{P}$ into the set of all closed densely defined linear operators in a Hilbert space $\boldsymbol{H}$, where the linearity is relative to the *strong operations* in the partial algebra of all closed densely defined operators in $\boldsymbol{H}$; the *strong sum* of two such operators is the closure of the usual sum, and is defined to exist only when this closure exists and is densely defined; and similarly for strong multiplication by real numbers. The equivalence relation in question is as follows: two such processes, Φ and Φ', with corresponding Hilbert spaces $\boldsymbol{H}$ and $\boldsymbol{H}'$ are equivalent if and only if there exists a unitary transformation U of $\boldsymbol{H}$ onto $\boldsymbol{H}'$ such that for all $x \in \boldsymbol{P}$, $U\Phi(x)U^{-1} = \Phi'(x)$. The notion of *weak operational process* is defined in the same way, relative to a given dense linear domain $\boldsymbol{D}$ in $\boldsymbol{H}$, except that the values $\Phi(x)$ are bilinear forms on $\boldsymbol{D} \times \boldsymbol{D}$. Any member of the equivalence class defining a process may be called a *concrete process*.

A *quantum process* is an operational process with an additional important element of structure called a "vacuum vector". More specifically, a quantum process in M with probe space $\boldsymbol{P}$ is an equivalence class of structures $(\Phi, \boldsymbol{H}, v)$ such that $(\Phi, \boldsymbol{H})$ is a concrete operational process, and v is a unit vector in $\boldsymbol{H}$ which is cyclic for the ring of operators determined by the $\Phi(x)$; two such structures are defined as equivalent in case there exists a unitary operator U which has the same property as earlier, and in addition the property that $Uv = v$. It will suffice to define the foregoing ring of operators (where I use the term "ring" in the original sense of MURRAY and VON NEUMANN) for the more general case of a weak operational process. Specifically, the ring of operators generated by any given set of bilinear forms on domains $\boldsymbol{D} \times \boldsymbol{D}$, where the $\boldsymbol{D}$ are given dense domains in $\boldsymbol{H}$, is the set of all bounded linear operators on $\boldsymbol{H}$ which commute with all unitary operators which leave invariant the forms in question.

The (direct product) combinations of essentially three distinct types of quantum processes provide mathematical representations of most of the quantum fields considered in heuristic practice; and these three types are also distinguished in a mathematically natural fashion. They are as follows.

(1) The abelian process, characterized as that for which the operator ring $\boldsymbol{R}$ determined by the $\Phi(x)$ is abelian. A (strict, i.e. non-weak) abelian process is mathematically essentially the same as a generalized stochastic process.

(2) The next simplest assumption to that of the essential commutativity of the $\Phi(x)$ is that any commutator of any two of the $\Phi(x)$ is essentially a scalar operator, not necessarily zero. Non-trivial operators of this type cannot be bounded, and in order to suppress irrelevant pathology connected with unbounded operators as well as for other purposes, it is advantageous to express the assumption in the form applicable to a self-adjoint process, i.e. one such that each $\Phi(x)$ is self-adjoint:

$$W(x)\,W(y) = e^{(i/2)A(x,y)}\,W(x+y),$$

where $W(x) = e^{i\Phi(x)}$, and $A(x, y)$ is a (necessarily) antisymmetric real bilinear form in the vectors x and y in $\boldsymbol{P}$. This may be called a *symmetric* process; when the form $A(\cdot\,,\cdot)$ is non-singular (as in Theorem 2 below) or effectively non-singular (as in Theorem 1 below) it is then totally distinct from the abelian process. It follows from Stone's theorem and the linearity of $\Phi(\cdot)$ that

$$[\Phi(x), \Phi(y)] \subset i\,A(x, y)\,I;$$

the latter relation is an abstract mathematical form of what are called the "canonical commutation relations" in much of the physical literature, and the physically associated quantum field is often referred to as "satisfying Bose-Einstein statistics", etc.

(3) Another generalization of the abelian case is that in which the anti-commutator of any two of the $\Phi(x)$ is a scalar operator:

$$\Phi(x)\,\Phi(y) + \Phi(y)\,\Phi(x) = S(x, y),$$

where $S(x, y)$ is a (necessarily) symmetric real bilinear form in the vectors x and y in $\boldsymbol{P}$. Such a process may be called *anti-symmetric*. In the physical literature, the indicated relations are referred to as the "canonical anti-commutation relations", and an associated quantum field is said to satisfy "Fermi-Dirac statistics", etc.

The algebraic differences between the anti-symmetric and symmetric processes are parallel to those between the orthogonal and symplectic groups, but are involved also with analytical complications, arising from the unboundedness of the operators $\Phi(x)$ in the symmetric case. In the most important case in which the symmetric form $S(\cdot\,,\cdot)$

is positive definite, the process operators $\Phi(x)$ are bounded. There is nevertheless a remarkable analogy between the results, if not all of the methods, for the anti-symmetric case and the symmetric case. Since in the time at my disposal I can at most give a suggestion of the essential ideas, I shall confine the further treatment to the symmetric case, which is the most generally familiar one.

3. What is a linear field?

As an illustrative example, and for its intrinsic interest, let me describe the simplest nontrivial quantum field which arises in theoretical physics, the so-called "neutral scalar free field". First, let me give an intrinsic characterization based on mathematical versions of conventional theoretical physical postulates. Let M denote 4-dimensional space-time; let $\boldsymbol{P}$ denote the space of all infinitely differentiable real-valued functions on M of compact support. Let T denote the linear differential operator $\square - m^2$, where $\square = -(\partial/\partial t)^2 + \Delta$, and m is a real constant, acting on functions on space-time; and let D denote the (Schwartz-) distribution solution of the equation $TD=0$ having the Cauchy data at time $t=0$: $D(x, 0)=0$; $(\partial/\partial t)\, D(\vec{x}, t)|_{t=0} = \delta(\vec{x})$, where $\vec{x}$ denotes the space variable. (Alternatively, D is the difference of the elementary solutions for the differential operator T which are supported respectively by the forward and backward cones.) This definition of D may seem slightly ad hoc, but $D(x-y)$ may be characterized as essentially the unique non-zero generalized function F of x and y which satisfies the differential equation $TF=0$ as a function of x, is anti-symmetric in x and y (and so satisfies the equation also as a function of y), is invariant under the Poincaré group (as is the operator T), and satisfies a very mild regularity condition. Let A denote the (degenerate) anti-symmetric bilinear form on $\boldsymbol{P}$ given by the equation

$$A(f, g) = \iint D(x-y)\, f(x)\, g(y)\, dx\, dy.$$

(A may be directly characterized as the unique anti-symmetric form on $\boldsymbol{P}$ which is invariant under the Poincaré group, satisfies the equation $A(Tf, g)=0$ for arbitrary f and g in $\boldsymbol{P}$, and is a continuous function of f and g in a certain weak topology.)

Theorem 1. *There exists a unique structure* $(\Phi, \boldsymbol{H}, v, \Gamma)$ *such that* $(\Phi, \boldsymbol{H}, v)$ *is a symmetric quantum process over* $(m, \boldsymbol{P}, A)$ *and* Γ *is a continuous one-parameter unitary group on* $\boldsymbol{H}$, *such that*:

(1) Φ *satisfies the differential equation, in the sense that*

$$\Phi(Tf) = 0, \quad f \text{ arbitrary in } \boldsymbol{P};$$

(2) *$\Gamma(t)v = v$ for all t, the infinitesimal generator of Γ given by Stone's theorem is non-negative, and Γ intertwines appropriately with Φ in the sense that*

$$\Gamma(t)\,\Phi(f)\,\Gamma(t)^{-1} = \Phi(f_t),$$

where $f_{t'}(\vec{x}, t) = f(\vec{x}, t+t')$ (this means that $\Gamma(t)$ acts as the temporal displacement operator, and that the energy is positive);

(3) *The mapping $f \to \exp(i\Phi(f))$ is continuous in a certain relatively weak sense, from $\boldsymbol{P}$ to the unitary operators on $\boldsymbol{H}$.*

I remark that the same is true if Γ is required to be a continuous unitary representation of the orthochronous Poincaré group (or of the full Poincaré group, if anti-unitary operators are admitted); the positivity condition then applies only to all time-like displacements, i.e. displacements conjugate within the Poincaré group to the indicated temporal displacement. But this Poincaré invariance is simply a corollary to the theorem in the light of the invariance of the operator T; the theorem really applies equally well to an extensive class of temporally-invariant operators which are not Poincaré-invariant, or even act in totally different spaces from conventional euclidean space.

Adopting distribution notation, the theorem states mainly that, apart from regularity features of an anticipated type, there are unique operators $\phi(x)$ for $x \in M$ (actually, of course $\phi(x)$ is not a well-defined operator, but only its average $\int \phi(x) f(x)\,dx = \Phi(f)$), and a cyclic vector v in the Hilbert space $\boldsymbol{H}$ on which these operators act, such that

$$(\Box - m^2)\,\phi(x) = 0;$$

$$[\phi(x), \phi(y)] = iD(x-y);$$

v defines a state which is both temporally invariant and of positive energy.

The last condition means that if F is any bounded function of a finite number of the $\Phi(f)$, and if F^t denotes the corresponding function of the $\Phi(f_t)$, then $\langle F^t v, v\rangle = \langle F v, v\rangle$; and if G is any similar bounded function, then $\langle F^t G v, v\rangle$ as a function of t has positive spectrum (i.e. its Fourier transform vanishes for negative values of the dual variable to t, the so-called "frequency"). The one-parameter unitary group $\Gamma(t)$ described earlier can be constructed from such a system $(\Phi, \boldsymbol{H}, v)$, and will have the earlier indicated properties.

The structure $(\Phi, \boldsymbol{H}, v, \Gamma)$ can be explicitly described in three essentially different known representations, all unitarily equivalent, but providing diagonalizations of differing sets of operators. The most natural representation in connection with the treatment of non-linear local interactions is that in which the (mutually commuting) operators

$\phi(\vec{x}, t_0)$, the time t_0 being fixed, are simultaneously diagonalized; because these operators determine a maximal abelian algebra of operators on $\boldsymbol{H}$, this representation is essentially unique. At various times I have called it the "real wave representation", because it makes explicit the wave properties of the free field, complementarily to its particle properties, clearly visible in the first representation given for free fields, developed in non-relativisitic form by COOK, following the heuristic work of FOCK; I have also called it the "renormalized Schrödinger representation", because it involves an infinite-dimensional analog to the familiar Schrödinger representation for operators satisfying the canonical commutation relations, in which the only formal change is a gross alteration of the euclidean volume element which can be described as a renormalization. Now it is not immediately obvious from Theorem 1 that it is legitimate to consider the free field at a fixed time, even after smoothing in space, for it asserts the existence of a bona fide operator $\Phi(f)$ only when f is a quite regular function of both space and time, i.e. after smoothing in space-time. This is however the case, and the only reasons for stating the theorem in terms of space-time averages is that this is very frequently done in the literature, and makes properties of relativistic invariance more manifest; and that the relation between the space-time averages and the space averages at fixed times is in non-linear cases a very sticky point, which should benefit therefore by illustration in the comparatively simple case of a free field.

Rather than merely restate the Theorem in terms of space averages of the quantum process $\phi(x)$ at fixed times, I shall take advantage of the opportunity to give a much more general theorem, which serves also to indicate how relatively little the geometrical structure of Minkowski space has to do with the general idea of quantization; the state of the underlying "classical" system at a fixed time need not be specified by functions on space (the "Cauchy data") for the equation $T\psi=0$, where ψ is an ordinary real-valued function, but by vectors in general Hilbert spaces. It is only with the consideration of non-linear interactions that multiplicative properties of these Hilbert spaces, arising from representations of them in terms of numerical function spaces, become really material.

Theorem 2. *Let B denote a non-negative self-adjoint operator on a real Hilbert space $\boldsymbol{G}$, having only 0 in its null space. Let $\boldsymbol{H}$ denote the Hilbert space completion of $\boldsymbol{D}_C \oplus \boldsymbol{D}_{C^{-1}}$, where $C = B^{\frac{1}{2}}$, and $\boldsymbol{D}_T$ for any operator T indicates the domain of T as a Hilbert space, relative to the inner product: $\langle x, y\rangle_T = \langle Tx, Ty\rangle$,* and *$\langle\cdot\,,\cdot\rangle$ denotes the inner product in $\boldsymbol{G}$. Let A denote the (non-singular) anti-symmetric form*

$$A(f_1 \oplus g_1, f_2 \oplus g_2) = \langle C f_1, C^{-1} g_2\rangle - \langle C f_2, C^{-1} g_1\rangle$$

on $\boldsymbol{H}$. *Let* $0(\cdot)$ *denote the continuous one-parameter unitary group on* $\boldsymbol{H}$ *whose matrix relative to the indicated decomposition is*

$$0(t) = \begin{pmatrix} \cos(tB) & \dfrac{\sin(tB)}{B} \\ -B\sin(tB) & \cos(tB) \end{pmatrix}.$$

Then there exists a unique symmetric quantum process $(\psi, \boldsymbol{K}, v)$ *and positive-generator one-parameter unitary group* $\Gamma(t)$ *on* $\boldsymbol{K}$, *with probe space* $\boldsymbol{H}$ *and relative to the given form* A, *such that* $\Gamma(t)v = v$,

$$\Gamma(t)\,\psi(z)\,\Gamma(t)^{-1} = \psi\big(0(t)\,z\big);$$

and denoting $\psi\big(0(t)\,(0\oplus f)\big)$ *as* $\Phi(f,t)$ *and* $\psi\big(0(t)\,(g\oplus 0)\big)$ *as* $\dot{\Phi}(g,t)$, *the following relations hold*:

$(\partial/\partial t)\,\Phi(f,t) = \dot{\Phi}(f,t)$ *weakly on a dense domain* $(f \in \boldsymbol{D}_C \wedge \boldsymbol{D}_{C^{-1}})$;

$[\Phi(f,t), \dot{\Phi}(g,t)] = \langle f, g\rangle$; $[\Phi(f,t), \Phi(g,t)] = 0 = [\dot{\Phi}(f,t), \dot{\Phi}(g,t)]$ *(on dense domains; indeed the Weyl relations hold)*;

the differential equation $u''(t) + B^2 u(t) = 0$ *is satisfied by* $\Phi(f,t)$ *as a distribution, i.e. the equation*

$$(\partial/\partial t)^2\,\Phi(f,t) + \Phi(B^2 f, t) = 0$$

holds weakly on a dense domain.

This result would for example ennable one to give a conceptually simple and mathematically rigorous meaning to the so-called quantization of MAXWELL's equation; or of such general equations as those of the form $\square\phi = V(x)\,\Phi + F(\Delta)\,\Phi$, where $V(x)$ is an arbitrary bounded measurable function on space, and $F(\Delta)$ denotes an arbitrary non-negative self-adjoint function of the laplacian. In all these cases, the quantum process $\phi(\vec{x}, t)$, as an operator-valued distribution in space, at a fixed time t, or in space-time, exists in a highly regular sense, and satisfies the prescribed differential equations and commutation relations in strong senses. The actual operators $\Phi(f,t)$ and $\dot{\Phi}(g,t)$ may be expressed quite explicitly as simple operators in the space $L_2(\boldsymbol{H}')$ of all square-integrable functionals over the space $\boldsymbol{H}'$ in $\boldsymbol{H}$ consisting of pairs $f\oplus g$ such that $f \equiv 0$, relative to a certain weak probability measure in $\boldsymbol{H}'$ known as the isonormal distribution (cf. [4]). In this representation the $\Phi(f,t)$ are diagonalized, i.e. consist of multiplication operators, while the $\dot{\Phi}(g,t)$ are essentially generators of vector displacements, through vectors related to g, in $\boldsymbol{H}'$.

The difference in appearance between the A-forms involved in Theorems 1 and 2 is largely due to the degeneracy of the form A in Theorem 1 and its non-degeneracy in Theorem 2. If in Theorem 1, the probe space $\boldsymbol{P}$ is replaced by its quotient space $\boldsymbol{P}'$ modulo the null-space

$\boldsymbol{P}_0$ of all vectors z such that $A(x, z) = 0$ for all $x \in \boldsymbol{P}$, little is lost, in as much as $\Phi(f) = \Phi(f')$ if $f \equiv f'$ modulo $\boldsymbol{P}_0$. The resulting space $\boldsymbol{P}'$ is, in the special case considered in Theorem 1, naturally isomorphic to a dense subspace of the probe space $\boldsymbol{H}$ of Theorem 2.

4. What is a non-linear field?

The theory of quantum processes associated with a given suitable linear partial differential equation is quite well founded and fairly well developed, but the situation is altogether different as soon as a local non-linearity is involved. The fundamental belief that the dynamics of physical fields are appropriately described by local partial differential equations has led to the study of such equations from a quantum point of view; and, as is well known, in the case of an equation postulated to describe the interaction between photons and electrons, to significant empirical confirmation. The problem of giving a definite mathematical meaning to such so-called "quantized" non-linear partial differential equations has however resisted solution until recently.

The difficulty is that a non-linear operation on a weakly-defined process is involved. Consider for example the differential equation $\square \phi = p(\phi)$, where $p(\phi)$ is a polynomial in ϕ; as long as ϕ is a generalized function, the right-hand side has no *a priori* meaning. On the other hand, close investigation shows that even in the simplest non-trivial cases, such as the neutral scalar field described earlier, the quantum process ϕ is not a strict process, in the sense that there is no bona fide function $\phi(x)$ such that $\Phi(f) = \int \phi(x) f(x)\, dx$; although it becomes quite close to a strict process as the number of space dimensions decreases to one. For a strict process, of course, there is no difficulty in defining a local non-linear function, e.g. the square; but this definition cannot be made directly in terms of the process-average Φ, in a fashion which is applicable to weak processes. In the case of the hypothetical so-called "interacting" process satisfying the indicated non-linear equation there is every reason to believe that the regularity properties would be at best not improved.

In the case of a symmetric process there is however a method for introducing a notion of quasi-power which is not applicable to an ordinary distribution or general type of operational process. In order for this method to be applicable to the definition of the right-hand member of an evolutionary partial differential equation, it appears to be necessary that it deal with the quasi-power at a fixed time; that is to say, smoothing in space does not essentially change the meaning of the equation, but smoothing in time destroys, apparently, this meaning. There is no known way to transform a non-linear equation of evolution which is local in the time into one which is expressed in terms of temporal averages of the solution. Conversely, if the non-linear term can be given mathematical meaning as a generalized function in space at any fixed time, the differential equation in question acquires a definite meaning and appears accessible to mathematical analysis as a Cauchy problem, although a relatively sophisticated one.

Before indicating the nature of this definition for powers of a symmetric quantum process at a fixed time, let me remark parenthetically that although the idea of quantum fields at fixed times has always been, and is currently, extensively used, the question of whether they are physically or mathematically meaningful has been somewhat controversial. In a reaction against the purely formal treatment of non-linear functions of quantum fields, axiomatic schools developed around 15 years ago which insisted on space-time smoothing and the consequent abandonment of the canonical commutation relations for interacting fields, as well as, of course, of local non-linear partial differential equations as a means of describing the dynamics of quantum fields. My own mathematical work on quantum fields has emphasized the utilization of fields at fixed times, and in the past few years some of the axiomatic people having been moving in this direction. Nevertheless, the subject remains a disputed one, and in the absence of a solid mathematical foundation (or for that matter, of experimental methods for measurement of interacting quantum fields at fixed times), it was not resoluble. As we shall see, the non-linear functions of the field at a fixed time are well-defined in a conceptually simple mathematical sense, but are in general rather singular objects from a differential equation or generalized function standpoint.

For simplicity, let us start with the case of squares of fields; the cases of higher powers will be similar. Recalling the canonical commutation relations for a field ϕ at a fixed time t:

$$[\phi(\vec{x}, t), \phi(\vec{x}', t)] = 0 = [\dot{\phi}(\vec{x}, t), \dot{\phi}(\vec{x}', t)] = 0;$$
$$[\phi(x, t), \dot{\phi}(\vec{x}', t)] = i\,\delta(\vec{x} - \vec{x}'),$$

we may note that $\phi(\vec{y}, t)^2$, although undefined, has simple explicit formal commutation relations with the $\phi(\vec{x}, t)$:

$$[\phi(\vec{y}, t)^2, \phi(x, t)] = 0, \quad [\phi(\vec{y}, t)^2, \dot{\phi}(\vec{x}, t)] = 2i\phi(\vec{y}, t)\,\delta(\vec{y} - \vec{x}).$$

If we regard $\phi(\vec{y}, t)^2$, or more precisely, its space-averaged form, say $\Phi^{(2)}(f, t) = \int \phi(\vec{y}, t)^2 f(\vec{y})\, d\vec{y}$, as an unknown quantity Z described by the space-smoothed forms of these equations, we obtain two mathematically rather unexceptionable equations:

$$[Z, \Phi(f, t)] = 0, \quad [Z, \dot{\Phi}(g, t)] = 2i\Phi(fg, t);$$

unbounded operators are involved here, so that these equations are to hold only on suitably dense domains. Now the $\Phi(f, t)$ and $\dot{\Phi}(g, t)$ form an essentially irreducible set of operators, so that these putative equations, if they have a solution at all, should have a unique solution modulo additive scalar operators. In this way the heuristic question as to the existence of a mathematically meaningful square of the field may be reduced, in a fashion which is by the standards of theoretical physics quite conservative, to a definite and conceptually simple mathematical problem.

It is preferable for the usual reasons to replace the foregoing relations by one involving bounded operators, and this is indeed possible. Setting $U(f) = e^{i\Phi(f,t)}$, $V(g) = e^{i\dot{\Phi}(g,t)}$, and $W(t) = e^{itZ}$, an argument of the formal type appropriate at this level shows that

$$W(s)\,U(f)\,W(s)^{-1} = U(f),$$
$$W(s)\,V(g)\,W(s)^{-1} = \exp\,\overline{[i\dot{\Phi}(g,t) - 2s\,\Phi(fg,t)]}\,.$$
$$= U(-2sfg)\,V(g)\,e^{2is\int fg^2 d\vec{x}}.$$

The putative transformation of the $U(f)$ and $V(g)$ by $W(s)$ is known as a canonical transformation; evidently this transformation is well-defined, independently of whether $W(s)$ exists or not; when a unitary operator playing the role of $W(s)$ in the foregoing equation does exist, the transformation is said to be *unitarily implementable*. Necessary and sufficient conditions for this to be the case were treated more than a decade ago, and the present case falls nicely under the general theory, which serves also to indicate the distinctive role played by Hilbert space, not only as a representation space for the process operators, but also in connection with the underlying probe space.

To summarize briefly, if $\boldsymbol{N}$ is any complex Hilbert space, a "Weyl system" over $\boldsymbol{N}$ is a continuous mapping $z \to W(z)$ from $\boldsymbol{N}$ to the unitary operators on a Hilbert space $\boldsymbol{K}$ such that $W(z)\,W(z') = e^{(i/2)\,Im\langle z, z'\rangle}\,W(z+z')$, for arbitrary z and z' in $\boldsymbol{N}$. The "free Weyl system", closely related to the free processes described earlier but invariantly attached to the Hilbert space $\boldsymbol{N}$, may be described briefly as the pair (W, v), where W is a Weyl system and v is a vector in $\boldsymbol{N}$ which is cyclic for the $W(z)$ and such that $\langle W(z)\,v, v\rangle = \exp\,[-\|z\|^2/4]$. If T is a symplectic transformation on $\boldsymbol{N}$, i.e. a continuous real-linear transformation which leaves invariant the anti-symmetric form $Im\langle z, z'\rangle$, and W is any Weyl system over $\boldsymbol{N}$, then it is easily seen that if W_T is defined by the equation $W_T(z) = W(Tz)$, then W_T is also a Weyl system. The unitary implementability question raised above is precisely a special case of the question of the unitary implementability of the canonical transformation $W(z) \to W_T(z)$; i.e. the question of when there exists a unitary operator S

such that $SW(z)S^{-1}=W(Tz)$. Such questions have now been extensively studied in the case of the free processes, and according to a result in SHALE's Chicago thesis, there exists such an operator S if and only if $T^*T=I+H$, where H is a Hilbert-Schmidt operator. In the present context, T is easily computable as an integral operator, and the usual square-integrability criterion for a Schmidt class integral operator shows that: a unitary operator $W(s)$ of the indicated type exists, for sufficiently smooth functions f and g (say of class C^∞ and of compact support) if and only if the number of space dimensions is 1. Since essentially only formal algebra was involved in the foregoing reduction, this eliminates any reasonable possibility that the square of the free field might exist as a well-defined strong operational process in space, at a fixed time, except when the number of space dimensions is 1.

In the case when the number of space dimensions $n=1$, we could go on and establish that $W(s)$ defined a projective one-parameter group and obtain the square of the field as its generator. But a different method is better suited to the treatment of cubes and higher powers. I shall indicate this method and describe some implications for the case $n=1$, following which I shall indicate the modifications needed to deal with the case $n=3$ (or other higher values of n).

5. Local polynomials in free quantum fields

In order to construct explicitly cubes, higher powers, and other polynomials in quantum processes, one needs a systematic procedure for avoiding the infinite terms which arise if one proceeds too rashly towards this end. The algebraic basis for this procedure emerges from the considerations involved in the foregoing treatment of squares, and may be briefly indicated as follows.

Let $\boldsymbol{P}$ be a real linear vector space on which is given a non-degenerate anti-symmetric bilinear form A; let $\boldsymbol{E}$ denote the essentially unique algebra generated by $\boldsymbol{P}$ and a unit e such that $[x, y]=A(x, y)\,e$, for all x and y in $\boldsymbol{P}$; let F denote a given linear functional on $\boldsymbol{E}$. If $z_1, \ldots, z_r$ are any given elements of $\boldsymbol{P}$, their "normal product" relative to F, denoted $:z_1 z_2, \ldots, z_r:$, is defined recursively by the equations (assuming that $F(e)\neq 0$)

$$[:z_1 \ldots z_r:, z'] = \sum_k :z_1 \ldots \hat{z}_k \ldots z_r: [z_k, z'],$$

$$F(:z_1 \ldots z_r:)=0, \qquad :e:=e.$$

It can be shown that such normal products exist and are unique. The method of proof leads to the introduction of what may be called quantized differential forms, similar somewhat to conventional differential forms except that operators rather than functions form the 0-forms, and useful for rationalizing some of the objects connected with quantum fields which relate to operators in a fashion analogous to the way in which differential forms relate to functions; but details will have to be omitted.

Now let us return to the problem of dealing with powers of the neutral scalar field, at a fixed time. The finite linear combinations of the $\Phi(f, t)$ and $\dot{\Phi}(g, t)$, t being held fixed, form a real linear vector space $\boldsymbol{P}$ such that the commutator of any two elements is a scalar. The vector v previously indicated is in the domain of any polynomial T in the elements in $\boldsymbol{P}$, and serves to define a linear functional $F(T) = \langle Tv, v\rangle$. The cited result then leads to a definition for the normal product $:z_1 \ldots z_r:$, where the z_t are in $\boldsymbol{L}$. This ennables one to define the n'th power, say $\Phi^{(n)}(f)$ of the process $\Phi(\cdot\,, t)$, by the equation

$$\Phi^{(n)}(f) = \lim_{g\to\delta} \int \cdots \int : \phi(x_1, t)\, \phi(x_2, t) \ldots \phi(x_n, t) : g(x_1 - x_2) \ldots g(x_{n-1} - x_n) \cdot f(x_n)\, d x_1\, d x_2 \ldots d x_n;$$

the integrand may appear singular at first glance, but may be given a meaning in a straightforward way, and it may be shown, partly by general theory of the normal products, and partly by computation, that the limit indicated actually exists, in any of various appropriate senses. The simplest is that of the convergence in $L_2(\boldsymbol{H}')$ of the functionals, multiplication by which represent, in the real wave representation, the operators involved in the foregoing equation. The point of the normal product is that without it the limit either doesn't exist or is infinite almost everywhere (cf. [14]).

The powers attained can be characterized in a fashion similar to squares. More specifically,

$$U(f)\, \Phi^{(r)}(h)\, U(f)^{-1} = \Phi^{(r)}(h)$$

is a reflection of the circumstance that all the $e^{i\Phi(f,t)}$ commute (t fixed), while the equation

$$V(f)\, \Phi^{(r)}(h)\, V(f)^{-1} = (\text{closure of})\ \Phi^{(r)}(h) + r\,\Phi^{(r-1)}(fh) + \cdots + \binom{r}{s} \Phi^{(r-s)}(fh^s) + \cdots + \left(\int f h^r\, d\vec{x}\right) I$$

is based formally simply on the binomial theorem, adapted to normal products, although the analytical justification requires some technical sophistication. In particular, the fact that the unbounded self-adjoint operators $\Phi^{(r)}(f)$ all commute, in the strong sense that their spectral projections do so, is involved here, and a by product of the analysis is the simultaneous spectral resolution of all these operators, in the real wave representation indicated earlier. The irreducibility of the $U(f)$

and $V(g)$ for all f and g shows that $\Phi^{(r)}(h)$ is uniquely characterized by the relations just indicated, in a recursive sense; i.e., $\Phi^{(2)}(\cdot)$ is determined directly, then $\Phi^{(3)}(\cdot)$ relative to the determination of $\Phi^{(2)}(\cdot)$ and so forth.

The foregoing construction shows some properties of the products $\Phi^{(r)}(\cdot)$ indicating that they are not a peculiar ad hoc creation, but have some foundational features befitting a local function of the free field. In the first place, they are indeed local, in the sense that $\Phi^{(r)}(g)$ is for any g affiliated with the ring of operators (in the Murray-von Neumann sense) determined by the $\Phi(h)$, as the supports of h range over an open set containing the range of g. In the second place, they are invariant under the euclidean group.

Remarkably, but in retrospect perhaps not surprisingly, the same pseudo-products have already appeared, in a much different and quite primitive form, in the theoretical physical literature. Around two decades ago, G. C. WICK studied a procedure for standardizing the removal of infinities from products of field operators, and derived a formula, known in the physical literature as WICK'S Theorem, which greatly facillitated field-theoretic computations when in terms of the "Wick products". The mathematical nature of these objects was somewhat obscure (as can be seen from the treatment given in the book of BOGOLIOUBOFF and SHIRKOV, which is actually one of the clearest in the theoretical physical literature). A few years ago, GÅRDING and WIGHTMAN [2] treated mathematically a version of Wick products in four space-time dimensions, obtaining densely defined operators by averaging over space-time; but without obtaining self-adjointness properties, or dealing with the products at a fixed time, or quite making an explicit identification with the conventional Wick formalism. (The latter was at the time mathematically somewhat nebulous, and for the definition of the Wick product they used an explicit formula ascribed to CAIENELLO.) More recently, JAFFE [7] studied Wick products at fixed times for the scalar field in two space-time dimensions. My own work proceeded from an observation I made years ago, that the fractional derivative of the Wiener Brownian motion process or of similar processes, of order 1/2, although not conventional stochastic processes but only stochastic distributions, could be squared in non-trivial finite fashion by what may be described, ex post facto, as an adaptation of "centering" in probability theory; and that higher powers could be treated by generalization of the centering process. These generalized Brownian motion processes are actually the same, as shown by the real wave representation, as the scalar field at a fixed time. The stochastic approach, although lacking transparence from the particle representation viewpoint which is the most familiar one in the theory of free fields, leads directly to the simultaneous spectral resolution of all the Wick products, at a fixed time, as well as to proofs of their self-adjointness, locality, etc.; and is, in my opinion, indubitably better adapted to the treatment of non-linear local interactions than the more elementary approach through the particle representation.

The identification of the stochastically-defined Wick product with the classical one is made by showing that it satisfies a known recursion relation formally characterizing the Wick products; in particular, the "theorem of Wick" acquires mathematical status as a simple but essential algebraic formula. The interrelations between the characteristics of the Wick products in the particle and real wave representations becomes crucial in the study of non-linear interactions, inasmuch as the "free" dynamics is more simply expressed in the particle than the real wave representation, while it is very much the other way around in the case of the non-linear local terms involved additionally in the "interacting" dynamics.

These reassuring results have been for the case of two space-time dimensions. What then of the case of four space-time dimensions? — this is the physically really interesting case, but we have seen that there is no chance that even the square of the field exists as a (strict) operational process at fixed times. Actually, the foregoing analysis for a two-dimensional space-time has a form which adapts quite cogently to the cases of higher-dimensional space-times. The analysis is largely in terms of the unitary transformation properties of the Wick products under specified unitary operators; and it makes just as good sense to treat the unitary transformation properties of bilinear forms as it does to treat the same properties for operators.

As a natural and convenient domain in the Hilbert space for the process operators, let us take the infinitely differentiable vectors for the one-parameter group of unitary operators $\Gamma(t)$ generated by the free energy operator. This domain is invariant under the $U(f)$ and $V(g)$ indicated earlier, and thereby the square of the free field at a fixed time may be characterized in essentially the same way as before, except that it is to be a continuous bilinear form on the indicated domain in the usual topology on such a domain. The uniqueness of the square then depends on an extension of SCHUR's lemma to show that the $U(f)$ and $V(g)$ leave no such bilinear form invariant. Next, the locality of the Wick powers can be treated in essentially the same way through the definition indicated earlier of the ring of operators determined by a set of a bilinear forms — a definition which extends the conventional one for bona fide operators. Having defined and treated squares, the higher powers may be treated recursively in essentially the same fashion as in the case of a two-dimensional space-time.

The Wick powers are, of course, structures associated with a free field, and the question remains of the significance and treatment of non-linear quantized field equations, which involve presumptively processes which are not "free".

6. The meaning of non-linear quantized field equations

In attempting to give meaning to non-linear equations involving generalized functions, one should bear in mind the ultimate purpose of the equations, for it often turns out that definite mathematical meaning can be given only to those features of the formalism which have essential relevance to this ultimate purpose. In quantum theory, there are two main objects which are relatively directly correlative with experiment: the energy operator, i.e. the self-adjoint generator of the one-parameter group of temporal displacements which is given by STONE's theorem; and the S-operator (or S-matrix), which can be described as a measure of the effect visible at time $+\infty$ of the interaction on a given non-interacting situation at time $-\infty$, relative to the hypothetical motion which the non-interacting system would have undergone from time $-\infty$ to time $+\infty$, had not the interaction been present. The S-operator is the more important object in quantum field theory, and indeed suffices in one way or another for virtually all specific purposes. But the energy is more familiar, so I shall consider that first.

To make a long story short, the dynamics of quantum fields requires a more sophisticated phenomenology than the primitive one of forty years ago, in which an observable is an hermitian operator, a state is determined by a vector or density matrix in the underlying Hilbert space, and the dynamics is given by a one-parameter group of unitary operators. An apparently adequate, and currently widely investigated, phenomenology is that based on the notion that the conceptual observables are elements of an abstract C^*-algebra (by abstract I mean an algebraic isomorphism equivalence class of concrete C^*-algebras, acting on Hilbert spaces); that the states are suitably regular positive normalized linear functionals on the algebra on the algebra; and that the dynamics is given by a one-parameter group of automorphisms of the algebra. A physically crucial object is the "physical vacuum", which is definable succinctly as a state E which is invariant under the cited one parameter group $A \to A^t$ — as a result of which $E(A^t A^*)$ is necessarily a positive definite function of t — having the distinctive property that the spectrum of this function is non-negative. One expects *mathematical* systems of the indicated type mostly to be unstable and not have vacuums, but if they correspond to reality they must in most cases have sufficient stability so that any appropriate mathematical model must admit a vacuum. Having the vacuum, the energy is determined as the self-adjoint generator of the one-parameter unitary group inducing the automorphism group in question in the representation canonically associated with the vacuum state.

Now the equations of motion of a non-linear quantum field suffices formally to determine a canonical (= anti-symmetric and/or symmetric) quantum process at any time t', given the process at time t. But the manner in which the equations determine the vacuum is considerably more subtle. The situation can be illustrated quite well through the consideration of a hypothetical relativistic equation such as

$$\Box\phi = p(\phi),$$

p being a given polynomial. At any fixed time t, the process whose kernel is formally $(\phi(\cdot,t), \dot\phi(\cdot,t))$ (i.e., more exactly, the process at the time t, say ψ_t, is defined on pairs of functions f and g which are, say, infinitely differentiable of compact support on space, and $\psi((f,g))$ is the closure of $\int\phi(\vec{x},t)\,f(\vec{x})\,d\vec{x} + \int\dot\phi(\vec{x},t)\,g(\vec{x})\,d\vec{x}$) involves in its very definition the vacuum. On the other hand, the definition of the vacuum involves a given solution of the equation, or equivalently, given process $(\phi(\cdot,t), \dot\phi(\vec{x},t))$. The relation between the ("interacting") quantum process ϕ and its vacuum is thus a highly implicit one, from a mathematical viewpoint, despite the simple heuristic description of the vacuum in theoretical physics as the "lowest eigenstate of the hamiltonian"; *ab initio*, the hamiltonian is not a bona fide operator in Hilbert space, but becomes one only ex post facto, after the vacuum has been determined as an expectation-value form (i.e. the form $T \to \langle Tv, v\rangle$, where v is the "vacuum state vector") so that in quantum field theory the heuristic definition is merely suggestive.

A certain novelty in the mathematical formulation of the quantized partial differential equations, relative to classical formulations regarding hyperbolic equations, is thus apparently inevitable, although there is a resemblence to a highly floating type of boundary value problem. This formulation, in its simplest form, in which the process ϕ is postulated to consist at fixed times of operators in Hilbert space, after suitable averaging over space, — a postulate which, as earlier indicated, probably restricts relativistic applications to the case of two-dimensional space-time, but which can be relaxed in a way which permits the extension of the formulation to higher dimensions, — is as follows.

I begin with the most essential point of the characterization of the non-linear functions of a given canonical process. Let $(\Phi(\cdot), \dot\Phi(\cdot))$ be a given anti-symmetric process with (for specificity) probe space $\boldsymbol{C}$, consisting of the infinitely differentiable functions of compact support in space; this means that each of Φ and $\dot\Phi$ is linear from $\boldsymbol{C}$ to the self-adjoint operators on a Hilbert space $\boldsymbol{H}$, and that the usual Weyl relations are satisfied. (At this point there is no dynamics in the situation, and the notation $\dot{\Phi}$ is purely in anticipation of later applications.) There it is no essential loss of generality, and will be assumed for simplicity, that the ring of operators determined by the Φ and $\dot{\Phi}$ is a factor. (The conventional heuristic assumption corresponds to the stronger mathematical assumption that this ring includes all bounded operators.) It is also assumed that there is given a vacuum vector v, as earlier. Let $\Phi^{(0)}(f)$ be defined as $\int f$, and suppose that $\Phi^{(j)}(\cdot)$ has been defined for $j<n$ in such a fashion that:

(1) $\Phi^{(k)}(f)$ is affiliated with the ring of operators $\boldsymbol{R}_\phi$ determined by the $\Phi(g)$, for all $k\leqq j$ and all f in the domain of $\Phi^{(k)}$;

(2) if $k\leqq j$ and $k>0$, then

$$e^{i\dot{\Phi}(h)}\,\Phi^{(k)}(f)\,e^{-i\dot{\Phi}(h)}=\text{closure of}\sum_{r=0}^{k}\Phi^{(k-r)}(fh^r)\binom{k}{r}$$

for all $h\in\boldsymbol{C}$; and conversely, if there exists for any given $f\in\boldsymbol{C}$ a self-adjoint operator T affiliated with $\boldsymbol{R}_\phi$ such that $v\in D_T$, $\langle Tv, v\rangle=0$, and

$$e^{i\dot{\Phi}(h)}\,Te^{-i\dot{\Phi}(h)}=\text{closure of } T+\sum_{r=1}^{k}\Phi^{(k-r)}(fh^r)\binom{k}{r}$$

for all $g\in\boldsymbol{C}$, then f is in the domain of $\Phi^{(k)}$ and the foregoing relation holds. (Note that the indicated closure automatically exists and is self-adjoint.)

Then $\Phi^{(n)}$ is defined to have in its domain all $f\in\boldsymbol{C}$ such that there exists an operator T having the indicated properties, with k replaced by n, and $\Phi^{(n)}(f)$ is defined as this operator T; by the factorial assumption on the operator ring determined by Φ and $\dot{\Phi}$, this operator is unique, if it exists at all.

With this definition, the n-th power of a symmetric process always exists as a well-defined process, whose domain is however not necessarily non-trivial; and I may proceed to the question of the meaning of a solution to a quantized equation such as $\Box\phi=p(\phi)$, p being a given polynomial of degree r. A solution of the equation is definable as a system $\left(\Phi(\cdot,t), \dot{\Phi}(\cdot,t), \boldsymbol{H}, v, \Gamma(\cdot)\right)$ such that

(1) For each t (or, in view of later assumptions, for $t=0$), the system $\left(\Phi(\cdot,t), \dot{\Phi}(\cdot,t), \boldsymbol{H}, v\right)$ forms a symmetric process with probe space $\boldsymbol{C}$, such that the operator ring determined by the $\Phi(f,t)$ and $\dot{\Phi}(g,t)$ as f and g range over $\boldsymbol{C}$ is a factor; and such that the s-th power $\Phi^{(s)}(\cdot,t)$ includes $\boldsymbol{C}$ in its domain, for $s\leqq r$;

(2) $\Gamma(\cdot)$ is a continuous one-parameter unitary group on $\boldsymbol{H}$ with non-negative generator, and such that

$$\Gamma(s)\,\Phi(f,t)\,\Gamma(s)^{-1}=\Phi(f,t+s); \qquad \Gamma(s)\,\dot{\Phi}(f,t)\,\Gamma(s)^{-1}=\dot{\Phi}(f,t+s);$$
$$\Gamma(t)\,v=v\,(s,t\in\boldsymbol{R}^1;\,f\in\boldsymbol{C});$$

(3) There exists a dense domain $\boldsymbol{D}$ in $\boldsymbol{H}$ such that for $w, w'\in\boldsymbol{D}$,

$$(\partial/\partial t)\,\langle\Phi(f,t)\,w,w'\rangle=\langle\dot{\Phi}(f,t)\,w,w'\rangle$$
$$(\partial/\partial t)\,\langle\dot{\Phi}(f,t)\,w,w'\rangle=-\langle\Phi(B^2f,t)\,w,w'\rangle-\langle\Phi^{(p)}(f,t)\,w,w'\rangle,$$
$$B=(m^2I-\Delta)^{\frac{1}{2}},$$

where $\Phi^{(p)}=\sum a_j\,\Phi^{(j)}$ if $p(\lambda)=\sum a_j\,\lambda^j$.

These are minimal conditions as regards regularity, etc., for a solution which is an operational process in space at each fixed time; it would be analytically inconvenient not to have somewhat more, and probably difficult in practice to establish the existence of a solution without establishing some additional regularity. The simplest and most natural assumption in this regard would be that of the invariance of $\boldsymbol{D}$ under the infinitesimal generator H of $\Gamma(\cdot)$, and under the operators $\Phi(f,t)$ and $\dot{\Phi}(g,t)$, as well as under the unitary groups generated by these operators. The dynamical equations given in (3) are presented as partial differential equations, but may equally well be given in pure operator-theoretic form; with a slight but natural increase in the regularity required, this form is:

(3′) There exists a dense domain $\boldsymbol{D}$ in $\boldsymbol{H}$, invariant under the generator H of $\Gamma(\cdot)$, under the $\Phi(f,0)$ and $\dot{\Phi}(f,0)$, and under the one-parameter unitary groups they generate, such that

$$\text{closure of } [\Phi(f,0),H]=i\,\dot{\Phi}(f,0)$$
$$\text{closure of } [\dot{\Phi}(f,0),H]=\text{closure of } -i\,\Phi(B^2f,0)-i\,\Phi^{(r)}(f,0).$$

A similar formulation of the quantized equations could be given for other types of relativistic differential equations and/or for symmetric quantum processes. The existence and uniqueness of solutions is of course another matter. For the free field equations, existence derives, as indicated earlier, from the adaptation of the Fock-Cook construction to the relativistic situation. Uniqueness is shown, in a slightly different formulation of the problem in [10] in the case of symmetric processes and in forthcoming work of WEINLESS [17] in the case of anti-symmetric processes. In general, one anticipates a high degree of uniqueness, which is essentially a matter of the uniqueness of the vacuum as an expectation functional, but present results are limited to linear fields.

7. The S-operator and the interaction representation

Both because of the difficulty of establishing existence of solutions of the foregoing equations for non-linear equations in two-dimensional space-time, and the analytical complications in higher space-time dimensions, it is advantageous to deal with the equations in a different representation, which is at the same time closer to the important physical notion of S-operator (or S-matrix, as it is often referred to by physicists). The putative solution of the foregoing equations is called the "Heisenberg field"; formally equivalent to these equations, but in the absence of a comprehensive mathematical theory rigorously essentially distinct from them are equations whose putative solution is sometimes called the "interaction-representation field". These equations derive from the ones given by the method of variation of constants, which was initiated in the present connection by Schwinger and Tomonaga, independently, who noted that the resulting equations, although explicitly time-dependent, had an advantage in that the commutation relations between process operators at different times can be given explicitly, unlike the commutators for the Heisenberg fields, being in fact the free-field commutation relations.

The temporal propagation in this "interaction representation" is given formally by the solution of a differential equation of the form

$$u' = i H(t) u, \tag{1}$$

where $u(\cdot)$ is a vector valued function on R^1, and each $H(t)$ satisfies a formal self-adjointness condition, and indeed has the special heuristic form

$$H(t) = e^{itH_0} H_1 e^{-itH_0},$$

where H_0 and H_1 are, heuristically, given formally self-adjoint operators. From a mathematical viewpoint there are however two serious complications, even in the relatively simple case of two space-time dimensions. First, H_1 is not really an operator in Hilbert space, but is rather what I have identified as a "quantized differential form". Just as a closed differential form of degree 1 is locally identifiable with a function modulo constants, but is not globally so identifiable when it is inexact, so H_1 is identifiable locally (in space) with a bona fide operator, modulo a constant operator which may be eliminated by normalizing so that the vacuum expectation value vanishes, but is not globally equivalent to any fixed operator.

In more conventional terms, H_1 is given heuristically by an expression of the form $\int q(\phi_0(\vec{x}, 0))\, d\vec{x}$, where q is a polynomial such that $q' = p$; this definitely fails to exist as a self-adjoint operator, by a unitary implementability argument for the putative one-parameter group it generates of the type indicated earlier in connection with squares. The way around this difficulty is the use of the locality of the interactions and the hyperbolicity of the underlying equation in conjunction with the possibility of ascribing a natural mathematical meaning as operators H_1^f to the expressions $\int q(\phi_0(\vec{x}, 0) f(\vec{x}) d\vec{x}$, when f is an adequately regular function of compact support (cf. [13]). The H_1^f for differing f define the same propagation within a given bounded region of space-time, provided all the f are 1 on a sufficiently large region in space, and the corresponding equations to (1) with H_1 replaced by H_1^f admit strong solutions at all.

The second difficulty is that while the H_1^f are bona-fide self-adjoint operators in Hilbert space, their transforms $H_1^f(t) = e^{itH_0} H_1 e^{-itH_0}$ have such variable domains that the existing general theory for equations of the indicated type is quite inapplicable. It can nevertheless be shown (as follows from [13]) that the Eq. (1) with H_1 replaced by H_1^f always admits solutions. The uniqueness and regularity properties of these solutions may depend on special properties of the polynomial p describing the interaction. In one of the simplest non-trivial cases, that in which $p(l) = l^3$, some of the relevant questions have been studied by NELSON [9] and GLIMM and JAFFE [3].

Added in proof. In forthcoming work, I have treated the case of an arbitrary polynomial p of the form $p = q'$, where q is bounded from below [abstracted in Bull. Am. Math. Soc. **75**, 1390—1395 (1969)].

In four-dimensional space-time, the situation is further complicated by the circumstance that the H_1^f are generalized operators (or equivalently, densely-defined bilinear forms), as earlier indicated. The resolution of this complication, at least to the point of leading to a rigorous and natural formulation of the fundamental dynamical equations, depends on two related special features. First, the two-dimensional theory was largely developed in terms of considerations of unitary transformation properties; and, as already noted, these considerations apply virtually equally well to bilinear forms. Second, the bilinear forms in question are not at all arbitrary but have the important regularity property of being invariant under a set of unitary operators whose commutor is a finite operator ring (in the case of the "scalar" field being used for illustrative purposes here, a maximal abelian ring; "finite" refers to the concept introduced by MURRAY and VON NEUMANN). This is a very strong property, which in the case of a densely defined symmetric operator enforces its essential self-adjointness (cf. e.g. [12]). The forms in question are thus a natural extension of those corresponding to self-adjoint operators, and indeed can, in a sense, be diagonalized, the corresponding operator being represented as multiplication by a generalized function. This aspect of the situation is related to a theory of generalized vectors in Hilbert space associated with any given ring of operators, which appears to be a natural framework for the treatment of singular perturbations of self-adjoint operators. For brevity, and since I have already spoken some time ago about these general matters, I shall here merely conclude with a presentation in essentially conventional mathematical terms of the resulting formulation of the quantized dynamical equations in any number of space dimensions.

Consider the quantized equation $\Box\phi = p(\phi)$, where $p'(0) > 0$, for the tangential equation $\Box\phi = p'(0)\phi$ has no vacuum when $p'(0) < 0$, as shown by WEINLESS [17]; and it is convenient to assume also $p(0) = 0$. The primary objects, both from mathematical and physical standpoints, are not so much the process operators ϕ as the operators $S(t', t)$, which I shall call propagators, which transform the system from one time t to another time t'. For the reasons indicated earlier, it is virtually essential, or at any rate quite convenient, to treat these operators (which do not exist as unitary operators even in simple cases, but only as automorphisms of the C^*-algebra $\boldsymbol{W}$ described in [13] as the "space-finite Weyl algebra") in terms of the operators $S(t', t, f)$ which are corresponding propagators for the equation $\Box\phi = p'(0)\phi + f(p(\phi) - p'(0)\phi)$, where $f \in \boldsymbol{C}$. Following these parenthetical indications, I can make the

Definition. A (quantized) propagator for the equation $\Box\phi = p(\phi)$, where p is a given polynomial of the above-indicated type, is a strongly continuous function $S(t', t, f)$ on $R^1 \times R^1 \times \boldsymbol{C}$ whose values are unitary operators on the free-field Hilbert space $\boldsymbol{H}$ for the free field $(\phi_0, \boldsymbol{H}, v, \Gamma(\cdot))$ determined by the equation $\Box\phi_0 = p'(0)\phi_0$ by the characterization given earlier, satisfying the conditions:

(1) $S(t'', t', f)\, S(t', t, f) = S(t'', t, f)$ $(t, t', t'' \in R^1; f \in \boldsymbol{C})$;

(2) $\Gamma(t)\, S(t', t'', f)\, \Gamma(t)^{-1} = S(t+t', t+t'', f)$;

(3) If $h \in \boldsymbol{C}$ is supported by the sphere S, and if $f, g \in \boldsymbol{C}$ and $f(x) = g(x) = 1$ for $x \in S + Q_a$, where Q_a denotes the open sphere of radius a around the origin in space, then $S(t, 0, f)\, S(t, 0, g)^{-1}$ commutes with both $\Phi_0(h, 0)$ and $\dot{\Phi}_0(h, 0)$, provided $|t| < a$.

(4) $\dfrac{\partial S(t, t', f)}{\partial t} = H_1^f(t)\, S(t, t')$; $S(t, t) = I$, weakly on a dense domain $\boldsymbol{D}$ invariant under the $S(t, t', f)$, where $H_1^f(t)$ is the generalized operator corresponding to the bilinear form B representing $\int :q(\phi(x, t)): f(x)\, dx$. (More specifically, $(\partial/\partial t) \langle S(t, t')\, w, w' \rangle = B(S(t, t')\, w, w')$ for arbitrary w and w' in $\boldsymbol{D}$.)

Actually, the basic condition here is the differential Eq. (4). The other conditions are formal consequences of (4), and are to some extent unnecessarily strong in a definition, but serve to indicate some important properties of the temporal propagation. Condition (3) is a generalization of a familiar region of influence result in the theory of hyperbolic equations, and reflects the locality of the Wick product operators $H_1^f(t)$. By virtue of (3) the transform by $S(t, t', f)$ of any element X of the operator ring determined by the $\Phi(f, 0)$ and $\dot{\Phi}(g, 0)$ as f and g range over the elements of $\boldsymbol{C}$ vanishing outside of a given region R, i.e. $S(t, t', f) X S(t, t', f)^{-1}$, is independent of f, for f sufficiently close to the unit function, and there results a well-defined automorphism $\sum(t, t')$ of the space-finite Weyl algebra $\boldsymbol{W}$ earlier indicated, satisfying the basic desiderata:

$$\Sigma(t'', t')\,\Sigma(t', t) = \Sigma(t'', t); \qquad \Gamma(t'')\,\Sigma(t', t)\,\Gamma(t'')^{-1} = \Sigma(t' + t'', t + t'').$$

These automorphisms are not in general, and probably never except in trivial cases, unitarily implementable, in spite of the irreducibility of the algebra $\boldsymbol{W}$; and this is one reason why the use of the functions f appears foundationally essential, and not merely technically advantageous. However, the putative limit $\Sigma = \lim_{t' \to +\infty,\, t \to -\infty} \Sigma(t', t)$ will be unitarily implementable, for an extensive class of quantum processes, according to a line of thought which in its earliest form is represented by a well-known paper by YANG and FELDMAN on scattering theory, according to which the S-operator may be defined essentially as the unitary operator in question. It may be noted that the partially parallel problem of the existence of an analogous S-operator for a certain class of similar classical non-linear relativistic partial differential equations has recently attained an affirmative solution.

8. Conclusion

The theory and application of normal products of canonical quantum processes thus provides a mathematical foundation for the formulation and treatment of the dynamics of non-linear local quantum fields. The basically simple conceptual characterization and properties of these products at fixed times remove them from the position of ad hoc devices for removing infinite terms in order to obtain numerical results in agreement with experiment, to a position of natural and intrinsically interesting, if mathematically not yet familiar phenomena. They are similarly foundationally relevant to differential equations which are systems, or whose solutions are anti-symmetric rather than symmetric processes. I contemplate also similar applications to local interactions which are not defined by partial differential equations; these involve free fields whose classical counterpart is not the solution manifold of a differential equation, but a direct integral of such manifolds.

In any event, what are now established are formulations and methods of treatment for the fundamental dynamical equations of typical nonlinear relativistic quantized fields, which seem conservative and relevant from both contemporary mathematical and physical positions. Indeed, I should judge that there are now sufficiently many results in a positive direction for valid optimism regarding the resolubility, within the next few years, of some of the basic existential questions concerning the mathematical framework for physical quantum field theory.

References

1. FELDMAN, J.: On the Schrödinger and heat equations for bad potentials. Trans. Am. Math. Soc. **108**, 251—264 (1963).
2. GÅRDING, L., and A. S. WIGHTMAN: Fields as operator-valued distributions in relativistic quantum theory. Arkiv Fysik **28**, 129 (1964).
3. GLIMM, J., and A. JAFFE: $A\lambda\phi^4$ quantum field theory without cut-offs. I Preprint, May, 1968; publ. Phys. Rev. (1968).
4. GOODMAN, R., and I. SEGAL: Anti-locality of certain Lorentz-invariant operators. J. Math. Mech. **14**, 629—638 (1965).
5. GROSS, L.: Integration and non-linear transformations in Hilbert space. Trans. Am. Math. Soc. **94**, 404—440 (1960).
6. — Potential theory in Hilbert space. J. Funct. Anal. **1**, 123—181 (1967).
7. JAFFE, A.: Wick polynomials at fixed times. J. Math. Phys. **7**, 1250—1255 (1966).
8. NELSON, E.: Feynman integrals and the Schrödinger equation. J. Math. Phys. **5**, 332—343 (1964).
9. — A quartic interaction in two dimensions, p. 69—78, in: Prof. Conf. Math. Th. El. Parts. Cambridge, Mass.: M. I. T. Press 1966.
10. SEGAL, I.: Mathematical characterization of the physical vacuum. Illinois J. Math. **6**, 500—523 (1962).
11. — Interprétation et solution d'équations non linéaires quantifiées. Compt. Rend. **259**, 301—303 (1964).
12. — An extension of a theorem of O'Raifeartaigh. J. Funct. Anal. **1**, 1—21 (1967).
13. — Notes toward the construction of non-linear relativistic quantum fields. I. Proc. Nat. Acad. Sci. US **57**, 1178—1183 (1967).
14. — Non-linear functions of weak processes. J. Funct. Anal. **4**, 404—456 (1969).
15. — Quantized differential forms. Topology, 141—171 (1968).
16. SHALE, D.: Linear symmetries of free boson fields. Trans. Am. Math. Soc. **103**, 149—167 (1962).
17. WEINLESS, M.: Uniqueness of the vacuum for linear quantum fields. Ph. D. thesis, M. I. T., 1968; publ. J. Funct. Anal. (1969).

On the Analogue of the Modular Group in Characteristic p

By André Weil

The Institute for Advanced Study, Princeton

1. It is well-known that the classical concept of modular forms may be introduced as follows. Write G for one of the two algebraic groups $GL(2)$, $SL(2)$; take for k the field $\mathbf{Q}$ of rational numbers, $\mathbf{R}$ being then the completion k_∞ of k at its infinite place; let Γ be the subgroup $G_{\mathbf{Z}}$ of $G_{\mathbf{R}}$ (i.e. the group of the matrices in $M_2(\mathbf{Z})$ with the determinant ± 1 if $G = GL(2)$, and $+1$ if $G = SL(2)$). On $G_{\mathbf{R}}$, consider the complex-valued functions which are left-invariant under Γ (or at any rate under some congruence subgroup of Γ), behave in a prescribed manner under a translation belonging to the center of $G_{\mathbf{R}}$, and behave in a prescribed manner under the right translations belonging to the usual maximal compact subgroup of $G_{\mathbf{R}}$ and under the Casimir operator for $G_{\mathbf{R}}$; the two latter conditions ensure that this determines in the upper half-plane a modular form of prescribed degree which is an eigenfunction for the Beltrami operator (in particular, if the corresponding eigenvalue is 0, it is holomorphic, or at any rate the sum of a holomorphic and an antiholomorphic function). If we write $\mathbf{A}$ for the ring of adeles of k and $G_{\mathbf{A}}$ for the adelized group G, one can then, from such a function, derive a function on $G_{\mathbf{A}}$, left-invariant under G_k, and behaving in a prescribed manner under the center, under right-translations belonging to the usual maximal compact subgroup of $G_{\mathbf{A}}$ (or at any rate under a subgroup of finite index of that group), and under the Casimir operator for G_∞. Such an interpretation is useful if one wishes to extend the classical theory to arbitrary number-fields or function-fields.

On the present occasion, we wish to take for k the field of rational functions in one indeterminate over a finite field. Then, just as in the case $k = \mathbf{Q}$, it is unnecessary to use the adele language, and the relevant concepts can be described in a completely elementary manner.

2. Accordingly, we put $k = \mathbf{F}(T)$, where $\mathbf{F}$ is a finite field and T an indeterminate over $\mathbf{F}$; we call q the number of elements of $\mathbf{F}$. As usual, we write ∞ for the place of k for which $|T|_\infty > 1$; then the completion k_∞ of k at that place is the field of formal power-series $x = \sum \alpha_i T^{-i}$ in T^{-1}, where $\alpha_i \in \mathbf{F}$ for all $i \in \mathbf{Z}$ and there is n such that $\alpha_i = 0$ for $i < n$;

if at the same time $\alpha_n \neq 0$, we have then $|x|_\infty = q^{-n}$. The elements x of k_∞ for which $|x|_\infty \leq 1$, i.e. $\alpha_i = 0$ for all $i < 0$, make up the maximal compact subring r_∞ of k_∞; its maximal ideal is $T^{-1} \cdot r_\infty$, and k_∞, as an additive group, is the direct sum of that ideal and of the ring $\boldsymbol{F}[T]$ of the polynomials in T with coefficients in $\boldsymbol{F}$. If $P \in \boldsymbol{F}[T]$ and $P \neq 0$, then $|P|_\infty = q^{deg(P)}$.

We will confine ourselves to the group $GL(2)$; to simplify notations, we write G (instead of G_∞) for the group $GL(2, k_\infty)$, i.e. for the group of invertible matrices $\begin{pmatrix} x & y \\ z & t \end{pmatrix}$ in the ring $M_2(k_\infty)$ of the matrices of size 2 over k_∞. The matrices of that form for which $z = 0$ (the "triangular" matrices) make up a subgroup B of G; those for which $z = 0$, $t = 1$ and $x = T^n$ with $n \in \boldsymbol{Z}$ make up a subgroup of B which we denote by B_1. We write $\mathfrak{K}$ for the group of the invertible matrices in $M_2(r_\infty)$ (i.e. the matrices in $M_2(r_\infty)$ whose determinant is in $r_\infty^\times$), and $\mathfrak{Z}$ for the center of G; $\mathfrak{K}$ is a maximal compact subgroup of G, and we have $G = B \cdot \mathfrak{K} = B_1 \cdot \mathfrak{K}\mathfrak{Z}$. We write Γ for the "modular group consisting of the invertible matrices in $M_2(\boldsymbol{F}[T])$, i.e. of the matrices $\begin{pmatrix} P & Q \\ R & S \end{pmatrix}$, with P, Q, R, S in $\boldsymbol{F}[T]$ and $PS - QR$ in $\boldsymbol{F}^\times$; this is a discrete subgroup of G. In obvious analogy with the classical theory, one can define congruence subgroups of Γ. In particular, take any unitary polynomial A in $\boldsymbol{F}[T]$ (i.e. one whose highest coefficient is 1); then we write Γ_A for the subgroup of Γ consisting of the matrices $\begin{pmatrix} P & Q \\ R & S \end{pmatrix}$ in Γ for which $R \equiv 0$ (mod. A), i.e. $R \in A \cdot \boldsymbol{F}[T]$; these are the most interesting congruence subgroups of Γ for us, because of their connection (discovered by HECKE in the classical case) with the theory of Mellin transforms.

3. Just as in the classical theory, Γ operates in a "properly discontinuous manner" on the space $G/\mathfrak{K}\mathfrak{Z}$ (which takes here the place of the upper half-plane). For every $n \in \boldsymbol{Z}$, write $\sigma_n = \begin{pmatrix} T^n & 0 \\ 0 & 1 \end{pmatrix}$. We begin by proving the following:

Every element g of G can be written as $g = \gamma \sigma_n g_0$ with $\gamma \in \Gamma$, $n \geq 0$, $g_0 \in \mathfrak{K}\mathfrak{Z}$; moreover, when g is given, the integer n in this formula is uniquely determined.

In substance, this is equivalent to the well-known classification of the projective line-bundles over a projective line (i.e. of the rational non-singular ruled surfaces) over the groundfield $\boldsymbol{F}$; that aspect of our problem will not be further considered here. On the other hand, the above result corresponds to the determination of the usual fundamental domain for the action of the classical modular group in the upper half-plane, and, as will be shown now, it can be proved similarly. For each non-zero vector (x, y), with x and y in k_∞, put $h(x, y) = \sup(|x|_\infty, |y|_\infty)$. Then h is invariant under $\mathfrak{K}$, i.e. we have $h((x, y) \cdot \mathfrak{k}) = h(x, y)$ for all $\mathfrak{k} \in \mathfrak{K}$. If g is any element of G, there is $C > 0$ such that

$$h((x, y) \cdot g^{-1}) \leqq C \cdot h(x, y),$$

or, what amounts to the same,

$$h(x, y) \leqq C \cdot h((x, y) \cdot g)$$

for all vectors (x, y). At the same time, for each $C' > 0$, there are only finitely many vectors (R, S) with R and S in $\boldsymbol{F}[T]$, such that $h(R, S) \leqq C'$. From these facts, it follows at once that, for a given $g \in G$, one can choose an element γ of Γ such that, if we put $\gamma^{-1} g = \begin{pmatrix} x & y \\ z & t \end{pmatrix}$, $h(z, t)$ has its smallest value h_0. Assume that γ has been so chosen, and put $\gamma^{-1} g = b_0 \mathfrak{k}$ with $b_0 = \begin{pmatrix} x_0 & y_0 \\ 0 & t_0 \end{pmatrix} \in B$ and $\mathfrak{k} \in \mathfrak{K}$, so that we have $h_0 = |t_0|_\infty$. Take any $\gamma' \in \Gamma$, and put $\gamma' \gamma^{-1} g = \begin{pmatrix} x' & y' \\ z' & t' \end{pmatrix}$; in view of our choice of γ, we have $h(z', t') \geqq h_0$. Writing $\gamma' b_0 = \begin{pmatrix} x'' & y'' \\ z'' & t'' \end{pmatrix}$, we have $(z', t') = (z'', t'') \cdot \mathfrak{k}$. Therefore:

$$h(z'', t'') = h(z', t') \geqq h_0 = |t_0|_\infty.$$

Now take $\gamma' = \begin{pmatrix} 0 & 1 \\ 1 & S \end{pmatrix}$, with $S \in \boldsymbol{F}[T]$; then $(z'', t'') = (x_0, y_0 + S t_0)$, so that we must have, for all S:

$$\sup(|x_0|_\infty, |y_0 + S t_0|_\infty) \geqq |t_0|_\infty.$$

As k_∞ is the direct sum of $\boldsymbol{F}[T]$ and $T^{-1} \cdot r_\infty$, we can write $y_0 t_0^{-1}$ in the form $S_0 + v$, with $S_0 \in \boldsymbol{F}[T]$ and $|v|_\infty < 1$. Taking now $S = -S_0$ in the above inequality, we have

$$|y_0 + S t_0|_\infty = |v t_0|_\infty < |t_0|_\infty$$

and therefore $|x_0|_\infty \geqq |t_0|_\infty$, so that we may write $x_0 t_0^{-1} = T^n u$ with $n \geqq 0$ and $|u|_\infty = 1$. Then:

$$g = \gamma \cdot \begin{pmatrix} 1 & S_0 \\ 0 & 1 \end{pmatrix} \cdot \sigma_n \cdot \begin{pmatrix} u & T^{-n} v \\ 0 & 1 \end{pmatrix} \cdot \mathfrak{k} \cdot \begin{pmatrix} t_0 & 0 \\ 0 & t_0 \end{pmatrix}.$$

In the right-hand side, the first two factors belong to Γ, the fourth and fifth ones to $\mathfrak{K}$, and the last one to $\mathfrak{Z}$. This proves the first part of our theorem, which we can also express by saying that the matrices σ_n for $n \geqq 0$ contain a full set of representatives for the double cosets $\Gamma \backslash G / \mathfrak{K}\mathfrak{Z}$ in G. Now we must show that, if two such matrices σ_n, σ_m are in the same double coset, then $n=m$. In fact, our assumption means that we have $\sigma_m = \gamma \sigma_n \mathfrak{k}\mathfrak{z}$ with $\gamma \in \Gamma$, $\mathfrak{k} \in \mathfrak{K}$, $\mathfrak{z} \in \mathfrak{Z}$. Write $\gamma = \begin{pmatrix} P & Q \\ R & S \end{pmatrix}$ and $\mathfrak{z} = T^{-i} u \cdot 1_2$ with $i \in \boldsymbol{Z}$, $|u|_\infty = 1$. Then the matrix

$$T^{-i} \cdot \begin{pmatrix} T^{-m} & 0 \\ 0 & 1 \end{pmatrix} \cdot \begin{pmatrix} P & Q \\ R & S \end{pmatrix} \cdot \begin{pmatrix} T^n & 0 \\ 0 & 1 \end{pmatrix} = \begin{pmatrix} T^{n-m-i} P & T^{-m-i} Q \\ T^{n-i} R & T^{-i} S \end{pmatrix}$$

must be in $\mathfrak{K}$, so that its determinant must be in $r_\infty^\times$; as the determinant $PS - QR$ of γ is in $\boldsymbol{F}^\times$, this gives $n - m = 2i$. As $T^{n-i} R$ and $T^{-i} S$ must be in r_∞, and the polynomials R and S are not both 0, we must have $i - n \geqq \deg(R)$ if $R \neq 0$, and otherwise $i \geqq \deg(S)$, hence in both cases $i \geqq 0$ and $n \geqq m$. Interchanging n and m, we get $n = m$, as was to be proved.

4. It is also useful to find out when an element g of G can be written in the above form in two different ways (this amounts to determining when an element γ of Γ, acting on $G/\mathfrak{K}\mathfrak{Z}$, can have a fixed point; here the analogy with the classical case would be misleading). Clearly it amounts to the same to find out when the matrix $\sigma_n^{-1} \gamma \sigma_n$, with $n \geqq 0$, $\gamma \in \Gamma$, can be in $\mathfrak{K}\mathfrak{Z}$. By considering its determinant, we see that, if it is in $\mathfrak{K}\mathfrak{Z}$, it is in $\mathfrak{K}$, and that it is in $\mathfrak{K}$ if and only if it is in $M_2(r_\infty)$. Writing $\gamma = \begin{pmatrix} P & Q \\ R & S \end{pmatrix}$, we see that this is so if and only if P and S are in $\boldsymbol{F}$, Q is at most of degree n (hence also in $\boldsymbol{F}$ if $n=0$), and R is 0 if $n>0$ and is in $\boldsymbol{F}$ if $n=0$. For $n=0$, this gives $\gamma \in GL(2, \boldsymbol{F})$; for $n>0$, it gives $\gamma = \begin{pmatrix} \alpha & Q \\ 0 & \beta \end{pmatrix}$ with α and β in $\boldsymbol{F}^\times$ and $\deg(Q) \leqq n$.

5. One may now consider those complex-valued functions on G which are right-invariant under $\mathfrak{K}\mathfrak{Z}$ and left-invariant under Γ or some congruence subgroup of Γ; an important example is given by the "Eisenstein series"

$$f(g) = |\det(g)|_\infty^{s/2} \sum_{R,S} a(R, S)\, h((R, S) \cdot g)^{-s};$$

here the summation is taken over all pairs of mutually prime polynomials R, S in $\boldsymbol{F}[T]$; the coefficients $a(R, S)$ depend only upon the congruence classes of R and S modulo some fixed polynomial A; and s is any complex number such that $Re(s) > 2$, so that the series is absolutely convergent (if this were not so, one could still attach a value to it by analytic continuation in the s-plane). These series will not be further considered here.

We restrict our attention to the "Hecke groups" Γ_A defined in no. 2; a complex-valued function on G, left-invariant under Γ_A and right-invariant under $\mathfrak{K}\mathfrak{Z}$, will be called an *automorphic function of level A on G*. A complete set of representatives of $\Gamma_A \backslash \Gamma$, i.e. of the left cosets $\Gamma_A \gamma$ in Γ, can be obtained as follows. In the set of all pairs (R, S) of mutually prime polynomials in $\boldsymbol{F}[T]$, consider the equivalence relation $(R, S) \sim (R', S')$ given by $RS' \equiv SR'$ (mod. A); write S/R mod. A for the equivalence class of the pair (R, S) for this relation, and D_A for the set of all such classes; for obvious reasons, the finite set D_A may be called the projective line modulo A (or, more accurately, the projective line over the ring $\boldsymbol{F}[T]/A \cdot \boldsymbol{F}[T]$). For each $\varrho \in D_A$, choose a representative (R_ϱ, S_ϱ), and two polynomials P_ϱ, Q_ϱ such that $P_\varrho S_\varrho - Q_\varrho R_\varrho = 1$; put $\gamma_\varrho = \begin{pmatrix} P_\varrho & Q_\varrho \\ R_\varrho & S_\varrho \end{pmatrix}$; then the elements γ_ϱ of Γ, for $\varrho \in D_A$, make up a complete set of representatives of $\Gamma_A \backslash \Gamma$. Consequently, in view of the theorem in no. 3, the elements $\gamma_\varrho \sigma_n$, for $\varrho \in D_A$, $n \geqq 0$, contain a complete set of representatives of the double cosets $\Gamma_A \backslash G / \mathfrak{K}\mathfrak{Z}$ in G. Therefore, if f is an automorphic function of level A on G, and if we put $f_n(\varrho) = f(\gamma_\varrho \sigma_n)$ for every $\varrho \in D_A$ and every $n \in \boldsymbol{Z}$, f is completely determined by the functions f_n on D_A for $n \geqq 0$.

In view of our results in no. 4, we have, for every $n \geqq 0$, $f(\gamma' \sigma_n) = f(\gamma \sigma_n)$ whenever $\gamma^{-1}\gamma'$ is of the form $\begin{pmatrix} 1 & Q \\ 0 & 1 \end{pmatrix}$ with $\deg(Q) \leqq n$; if we define $\varrho + Q$, for $\varrho = S/R$ mod. A, by $\varrho + Q = (S + QR)/R$ mod. A, this gives $f_n(\varrho) = f_n(\varrho + Q)$ for $\deg(Q) \leqq n$. Similarly, for $\varrho = S/R$ mod. A and $\alpha \in \boldsymbol{F}^\times$, define $\alpha\varrho = (\alpha S)/R$ mod. A and $\varrho^{-1} = R/S$ mod. A. As the matrix $\begin{pmatrix} 1 & 0 \\ 0 & \alpha \end{pmatrix}$ is in Γ and in $\mathfrak{K}$ and commutes with σ_n for all $n \in \boldsymbol{Z}$, we have $f_n(\alpha\varrho) = f_n(\varrho)$ for all $n \in \boldsymbol{Z}$ and all $\alpha \in \boldsymbol{F}^\times$. Finally, for $\gamma = \begin{pmatrix} P & Q \\ R & S \end{pmatrix} \in \Gamma$, put $\gamma' = \begin{pmatrix} Q & P \\ S & R \end{pmatrix}$; we have

$$\gamma \sigma_{-n} = \gamma' \sigma_n \cdot \begin{pmatrix} 0 & T^{-n} \\ T^{-n} & 0 \end{pmatrix}.$$

As the last matrix in the right-hand side is in $\mathfrak{K}\mathfrak{Z}$, this gives $f(\gamma\sigma_{-n}) = f(\gamma'\sigma_n)$, hence, for $\gamma = \gamma_\varrho$, $f_{-n}(\varrho) = f_n(\varrho^{-1})$ for all $n \in \boldsymbol{Z}$ and all $\varrho \in D_A$.

6. As in the classical case, one can attach "Hecke operators" to the places of the field k; we begin with the places other than ∞, since for these the analogy is more obvious. Let Π be a unitary prime polynomial in $\boldsymbol{F}[T]$, not a divisor of A; call π its degree. Consider all the matrices $\begin{pmatrix} P & Q \\ R & S \end{pmatrix}$ with coefficients P, Q, R, S in $\boldsymbol{F}[T]$, $R \equiv 0$ (mod. A),

whose determinant is of the form $\alpha\Pi$ with $\alpha\in\boldsymbol{F}^\times$. It is easily seen that these matrices make up a union of finitely many left cosets of Γ_A in $GL(2,k)$, and that a complete set of representatives for these cosets consists of $\begin{pmatrix}\Pi & 0\\ 0 & 1\end{pmatrix}$ and of the matrices $\begin{pmatrix}1 & M\\ 0 & \Pi\end{pmatrix}$ when one takes for M any complete set of representatives of the congruence classes modulo Π in $\boldsymbol{F}[T]$, for instance all the polynomials of degree $<\pi$. Now, for any function f on G, consider the function f_Π on G given by

$$f_\Pi(g)=f\left(\begin{pmatrix}\Pi & 0\\ 0 & 1\end{pmatrix}\cdot g\right)+\sum_M f\left(\begin{pmatrix}1 & M\\ 0 & \Pi\end{pmatrix}\cdot g\right),$$

the sum in the right-hand side being taken over all the polynomials of degree $<\pi$ in $\boldsymbol{F}[T]$. If f is right-invariant under $\mathfrak{K}\mathfrak{Z}$, so is f_Π; and one sees at once that, if f is left-invariant under Γ_A, f_Π has the same property. Therefore the mapping $f\to f_\Pi$ induces, on the space of automorphic functions of level A on G, an operator which we denote by H_Π. As in the classical case, it is easily seen that any two of these operators commute with each other.

From a birational point of view, there is no difference between the place ∞ of k and those attached to the prime polynomials Π; therefore we may expect to find a Hecke operator attached to ∞. Had we used the adele language, we would not need a special definition for this; here, as we have given a special role to play to the place ∞, we shall proceed as follows. For each function f on G, right-invariant under $\mathfrak{K}\mathfrak{Z}$, consider the function f_∞ given by

$$f_\infty(g)=\int_{\mathfrak{K}} f(g\mathfrak{k}\sigma_{-1})\,d\mathfrak{k}$$

where $d\mathfrak{k}$ is a suitably normalized Haar measure on the compact group $\mathfrak{K}$. Clearly f_∞ is right-invariant under $\mathfrak{K}\mathfrak{Z}$; if f is left-invariant under Γ_A, so is f_∞. We write H_∞ for the operator induced by the mapping $f\to f_\infty$ on the space of automorphic functions of level A on G; obviously it commutes with the operators H_Π.

In the integral defining f_∞, the integrand is constant on the right cosets modulo $\mathfrak{K}\cap\sigma_{-1}\mathfrak{K}\sigma_1$ in $\mathfrak{K}$; it is easily seen that a complete set of representatives for these cosets is given by $\begin{pmatrix}0 & 1\\ 1 & 0\end{pmatrix}$ and $\begin{pmatrix}1 & \xi\\ 0 & 1\end{pmatrix}$ for $\xi\in\boldsymbol{F}$. If we normalize $d\mathfrak{k}$ so that the measure of each such coset is 1, and if we observe that

$$\begin{pmatrix}0 & 1\\ 1 & 0\end{pmatrix}\cdot\sigma_{-1}=\sigma_1\cdot\begin{pmatrix}0 & 1\\ 1 & 0\end{pmatrix}\cdot T^{-1},$$

we get (since f was assumed to be right-invariant under $\mathfrak{K}\mathfrak{Z}$):

$$f_\infty(g) = f(g\sigma_1) + \sum_{\xi\in \boldsymbol{F}} f\left(g\cdot\begin{pmatrix} T^{-1} & \xi \\ 0 & 1\end{pmatrix}\right).$$

Assume now that f is automorphic of level A; as in no. 5, put $f_n(\varrho) = f(\gamma_\varrho\sigma_n)$ for $\varrho\in D_A$, $n\in\boldsymbol{Z}$, and write $(H_\infty f)_n$ for the function similarly derived from $H_\infty f$. Observe that we have

$$\sigma_n\cdot\begin{pmatrix} T^{-1} & \xi \\ 0 & 1\end{pmatrix} = \begin{pmatrix} 1 & \xi T^n \\ 0 & 1\end{pmatrix}\cdot\sigma_{n-1}.$$

Taking now $g = \gamma_\varrho\sigma_n$ with $n\geqq 0$ in the above formula for $f_\infty(g)$, we get:

$$(H_\infty f)_n(\varrho) = f_{n+1}(\varrho) + \sum_\xi f_{n-1}(\varrho + \xi T^n).$$

In particular, assume that f is an eigenfunction of H_∞; this corresponds, in the classical case, to prescribing that the function f on $GL(2, \boldsymbol{R})$ be an eigenfunction of the Casimir operator (or that the modular form determined by f be an eigenfunction of the Beltrami operator in the upper half-plane). Let λ be the eigenvalue of H_∞ to which f belongs. Then $H_\infty f = \lambda f$, and we have, for all $n\geqq 1$:

$$f_{n+1}(\varrho) = \lambda f_n(\varrho) - \sum_\xi f_{n-1}(\varrho + \xi T^n).$$

This shows that f_n is uniquely determined for all $n > 2$ when f_0 and f_1 are given on D_A. As D_A is a finite set, and as we have seen that f is uniquely determined by the values of f_n for all $n\geqq 0$, this shows that the automorphic functions of given level, belonging to a given eigenvalue of H_∞, make up a vector-space of finite dimension over $\boldsymbol{C}$.

7. As we have observed before, we have $G = B_1\cdot\mathfrak{K}\mathfrak{Z}$, where B_1 is as defined in no. 2; therefore a function f on G, right-invariant under $\mathfrak{K}\mathfrak{Z}$, is uniquely determined by the function φ induced by it on B_1, i.e. by the function

$$\varphi(n, y) = f\left(\begin{pmatrix} T^n & y \\ 0 & 1\end{pmatrix}\right)$$

for all $n\in\boldsymbol{Z}$ and $y\in k_\infty$.

Conversely, let φ be given; we wish to investigate whether there is an automorphic function of level A on G, inducing φ on B_1. For this to be so, it is obviously necessary that φ should be right-invariant under the group $B_1\cap\mathfrak{K}\mathfrak{Z}$; as this consists of the matrices $\begin{pmatrix} 1 & v \\ 0 & 1\end{pmatrix}$ with $v\in r_\infty$, we must have $\varphi(n, y + T^n v) = \varphi(n, y)$ for all $v\in r_\infty$. When that is so, φ can be extended in one and only one way to a function f on G, right-invariant under $\mathfrak{K}\mathfrak{Z}$; we have now to investigate whether f is left-invariant under Γ_A. Express first that it is left-invariant under $\begin{pmatrix}\alpha & 0\\ 0 & 1\end{pmatrix}$ for $\alpha\in\boldsymbol{F}^\times$; as this is also in $\mathfrak{K}$, and as we have

$$\begin{pmatrix}\alpha & 0\\ 0 & 1\end{pmatrix}\cdot\begin{pmatrix}T^n & y\\ 0 & 1\end{pmatrix}\cdot\begin{pmatrix}\alpha & 0\\ 0 & 1\end{pmatrix}^{-1}=\begin{pmatrix}T^n & \alpha y\\ 0 & 1\end{pmatrix},$$

we get $\varphi(n, y)=\varphi(n, \alpha y)$ for all n, y and $\alpha\in\boldsymbol{F}^\times$. Express now that f is left-invariant under $\begin{pmatrix}1 & Q\\ 0 & 1\end{pmatrix}$ for all $Q\in\boldsymbol{F}[T]$; this gives $\varphi(n, y)=\varphi(n, y+Q)$; in other words, for each n, the function $y\to\varphi(n, y)$ can be expanded in a Fourier series on the compact group $k_\infty/\boldsymbol{F}[T]$. Take a fixed non-trivial character ψ_0 of the additive group of $\boldsymbol{F}$; for every element $x=\sum\alpha_i T^{-i}$ of k_∞, put $\psi(x)=\psi_0(\alpha_1)$. Then the characters of k_∞, trivial on $\boldsymbol{F}[T]$, are those of the form $y\to\psi(Qy)$ with $Q\in\boldsymbol{F}[T]$. This gives:

$$\varphi(n, y)=\sum_Q c(n, Q)\psi(Qy),$$

where the sum is taken over all $Q\in\boldsymbol{F}[T]$. Writing that φ is invariant under $y\to\alpha y$ for all $\alpha\in\boldsymbol{F}^\times$, and under $y\to y+T^n v$ for all $v\in r_\infty$, we get $c(n, \alpha Q)=c(n, Q)$ for all n, Q and $\alpha\in\boldsymbol{F}^\times$, and $c(n, Q)=0$ unless $Q=0$ or $n+\deg(Q)\leqq -2$. In particular, for a given value of n, only finitely many terms in the Fourier series for φ can be other than 0, viz., the term with $Q=0$ if $n\geqq -1$, and otherwise that term and those for which $\deg(Q)\leqq -n-2$. We will say that φ and f are *B-cuspidal* when $c(n, 0)=0$ for all $n\in\boldsymbol{Z}$.

8. Let notations be as in no. 7, and let f_∞ be as defined in no. 6; as observed there, f_∞ is right-invariant under $\mathfrak{K}\mathfrak{Z}$ since f is so, and therefore we have $f_\infty=\lambda f$ if and only if f_∞ coincides with λf on B_1; in view of the formula given for f_∞ in no. 6, this can be written as follows:

$$\lambda\varphi(n, y)=\varphi(n+1, y)+\sum_{\xi\in\boldsymbol{F}}\varphi(n-1, y+\xi T^n).$$

In particular, if f is automorphic of level A, this expresses that it is an eigenfunction of H_∞ belonging to the eigenvalue λ. Replacing φ by its Fourier series, we get the equivalent condition

$$\lambda c(n, Q)=c(n+1, Q)+c(n-1, Q)\sum_\xi\psi(\xi T^n Q).$$

The sum in the last term in the right-hand side has the value 0 or q according as the polynomial Q contains a non-zero term in T^{-n-1} or not. For $Q=0$, this gives

$$\lambda c(n, 0)=c(n+1, 0)+qc(n-1, 0)$$

for all n. For $Q \neq 0$, put $d = \deg(Q)$, so that $c(n, Q) = 0$ for $n \geqq -d-1$. Then the above condition is trivially fulfilled for $n \geqq -d-1$; for $n \leqq -d-2$, it gives a recurrence relation which is easily solved by writing the formal power-series expansion

$$\left(1 - \frac{\lambda}{q} U + \frac{1}{q} U^2\right)^{-1} = \sum_{i=0}^{\infty} a_i U^i .$$

Then $c(n, Q) = a_{-n-d-2} c(-d-2, Q)$ for $n \leqq -d-2$ (and for all n if we put $a_i = 0$ for $i < 0$). When the coefficients $c(n, Q)$, for $Q \neq 0$, have that property, for some value of λ, we will say (for reasons which will become clearer presently) that they are *eulerian at* ∞.

On the other hand, take Π, and define f_Π, as in no. 6; write φ_Π for the function induced by f_Π on B_1. We have

$$\varphi_\Pi(n, y) = f\left(\begin{pmatrix} \Pi T^n & \Pi y \\ 0 & 1 \end{pmatrix}\right) + \sum_M f\left(\begin{pmatrix} T^n & y+M \\ 0 & \Pi \end{pmatrix}\right),$$

where the sum is taken over all the polynomials M of degree $< \pi = \deg(\Pi)$. As f is right-invariant under $\mathfrak{R}\mathfrak{Z}$, this can also be written as

$$\begin{aligned} \varphi_\Pi(n,y) &= f\left(\begin{pmatrix} T^{n+\pi} & \Pi y \\ 0 & 1 \end{pmatrix}\right) + \sum_M f\left(\begin{pmatrix} T^{n-\pi} & \Pi^{-1}(y+M) \\ 0 & 1 \end{pmatrix}\right) \\ &= \varphi(n+\pi, \Pi y) + \sum_M \varphi(n-\pi, \Pi^{-1}(y+M)) . \end{aligned}$$

Replacing φ by its Fourier series, we get

$$\varphi_\Pi(n, y) = \sum_Q c(n+\pi, Q) \psi(Q \Pi y) + \sum_Q c(n-\pi, Q) \psi(Q \Pi^{-1} y) \sum_M \psi(Q \Pi^{-1} M) .$$

Here the sum with respect to M has the value q^π or 0 according as $\psi(Q\Pi^{-1}M)$ is 1 for all M or not, i.e. according as $Q \equiv 0 \pmod{\Pi}$ or not. Consequently the Fourier coefficients of φ_Π are

$$c_\Pi(n, Q) = c(n+\pi, \Pi^{-1}Q) + q^\pi c(n-\pi, \Pi Q)$$

provided we agree that the first term in the right-hand side is 0 unless $Q \equiv 0 \pmod{\Pi}$.

We can now express, in terms of the coefficients $c(n, Q)$, the conditions for f to be an eigenfunction of the operator $f \to f_\Pi$; if ω is the corresponding eigenvalue, we get, for $Q = 0$:

$$\omega c(n, 0) = c(n+\pi, 0) + q^\pi c(n-\pi, 0)$$

for all n; on the other hand, taking for Q a polynomial prime to Π, and substituting $\Pi^i Q$ for Q in the above formula, with $i \geqq 0$, we get

$$\omega c(n, \Pi^i Q) = c(n+\pi, \Pi^{i-1} Q) + q^\pi c(n-\pi, \Pi^{i+1} Q) .$$

As before, this can be solved by writing the formal expansion

$$\left(1-\frac{\omega}{q^{\pi}}U+\frac{1}{q^{\pi}}U^2\right)^{-1}=\sum_{i=0}^{\infty}A_i U^i.$$

Then we have $c(n,\Pi^i Q)=A_i c(n+i\pi, Q)$ for all $i\geqq 0$. When that is so, we shall say that the coefficients $c(n,Q)$ are *eulerian at the place* Π.

9. The "eulerian" property can be expressed conveniently by introducing the formal power-series

$$P(U)=\sum_{n=0}^{\infty}\Big(\sum_{Q}{}' c(-n-2,Q)\Big)U^n$$

where $\sum'$ denotes the summation over all unitary polynomials in $\boldsymbol{F}[T]$. For each n, the sum $\sum'$ may be restricted to the polynomials of degree $\leqq n$, since all the other terms in it are 0; therefore P is well-defined as a formal power-series. If f is B-cuspidal, we have $c(n,0)=0$ for all n; as $c(n,\alpha Q)=c(n,Q)$ for all $\alpha\in\boldsymbol{F}^{\times}$, this gives

$$f\left(\begin{pmatrix}T^n & 0\\ 0 & 1\end{pmatrix}\right)=\varphi(n,0)=\sum_{Q}c(n,Q)=(q-1)\sum_{Q}{}'c(n,Q),$$

so that, in that case, the power-series P may be written

$$P(U)=(q-1)^{-1}\sum_{n=0}^{\infty}f\left(\begin{pmatrix}T^{-n-2} & 0\\ 0 & 1\end{pmatrix}\right)U^n.$$

As can be seen at once, the $c(n,Q)$ are eulerian at ∞ if P can be rewritten as

$$P(U)=\left(1-\frac{\lambda}{q}U+\frac{1}{q}U^2\right)^{-1}\sum_{Q}{}'c(-\deg(Q)-2,Q)\,U^{\deg(Q)}.$$

Here the series in the right-hand side may be regarded as corresponding, in the classical theory, to the Dirichlet series whose Mellin transform is the modular form defined by f, while the first factor corresponds to the gamma factor in the functional equation for that series.

On the other hand, notations being as in no. 8, the $c(n,Q)$ are eulerian at Π if and only if

$$P(U)=\left(1-\omega\left(\frac{U}{q}\right)^{\pi}+\left(\frac{U^2}{q}\right)^{\pi}\right)^{-1}\sum_{n=0}^{\infty}\Big(\sum_{Q}{}''c(-n-2,Q\Big)U^n$$

where $\sum''$ denotes the summation over all unitary polynomials Q prime to Π in $\boldsymbol{F}[T]$.

10. We come back to the question, raised in no. 7, whether the function f (derived from φ as explained there) is automorphic of level A.

If so, the function f' on G defined by

$$f'(g) = f\left(\begin{pmatrix} 0 & 1 \\ A & 0 \end{pmatrix} \cdot g\right)$$

is also automorphic of level A; one sees at once that, if f is an eigenfunction of one of the Hecke operators H_∞, H_Π, f' is an eigenfunction of the same operator, belonging to the same eigenvalue.

We assume now that f, and consequently f', are automorphic of level A; moreover, to simplify matters, we will assume that they are both B-cuspidal. Call φ' the function induced by f' on B_1, $c'(n, Q)$ its Fourier coefficients, and P' the formal power-series derived from these just as P was derived from $c(n, Q)$ in no. 9. As f' is assumed to be B-cuspidal, this gives

$$P'(U) = (q-1)^{-1} \sum_{n=0}^{\infty} f'\left(\begin{pmatrix} T^{-n-2} & 0 \\ 0 & 1 \end{pmatrix}\right) U^n = (q-1)^{-1} \sum_{n=0}^{\infty} f\left(\begin{pmatrix} 0 & 1 \\ A\,T^{-n-2} & 0 \end{pmatrix}\right) U^n.$$

Now we have, for $u \neq 0$:

$$\begin{pmatrix} 0 & 1 \\ A\,T^{-n-2} & 0 \end{pmatrix} = \begin{pmatrix} A^{-1} T^{n+2} u^{-1} & 0 \\ 0 & 1 \end{pmatrix} \cdot \begin{pmatrix} 0 & u \\ 1 & 0 \end{pmatrix} \cdot A\,T^{-n-2}.$$

If we put $a = \deg(A)$ and $u = T^a A^{-1}$, this gives, since u is then in $r_\infty^\times$, and since f is right-invariant under $\mathfrak{R}\mathfrak{Z}$:

$$f\left(\begin{pmatrix} 0 & 1 \\ A\,T^{-n-2} & 0 \end{pmatrix}\right) = f\left(\begin{pmatrix} T^{n+2-a} & 0 \\ 0 & 1 \end{pmatrix}\right).$$

Comparing now the power-series for P and for P', we get the "functional equation"

$$P(U) = U^{a-4} P'(U^{-1}),$$

which implies that both P and P' are polynomials of degree $a-4$. This corresponds to the fact, discovered by Hecke, that the Mellin transforms of modular forms of suitable type satisfy functional equations. Of course, here, we must have $P = P' = 0$ if $a < 4$.

11. As in the classical case, the above result can be significantly extended. Let M be a unitary polynomial in $\boldsymbol{F}[T]$; call μ its degree. Let χ be a character of the multiplicative group of polynomials prime to M modulo M (i.e. of the group $(\boldsymbol{F}[T]/M \cdot \boldsymbol{F}[T])^\times$), which we extend to all polynomials by writing $\chi(Q) = 0$ when Q is not prime to M. Assume that χ is primitive, i.e. that there is no divisor M' of M, other than M, such that $\chi(Q) = 1$ whenever Q is $\equiv 1 \pmod{M'}$ and prime to M. We will also assume that χ is trivial on $\boldsymbol{F}^\times$ (from a birationally invariant point of view, this guarantees that χ corresponds to a character

of $k_A^\times/k^\times$, trivial on $k_\infty^\times$). Together with the power-series P introduced in no. 9, we also consider, for all such characters χ whose conductor M is prime to the level A of f, the series

$$P(\chi, U) = \sum_{n=0}^{\infty} \Big(\sum_Q{}' c(-n-2, Q)\chi(Q)\Big) U^n$$

and the series $P'(\chi, U)$ similarly derived from f'. In order to establish the functional equations for these series, we observe that the characters of the additive group of $\boldsymbol{F}[T]$ modulo M are $Q \to \psi(M^{-1}NQ)$, where N runs through a complete set of representatives of the congruence classes modulo M in $\boldsymbol{F}[T]$, for instance the polynomials of degree $<\mu$ in $\boldsymbol{F}[T]$. Therefore, on that group, the character χ may be written as a linear combination of such characters. Just as in the classical theory, this is

$$\chi(Q) = q^{-\mu} g(\chi) \sum_N \bar{\chi}(N) \psi(M^{-1}NQ)$$

where $g(\chi)$ is the "Gaussian sum"

$$g(\chi) = \sum_{Q \bmod M} \chi(Q)\psi(-M^{-1}Q).$$

Consequently, we have

$$P(\chi, U) = (q-1)^{-1} q^{-\mu} g(\chi) \sum_n \sum_N \bar{\chi}(N) f\left(\begin{pmatrix} T^{-n-2} & -M^{-1}N \\ 0 & 1 \end{pmatrix}\right) U^n$$

and similarly

$$P'(\bar{\chi}, U) = (q-1)^{-1} q^{-\mu} g(\bar{\chi}) \sum_{n'} \sum_{N'} \chi(N') f\left(\begin{pmatrix} 0 & 1 \\ A\,T^{-n'-2} & -A\,M^{-1}N' \end{pmatrix}\right) U^{n'}.$$

As A is prime to M, the congruence $ANN' \equiv 1 \pmod{M}$ determines a one-to-one correspondence between representatives N, N' of the congruence classes prime to M modulo M. For such a pair N, N', write $ANN' = 1 - MX$, with $X \in \boldsymbol{F}[T]$. Then the matrix $\gamma = \begin{pmatrix} M & N \\ -AN' & X \end{pmatrix}$ is in Γ_A, and we have:

$$\begin{pmatrix} 0 & 1 \\ A\,T^{-n'-2} & -A\,M^{-1}N' \end{pmatrix} = \gamma \cdot \begin{pmatrix} T^{-n-2} & -M^{-1}N \\ 0 & 1 \end{pmatrix} \cdot \begin{pmatrix} 0 & u \\ v & 0 \end{pmatrix}$$

with $u = M^{-1}T^{n+2}$, $v = AMT^{-n'-2}$. The last factor in the right-hand side is in $\mathfrak{R}\mathfrak{Z}$ if $u^{-1}v$ is in $r_\infty^\times$, i.e. if $n + n' = a + 2\mu - 4$; therefore, when that is so, we have

$$f\left(\begin{pmatrix} T^{-n-2} & -M^{-1}N \\ 0 & 1 \end{pmatrix}\right) = f\left(\begin{pmatrix} 0 & 1 \\ A\,T^{-n'-2} & -A\,M^{-1}N \end{pmatrix}\right).$$

From this, it follows at once that $P(\chi, U)$ and $P'(\bar{\chi}, U)$ are polynomials of degree $a+2\mu-4$ and satisfy the functional equation

$$P(\chi, U) = \frac{g(\chi)}{g(\bar{\chi})}\,\chi(A)\,U^{a+2\mu-4}\,P'(\bar{\chi}, U^{-1}).$$

12. The analogy with the classical case [cf. my note in Math. Ann. **168**, 149—156 (1967)] suggests the following problem. Let the coefficients $c(n, Q)$ be given; assume that $c(n, \alpha Q) = c(n, Q)$ for all n, Q and $\alpha \in \boldsymbol{F}^{\times}$, $c(n, 0) = 0$ for all n, and $c(n, Q) = 0$ for $Q \neq 0$ and $n + \deg(Q) \leqq -2$. Define $P(U)$ as in no. 9, and $P(\chi, U)$ as in no. 11. Define φ on B_1 by

$$\varphi(n, y) = \sum_{Q} c(n, Q)\,\psi(Qy);$$

then, as we have seen in no. 7, φ is right-invariant under $B_1 \cap \mathfrak{K}\mathfrak{Z}$, and we can extend it to a function f on G, right-invariant under $\mathfrak{K}\mathfrak{Z}$. Similarly, let coefficients $c'(n, Q)$ be given, satisfying the same conditions as above; from these, derive power-series $P'(U)$, $P'(\chi, U)$ and functions φ', f' in the same manner. Now we ask: if P, P' and (for all primitive characters χ whose conductor M is prime to a fixed multiple B of A) the series $P(\chi, U)$, $P'(\bar{\chi}, U)$ satisfy the functional equations obtained in nos. 10—11, does it follow that f, f' are left-invariant under Γ_A and that $f'(g) = f\left(\begin{pmatrix} 0 & 1 \\ A & 0 \end{pmatrix} \cdot g\right)$?

The answer is affirmative, at any rate, if one adds the assumption that the coefficients $c(n, Q)$ are eulerian at almost all places of k; this is a special case of a more general result, valid for any number-field or function-field k. Perhaps, as the analogy with the classical case seems to suggest (cf. loc. cit.), it might be more appropriate to assume the eulerian property merely at ∞; perhaps even this could be dropped. As to these questions, I shall refrain from the temptation of offering "conjectures", which, in the absence of any evidence, would be mere guesswork; a more thorough investigation is clearly needed.

A Theorem on the Formal Multiplication of Trigonometric Series

By Antoni Zygmund
The University of Chicago

1. The classical theorem of Riemann asserts that if the two-way infinite sequence of numbers c_n tends to 0 as $n \to \pm\infty$, then the series

$$\sum_{-\infty}^{+\infty} c_n e^{inx} \tag{S}$$

integrated termwise twice has sum

$$F(x) = \tfrac{1}{2} c_0 x^2 - \sum_{-\infty}^{+\infty}{}' c_n n^{-2} e^{inx}$$

which is *smooth* at each x, that is

$$\frac{F(x+h) + F(x-h) - 2F(x)}{h} \to 0 \; (h \to 0). \tag{1.1}$$

As a matter of fact, the last relation holds uniformly in x.

If the sequence c_n is merely bounded the result is no longer true and we can only assert that the ratio in (1.1) remains bounded, uniformly in x, as $h \to 0$.

Since the property of smoothness is important in certain applications we may ask when (1.1) holds, for some points x at least, if we no longer assume that the $c_n \to 0$. The theorem which follows may be useful here.

Theorem 1. *Suppose that the coefficiens c_n of the series $S = \sum c_n e^{inx}$ are uniformly bounded and consider the formal product of S with an absolutely convergent trigonometric series*

$$\sum \gamma_n e^{inx} \quad \left(\sum |\gamma_n| < \infty\right) \tag{T}$$

If $\sum c'_n e^{inx}$ is this product, then clearly $c'_n = O(1)$. However the function

$$F_1(x) = \tfrac{1}{2} c'_0 x^2 - \sum_{-\infty}^{+\infty}{}' c'_n n^{-2} e^{inx}$$

obtained by integrating the series $\sum c'_n e^{inx}$ twice is smooth at each point x_0 where the sum $\lambda(x)$ of $\sum \gamma_n e^{inx}$ is zero.

We need the following lemma which is a slight generalisation of the Riemann theorem stated above.

Lemma. *Suppose that the sequence* $\{s_n\}_{n=1,2,\ldots}$ *is bounded and summable* $(C, 1)$ *to limit* s. *Then*

$$\lim_{h\to 0} \frac{2}{\pi} \sum_{n=1}^{\infty} s_n \frac{\sin^2 nh}{n^2 h} = s. \tag{1.2}$$

This lemma is known but we indicate briefly the proof. In view of the formula

$$\frac{1}{2} h + \sum_{n=1}^{\infty} \frac{\sin^2 nh}{n^2 h} = \frac{1}{2}\pi \qquad \left(0 < h \leqq \frac{1}{2}\pi\right)$$

we may assume that $s=0$, the only case that interests us. Let $h \to +0$, $N=[1/h]$, and let k be a large but fixed integer. We write the sum in (1.2) in the form

$$\sum = \sum_{n=1}^{kN} + \sum_{n=kN+1}^{\infty} = \sum{}' + \sum{}''.$$

Clearly, if $|s_n| \leqq M$ for all n,

$$\left|\sum{}''\right| \leqq M h^{-1} \sum_{kN+1}^{\infty} \frac{1}{n^2}$$

and so is small if k is large. On the other hand, if $s_1 + s_2 + \cdots + s_n = n\sigma_n$ summation by parts shows that

$$\begin{aligned} \left|\sum{}'\right| &\leqq \left| h \sum_{n=1}^{kN-1} n\sigma_n \Delta \frac{\sin^2 nh}{n^2 h^2} \right| + o(1) \\ &\leqq h \operatorname*{Max}_{n<kN} n|\sigma_n| \sum_{n=1}^{\infty} \left| \Delta \frac{\sin^2 nh}{n^2 h^2} \right| + o(1) \\ &\leqq h \cdot o(N)\, V + o(1) = o(1), \end{aligned}$$

where V denotes the total variation of the function $(\sin x)^2/x^2$ over the interval $(0, +\infty)$. This completes the proof of the lemma.

Passing to the proof of the theorem we may assume that $x_0 = 0$. The formal product $S' = \sum c'_n e^{inx}$ of the series $S = \sum c_n e^{inx}$ and $T = \sum \gamma_n e^{inx}$ has coefficients

$$c'_n = \sum_p c_p \gamma_{n-p}.$$

Since $c_n = o(1)$, $\sum |\gamma_n| < \infty$, we have $c'_n = o(1)$. We have to show that $F_1(h) + F_1(-h) - 2F_1(0) = o(h)$. Now,

$$F_1(2h) + F_1(-2h) - 2F_1(0) = o(h) + 4 \sum_{n\neq 0}{}' c'_n \frac{\sin^2 nh}{n^2 h}. \tag{1.3}$$

and if we show that under our hypotheses,

$$\sum_{k=-N}^{+N} c'_k = o(N),$$

an application of the lemma will show that the expression (1.3) is actually $o(h)$ and the theorem will be established.

Write

$$\Gamma_q = \sum_{p=q}^{+\infty} \gamma_p$$

Then

$$\sum_{n=-N}^{N} c'_n = \sum_{n=-N}^{N} \sum_{p=-\infty}^{+\infty} c_p \gamma_{n-p} = \sum_p c_p \{\Gamma_{-N-p} - \Gamma_{N-p+1}\}$$

and

$$\left|\sum_{-N}^{N} c'_n\right| \leqq M \sum_{p=-\infty}^{+\infty} |\Gamma_{-N-p} - \Gamma_{N-p+1}|$$

Let us split the last sum into two, extended over the ranges $|p| \leqq 2N$ and $|p| > 2N$. Denoting these sums respectively by $\sum'$ and $\sum''$ and observing that in view of the hypothesis $\lambda(0) = 0$ we have $\Gamma_s = o(1)$ for s tending both to $+\infty$ and $-\infty$ we clearly have

$$\sum{}' = o(N).$$

On the other hand,

$$\sum{}'' < \sum_{|p|>2N} \sum_{-N-p}^{N-p} |\gamma_q| \leqq (2N+1) \sum_{|q|>N} |\gamma_q| = o(N).$$

Thus $\sum' + \sum'' = o(N)$ and the theorem is established.

2. Let $\widetilde{F}(x)$ be the sum of the series obtained by integrating formally twice the series $\widetilde{S} = \sum(-i \operatorname{sign} n) c_n e^{inx}$ conjugate to S. Thus

$$\widetilde{F}(x) = i \sum{}' (\operatorname{sign} n) \frac{c_n}{n^2} e^{inx}.$$

Theorem 2. *Under the hypotheses of Theorem 1, the function $\widetilde{F}$ is likewise smooth at the points where $\lambda(x)$ is zero.*

The proof parallels that of Theorem 1. Assuming that $x_0 = 0$ it is enough to show that

$$\sum_{-N}^{+N} (\operatorname{sign} n) c'_n = \sum_{p=-\infty}^{+\infty} c_p \{\Gamma_{1-p} - \Gamma_{N-p+1}\} - \sum_{p=-\infty}^{+\infty} c_p \{\Gamma_{-N-p} - \Gamma_{-p}\} = o(N).$$

We omit the details.

It is clear that both in Theorems 1 and 2 the smothness is uniform over the set of the zeros of λ.

Also, if at some x_0 the value of λ is not zero, writing $\lambda(x)$ in the form $\{\lambda(x) - \lambda(x_0)\} + \lambda(x_0)$ we easily find that

$$\frac{F_1(x_0+h) + F_1(x_0-h) - 2F_1(x_0)}{h} - \lambda(x_0) \frac{F(x_0+h) + F(x_0-h) - 2F(x_0)}{h} \to 0$$

and likewise for $\widetilde{F}$.

Theorems 1 and 2 have connection with the theorems of Rajchman about the formal multiplication of trigonometric series. In Rajchman's theorems (see A. ZYGMUND, "Trigonometric series", Chapter IX) we assume that the "bad" series S has coefficients tending to 0 and we show that if the "good" series T has coefficients satisfying the condition $\sum |n \gamma_n| < \infty$, then the formal product ST converges to 0 at each point where T vanishes. In our case we assume less about both S and T, and of course get less in the conclusion.

The Influence of M. H. Stone on the Origins of Category Theory

By SAUNDERS MACLANE
The University of Chicago

This talk is a small piece of historical investigation, intended to be an example of history in the retrospective sense: Starting with some currently active ideas in category theory, it will examine their origins in particular in certain work of MARSHALL STONE. Hence this talk will not even mention many of STONE'S contributions (his theorem on one-parameter unitary groups, the Stone-Weierstrass theorem, or his results on spectra, on integration, or on convexity); instead we will examine the connection of just a few of his ideas with the subsequent development of category theory.

Our historical study focuses on the mathematical developments of the decade 1929—1939. This decade runs from the onset of the great depression to the beginning of the second World War, so its initial and terminal dates have some general historical significance. This decade also has special mathematical significance; it saw the rapid development of the methods of general analysis, modern algebra, and topology and witnessed an increasingly active interaction between these disciplines. The mathematical climate of the decade may be summarized by noting the most influential books published in this period. Such a list would surely include HERMANN WEYL'S Group Theory and Quantum Mechanics (1928), VAN DER WAERDEN'S Moderne Algebra (1930 and 1931), STONE'S Linear Transformations in Hilbert Space and Their Applications to Analysis (1932), BANACH'S Théorie des Opération linéaires (1932), ALEXANDROFF-HOPF'S Topology (1935), WEIL'S Spaces with Uniform Structure and General Topology (1938) and the first section (Summary of the theory of sets) of BOURBAKI'S Fundamental Structures of Analysis (1939).

The dominant tone of these investigations is very different from that of previous decades. The general attitude is well reflected by STONE'S statement, in his summary article [16] on the representation of Boolean algebras: "A cardinal principle of modern mathematical research may be stated as a maxim: 'One must always topologize.'" In the same article, STONE goes on to observe that this process of employing a suitable topology has given, for example, a deeper insight into the structure of Boolean rings. This emphasis on the use of all relevant mathematical methods to get at the deeper structure of the mathematical entities at hand is typical of STONE's influence on the development of Mathematics. The decade 1929—1939 emphasized the study of a variety of mathematical structures, and so set the stage for many future developments; in particular, for the ideas of category theory.

Category theory asks of each type of mathematical structure "What are the morphisms?" All the morphisms from A to B constitute the set hom (A, B), in the category of all objects $A, B, C \ldots$ of the given structure, and the axioms for a category (see [10] or [7]) specify the properties of composition of morphisms. This emphasis on morphisms as such came in the 1940's and not in the decade 1929—1939, when attention was focused rather on subobjects (monomorphisms) and quotients (epimorphisms). For example, VAN DER WAERDEN's Moderne Algebra, following the lead of EMMY NOETHER, studies homomorphisms $G \to H$ of groups, and of rings, but only such as map G *onto* H. The utility of considering the more general homomorphisms $G \to H$ of one group *into* another first became clear from the example of algebraic topology, where one was forced to study continuous maps $X \to Y$ of one topological space *into* another, and the corresponding homomorphisms on the homology groups. Indeed the notion of a functor as a morphism of categories is suggested by the decisive example of the homology functor H_n on the category of topological spaces to the category of abelian groups; it sends each space X to the corresponding nth (singular or Čech) homology group $H_n(X)$ and each morphism $f: X \to Y$ of spaces to the induced morphism $H_n(f): H_n(X) \to H_n(Y)$ of homology groups.

The general idea of a functor as a morphism of categories was foreshadowed by many other examples. One example much emphasized in the work of MARSHALL STONE is the ring $C(X)$ of all continuous real-valued functions $f: X \to R$ on the topological space X. We may regard C as a functor on the category of such spaces to the category of rings. Just as in the case of homology, the use of such functors emphasizes and formalizes the passage between topology and algebra, a passage fundamental to many of STONE's investigations.

STONE's work emphasized certain cases where such a passage provides an equivalence between topological and algebraic notions — a relation we now formulate as an equivalence of categories. For example, in Theorem 4, p. 383 of [15] he writes "The algebraic theory of Boolean rings is mathematically equivalent to the topological theory of Boolean spaces by virtue of the following relations ..." The relations which then follow involve the functor E which assigns to each Boolean ring A its prime ideal spectrum $E(A)$, which is a Boolean space (a locally compact totally disconnected Hausdorff space) and the functor B which reciprocally sends each such space S to the Boolean ring $B(S)$ of all compact open subsets of S. The cited theorem of Stone goes on to specify the functorial character of these two constructions E and B. To be sure, this is specified not in terms of the general action of E and B on morphisms, but in terms of their effect, more specifically, upon automorphisms, monomorphisms (subrings), and epimorphisms (via ideals). A previous theorem of Stone's had produced a homeomorphism $S \cong EB(S)$ for any Boolean space S and an isomorphism $A \cong BE(A)$ for any Boolean algebra A. Both these isomorphisms are natural ones in the technical sense of category theory. The presence of two such natural isomorphisms $I \cong EB$ and $I' \cong BE$, with I and I' identity functors, is exactly the assertion that the functors E and B provide an *equivalence of categories*; more exactly an equivalence of the category of Boolean algebras to the category of Boolean spaces. Stone's prescient emphasis on the careful formulation of this particular equivalence amounts to a clear recognition of the importance of the general notion, an importance which can now be illustrated by many different examples of equivalences of categories — simplicial sets and CW-complexes, or the useful equivalence between the category of *all* finite sets and the "skeletal" category of finite sets, in which there is just one finite set n for each natural number $n = 0, 1, 2, \ldots$. (With this example, one can clearly note that mathematics really doesn't use all those different finite sets.)

The categorical notions mentioned so far — morphism, functor, natural isomorphism, and equivalence of categories — belong to descriptive category theory. The cited instances of their use document the claim that category theory provides a handy language for the formulation of a large class of those general observations about mathematical structure which were first recognized by Stone and others in the decade of the 30's, and were then formulated by Eilenberg-MacLane in the 40's (see [5, 6]). More recent developments indicate that the deeper aspects of category theory are those involved in the concept of adjointness, due in its complementary aspects to Samuel (1948, [11]) and Kan (1958, [8]). We now indicate how Kan's notion of adjoint functor may have been adumbrated in the analytical study of adjoint transformations. It will appear that Stone's work provides both a formalism which inspired that for adjoint functors and an exhibition of several decisive examples of such adjointness.

The notion of an adjoint operator is an old one, appearing for differential equations in the work of Legendre and elsewhere, notably later in the work of Bôcher. In algebra, adjoints appear as conjugate transpose operators. For a Hilbert space H, the adjoint T^* of the linear operator T on H is defined by the familiar condition on the inner product

$$(T^* f, g) = (f, Tg), \tag{1}$$

required for all relevant vectors f, g in H. This definitive formulation is due to Stone in 1929 (see [12], p. 198); it was almost immediately put to extensive use by von Neumann ([19], 1932).

The definition of adjoint functor, as first stated by Kan ([8], 1958) is formally parallel, for functors $U:\boldsymbol{A}\to\boldsymbol{X}$ and $F:\boldsymbol{X}\to\boldsymbol{A}$ in the opposite directions between categories $\boldsymbol{A}$ and $\boldsymbol{X}$. Indeed, F is a *left adjoint* of U if there is an isomorphism

$$\hom_{\boldsymbol{A}}(FX, A) \cong \hom_{\boldsymbol{X}}(X, UA) \tag{2}$$

of hom-sets, defined for all objects X of $\boldsymbol{X}$ and A of $\boldsymbol{A}$, and natural in these objects. The formal analogy to the definition (1) of adjoint transformation is striking — and becomes even more so if we use the abbreviated notation (FX, A) for the set $\hom_{\boldsymbol{A}}(FX, A)$ of all morphisms in $\boldsymbol{A}$ from FX to A. On the factual side, we may illustrate by several examples of adjoint pairs of functors: If $\boldsymbol{A}$ is the category of all (real) vector spaces, $\boldsymbol{X}$ that of sets, and U the functor which assigns to each vector space A its underlying set of vectors (the "forgetful" functor, which forgets the vector space structure), then the corresponding left adjoint is the functor F which assigns to each set X the real vector space FX with basis X. The isomorphism (2) is then the familiar one, which states that a linear transformation $f:FX\to A$ on the vector space FX with basis X is completely determined once its values $f':X\to UA$ on the basis X are known.

In this example, we may regard FX as the "free" real vector space over the set X of "generators"; other constructions of "free" algebraic systems of various types lead in the same way to pairs of adjoint functors. For instance, the forgetful functor from groups to sets has as left adjoint the functor which assigns to each set X the free group FX with generators X, and the forgetful functor from algebras (over the reals) to real vector spaces has as left adjoint the functor T which sends each vector space V to the tensor algebra TV. For topological spaces with a base point, the suspension ΣX is left adjoint to the loop space ΩY, and this fact is used repeatedly in homotopy theory. The tensor product of vector spaces is characterized by the equation

$$\hom(V\otimes C, W) \cong \hom\big(V, \hom(C, W)\big),$$

natural for all vector spaces V, C, and W (over the same field); this equation states simply that the functor $-\otimes C$ is left adjoint to the functor hom$(C, -)$. Many other examples of the ubiquity of adjoint functors may be found in any one of the standard sources; for example [7] or [10].

The basic formal properties of adjoint functors are strikingly parallel to those of adjoint linear operators. First, when the adjoint (or the adjoint functor) exists, it is unique; for functors, this of course means unique up to a natural isomorphism. Second, the composition rule $(ST)^* = T^* S^*$ for adjoint operators has an exact analog governing the composition of adjoint functors. Third, the adjoint of a linear operator is linear, and correspondingly the adjoint of an additive functor is additive ([10], Theorem 13.1); more generally any left adjoint commutes with direct limits. Only in the existence assertions is there a notable difference; Freyd's adjoint functor theorem ([7], p. 84) appears to have no analog for adjoint operators. In all the other cited properties the parallel is so strong as to raise the evident question "Why?". However, I do not know any formal explanation, for example, any more general concept which would subsume both adjoint functors and adjoint operators with the just noted corresponding formal properties.

Even the properties of norms of operators have a parallel. In the definition (1) of an adjoint operator we may set $g = T^* f$ to calculate the norm $\|T^* f\|$ as

$$\|T^* f\|^2 = (f, T T^* f).$$

Similarly one may set $A = FX$ in the definition (2) of an adjoint functor. Under this adjunction isomorphism the identity map $1: FX \to FX$ then corresponds to a special morphism $m: X \to UFX$. This morphism m is universal from the object X to the functor U, in the sense that every morphism $g: X \to UA$ from X to a value of U factors uniquely through m, via a morphism $g': FX \to A$. Such universal constructions were first described by Samuel, and indeed the notion of adjointness may be characterized completely in terms of the universality of m (see [10], Theorems 7.1 and 8.3).

Marshall Stone's work not only set the stage for the general definition of adjoint functors by providing the clearly parallel definition of adjoint operator; his studies also gave some decisive examples of the construction of explicit adjoint functors. One such example arose directly in his pioneering and systematic study [14] of the relation between the "abstract" notion of a Boolean ring and the more "concrete" notion of an algebra of classes. It is well known that he raised (and solved) the question whether any Boolean ring could be realized as an algebra of classes, but one can note in addition that his solution amounted exactly to the construction of a pair of adjoint functors. Each algebra of classes may be regarded as a Boolean ring, and this by the evident forgetful functor (forget that the elements of the algebra are in fact classes). Stone explicitly constructed a left adjoint, which assigns to each Boolean ring A the algebra $E(A)$ of all subclasses of the class of all prime ideals of A. He also constructed the morphism $E:A \to E(A)$ of Boolean rings which sends each element a of the Boolean ring A to the class $E(a)$ of all those prime ideals p in A with a not in p. This morphism E is in fact the universal morphism of the Boolean ring A to (a value of) the forgetful functor from algebras of classes, and E is the left adjoint of the forgetful functor precisely because of this universality of E. Currently this universality is stated in terms of morphisms, but classically it can be stated in terms of the construction of suitable quotients (replacing epimorphisms). This universality of E is exactly so stated in the concluding Theorems 68 and 69 of STONE's paper [14] on the representation of a Boolean ring by an algebra of classes. As he remarks "We now complete the theory of representation by means of the following result" (the universality of E).

A more famous example of adjoint functors is the Stone-Čech compactification β. It is a functor.

$$\text{Completely regular spaces} \xrightarrow{\beta} \text{Compact Hausdoff spaces},$$

and is the left adjoint to the forgetful functor (opposite direction). The usual description of the compactification βX of the completely regular space X states that βX is compact, that the embedding $m:X \to \beta X$ is dense, and that every continuous function from X to the reals R extends in a unique way to a continuous functor $\beta X \to R$. This property of β was stated both by STONE and ČECH [4]. STONE went further, and observed (in [15], Theorem 88) that any continuous mapping of X into a compact Hausdorff space can be factored (uniquely) through βX. This observation amounts exactly to the assertion that the embedding $m:X \to \beta X$ is universal. In other words, STONE not only constructed β but established in full the properties which assert that β is a left adjoint.

For X completely regular, this universal map $X \to \beta X$ is a monomorphism (an embedding). One may also search for a compactification of any T_0 space, except that in this case the universal mapping will no longer be a monomorphism. This compactification is provided by composing β with the functor γ

$$T_0\text{-spaces} \to \text{completely regular spaces},$$

adjoint to the forgetful functor. Stone explicitly constructed this functor ([15], Theorem 77) and showed it, via universality, to be the desired adjoint ([15], Theorem 89).

We have now summarized some aspects of MARSHALL STONE's work on adjoint linear operators and on the representation of Boolean rings.

It has long been realized that Stone's investigations on the latter type have had a decisive influence on the subsequent study of rings of function What we have noted now is that these investigations also provided a sharp and completely formulated example of the construction of adjoint functors. This idea, central to the present usage of categories, was there in embryo in Stone's papers of 1936 and 1937.

It is of course not surprising that the abstraction presented in category theory should arise from a variety of concrete instances; this is a common aspect in the historical development of mathematics. What is surprising is that the clear recognition of the notion of adjoint functor waited for over twenty years, till Kan's paper in 1958. The notions of category and functor themselves were developed quite soon after Stone's work, in 1942 and 1945 ([5] and [6]). The related notion of universal construction also appeared explicitly in the work of Samuel [11] in 1948 and was extensively used by Bourbaki, though not in the efficient language of categories. (Indeed, it was some time after the general recognition of the importance of Kan's work that the relation between adjoint functors and universality was clearly realized.)

This situation is a striking instance of a historical question which can be raised about the time of appearance of many mathematical concepts: Given the available formalism of adjoint operators and the numerous examples of adjoint functors, why was the general notion so late in arising?

We have little expertise in answering such questions in the history of mathematics and any attempt at an answer can only be speculation (and is not to be regarded as part of my analysis of Stone's work). It is my own view that the climate of mathematical opinion in the decade 1946—1956 was not favorable to further conceptual development. Investigation of concepts as general as those of category theory were heartily discouraged, perhaps because it was felt that the scheme provided by Bourbaki's structures produced enough generality. It is to be noted that Kan, when developing adjoint functors, came at the time from a solitary position more or less outside active mathematical circles. It may even be that we should be on our guard lest the current very active mathematical life inhibit the development of ideas which fall outside the established directions of research.

There are other questions, not historical but mathematical, which bear on future possibilities and new directions. Stone's investigations emphasized the utility of some axiomatic methods in analysis. The current development of category theory raises the possibility of a different style of axiomatics: One axiomatizes (say) not a single Hilbert space, but the category of all Hilbert spaces. The model of such axioms on a whole category has been indicated by Lawvere, in his axioms [9] on the category of sets. The method has as yet been little exploited, but there are many possible cases at hand.

Category theory has had lively use in Topology, Homological Algebra, and Algebraic Geometry, but there have not as yet been decisive applications to General Topology or to Functional Analysis. We have observed that these fields, in the hands of Marshall Stone, did provide some starting points for categorists. There is a clear prospect that there may be future developments there, either using categories or some yet-to-be-discoreved parallel notions.

References

1. Alexandroff, P., u. H. Hopf: Topologie I. Berlin: Springer 1935.
2. Banach, S.: Théorie des opérations linéaires. Warsaw 1932.
3. Bourbaki, N.: Eléménts de mathématique. Première partie, Les structures fondamentales de l'analyse, Livre I, Théoriè des ensembles (fascicule de résultats). Act. Sci. et Ind., No. 846. Paris: Hermann 1939.
4. Čech, E.: On bicompact spaces. Ann. of. Math. **38**, 823–846 (1937).
5. Eilenberg, S., and S. MacLane: Natural isomorphisms in group theory. Proc. Nat. Acad. Sci. U.S. **28**, 537–543 (1942).
6. — — General theory of natural equivalences. Trans. AMS **58**, 231–294 (1945).
7. Freyd, P.: Abelian categories, an introduction to the theory of functors. New York: Harper & Row 1964.
8. Kan, D. M.: Adjoint functors. Trans. AMS **87**, 294–329 (1958).
9. Lawvere, F. W.: An elementary theory of the category of sets. Proc. Nat. Acad. Sci. U.S.A. **52**, 1506–1511 (1964).
10. MacLane, S.: Categorical algebra. Bull. Am. Math. Soc. **71**, 40–106 (1965).
11. Samuel, P.: On universal mappings and free topological groups. Bull. Am. Math. Soc. **54**, 591–598 (1948).
12. Stone, M. H.: Linear transformations in Hilbert space. I. Geometrical aspects. Proc. Nat. Acad. Sci. U.S.A **15**, 198–202 (1929).
13. — Linear transformations in Hilbert space and their applications to analysis. Am. Math. Soc. Coll. Publ., vol. XV. New York 1932.
14. — The theory of representations for Boolean algebras. Trans. AMS **40**, 37–111 (1936).
15. — Applications of the theory of Boolean rings to general topology. Trans. AMS **41**, 321–364 (1937).
16. — The representation of Boolean algebras. Bull. A.M.S. **44**, 807–816 (1938).
17. — A general theory of spectra. I. Proc. Nat. Acad. Sci. U.S **26**, 280–283 (1940).
18. Waerden, B. L. van der: Moderne Algebra. Berlin: Springer, vol. 1 (1930), vol. 2 (1931).
19. von Neumann, J.: Über adjungierte Funktionaloperatoren. Ann. Math. **33**, 249–310 (1932).
20. Weil, A.: Sur les espaces á structure uniforme et sur la topologie générale. Act. Sci. et Ind., No. 551. Paris 1938.
21. Weyl, H.: Gruppentheorie und Quantenmechanik. Leipzig 1928.

Remarks of Professor Stone

In his lecture, Professor MacLane asked some intriguing historical questions about the concept of adjoint functors or operators. I may perhaps be permitted to make a few comments by way of response to these questions. If one seeks to explain why the notion of an adjoint functor appeared in category theory as much as fifteen years after its inception, it seems to me that two reasons have to be suggested. In the first place, the concept had its origins and early development in analysis, in the theory of differential equations. It was not a vital part of the mathematical experience of the algebraically-oriented pioneers in category theory. The other historical factor that has to be cited is the interruption of mathematical research imposed by World War II and the many professional readjustments that followed it.

Without an opportunity to do some historical research on the point, I cannot now trace the concept of adjoint back to its origins. In the theory differential equations it is rather old. Bôcher in his "Leçons sur les mèthodes de Sturm" was already able to give a classic exposition of the concept and its applications to ordinary differential equations. In the late twenties the concept assumed a more abstract, algebraic form as a result of its relevance to the theory of operators in Hilbert space. The history of this development is interesting and instructive. I should like to sketch it briefly.

Stimulated by an interest in quantum mechanics, J. von Neumann began the work in operator theory which he was to continue as long as he lived. Most of the ideas essential for an abstract theory had already been developed by F. Riesz, who had established the spectral theory for bounded symmetric operators in a form very much like that now regarded as standard. Von Neumann saw the need for extending Riesz's treatment to nonbounded operators and found a clue to doing this in Carleman's highly original book on integral operators with singular kernels. The result was a paper von Neumann submitted for publication to the Mathematische Zeitschrift but later withdrew. The reason for the withdrawal was that in 1928 Erhard Schmidt and I independently saw the role which could be played in the theory by the concept of the adjoint operator and the importance which should be attached to self-adjoint operators. When von Neumann learned from Professor Schmidt of this observation, he was at once enabled to rewrite his paper in much more satisfactory and complete form, giving a full spectral theory for all closed symmetric operators as well as for the self-adjoint operators. This he did by abandoning Carleman's methods, which he had been able to apply only by use of a transfinite induction, and introducing the Cayley transform, which served to reduce the theory of non-bounded symmetric operators to that of bounded isometric operators. Incidentally, for permission to withdraw the paper without penalties when it was already in page proof the publisher exacted from Professor VON NEUMANN a promise to write for him a book on quantum mechanics. The book soon appeared and has become one of the classics of modern physics, particularly valued for its analysis of quantum statistics. My own interest in Hilbert space problems was aroused by reading the page proofs of VON NEUMANN'S paper when they were given to me by Professor CARATHEODORY, an editor at that time of the Mathematische Zeitschrift, just as he was finishing a visiting lectureship at Harvard in the spring of 1928. I saw almost at once that VON NEUMANN needed the concept of adjoint as a key to his problem and that a successful approach to the spectral theorem for self-adjoint operators could be made by the use of methods already familiar in the theory of differential equations. Looking back, I draw from this bit of mathematical history the lesson that a broad background is sometimes useful in quickly finding important clues to particular problems. Certainly VON NEUMANN would have seen for himself the relevance of the notions of adjoint and self-adjointness had his experience with differential integral operators then been as extensive as Professor SCHMIDT'S or my and own.

By analogy, I am equally certain that EILENBERG, MACLANE, and the other early pioneers in category theory would have hit quite early upon the idea of adjoint functor if they had been more fully aware of contemporary work in Hilbert space theory and its background in the theories of differential equations and topological groups. As it was, they drew mainly upon algebra and algebraic topology-especially the theory of abelian groups — for sources of the new theory. Of course, it will never take very long for such a central and essentially simple concept as that of adjoint functor to emerge from research carried on strictly inside a growing branch of mathematics. While I do not know the private history of KAN'S introduction of the adjoint functor, I suspect that it may have come about in just this manner. Even if such be the case, one could still ask with Professor MACLANE why, as a matter of the purely internal development of category theory, adjoint functors did not appear considerably earlier upon the cene. After all, the well-known dualities for abelian groups and for vector spaces could have provided clues without any need of venturing very far away from the initial concerns of category theory. If an answer is to be found here I think we have to look at another factor, the impact of the Second World War upon mathematicians and mathematics.

Before doing this, let me indulge in a few comments on the relation between category theory and other branches of mathematics, with an eye upon the future rather than upon the past. Category theory has scored real triumphs in algebra, algebraic topology, and algebraic geometry, and in so doing has established itself as a respectable branch of mathematics, worthy of intensive cultivation both for its own sake and for the sake of its applications. However, the relevance and the importance of category theory for some other branches of mathematics, such as analysis and set-theoretical or general topology is still to be demonstrated. In spite of a considerable amount of research intended to connect category theory with these branches of mathematics, we still have few if any indications that categories will serve as powerful tools in their further development. Professor MacLane has been very kind in pointing out the presence in my paper on "Applications of the Theory of Boolean rings to General Topology" of a number of mathematical phenomena which have since been recognized and studied in category theory. More recent work by others has shown that there are indeed many close links between general topology and category theory. Fundamentally, however, this still amounts to saying that many topological situations appear as special cases of general results in category theory. We do not yet have many new results in general topology made possible by essential applications of category theory. Possibly this is due to the observation, made in a recent paper of Steenrod, that the kinds of topological space that can be treated smoothly in terms of categories are not coextensive with the kinds that topologists have chosen as the subject of their investigations. The connections between measure theory and category theory have likewise been investigated, especially by Linton, with somewhat similar results. Some measure-theoretical phenomena can be neatly expressed in terms of category theory, but others seem to remain elusive. Indeed, some problems in measure theory suggested by the connections with category theory and essential to a full understanding of those connections can apparently be attacked only within measure theory and remain unsolved. It is possible that some of Laurent Schwartz's recent work on measure theory could be interpreted as effecting a rapprochement with category theory through a revision of the fundamental concepts of measure theory. If so, there is an obvious analogy with Steenrod's proposal in the case of topology. These two illustrations may serve to remind us that in those parts of mathematics where limit processes play an essential part, as in general topology, measure theory, and analysis taken as a whole, we often have to contend with situations bordering on the pathological and sometimes need to avoid entanglement in irresoluble pathologies by judicious limitation of the problems we try to attack. This clearly presents a certain contrast with the situation in algebra. Furthermore it suggests that however well category theory may be adapted to algebraic situations, it may still not fit the requirements of analysis and general topology in an entirely satisfactory manner. Whether the answer lies in changing our attitude towards analysis or in modifying category theory or in working out some sort of mutual adjustment between them, only future researches can tell.

Returning now to the influence of World War II upon mathematics in general and upon the development of category theory in particular, we need only recall the almost universal disruption of mathematical activity to realize how many potential advances must have been held up or even made forever impossible. It is unnecessary to recall in detail the breakdown of international communications, the diversion of mature mathematicians to war-time tasks, the suspension of normal teaching functions in the universities, the absorption of youth by the military services, or the tragic loss of life on the battle field and in concentration camps. All these evils had a heavy impact upon our science. Yet it seems to me equally important to dwell upn a less obvious influence of the war, but one that should not be underestimated. In the war's aftermath we had to deal with many problems of readjustment and reconstruction which undoubtedly took their toll of research. At the same time many of us returned to our normal pursuits as mathematicians with a certain restlessness of spirit and a new vision of a more intense and highly organized mathematical activity than we had previously known. Those of us who shared this sense of restlessness and of the opening up of new possibilities were not content to pick things up just as we had left them. Instead we looked for new beginnings and devoted much time and energy over a number of years to organizing a significantly higher level of mathematical activity than we had known before the war. In retrospect it seems to me plain that this upsurge for which we worked in the years just after the war was partly the result of the intellectual revolution in mathematics that had begun at the turn of the century and was in full swing by 1939, partly the result of new organizational trends in science which received a great impetus from experience gained in the prosecution of the war. If research suffered a little because of such new preoccupations — I think there is no doubt that it did — it was only for a short time. We all know that since then mathematical research has attained impressive new levels of quantity and quality.

The effects which we have tried to describe here can be traced in more specific terms with reference to developments in the field of which category theory is a major part. In the period just before the war a number of mathematicians began to look for general principles of the kind with which category theory is concerned. Some of them found leads in my paper on "Applications of the Theory of Boolean Rings to General Topology". One problem posed in this paper was that of characterizing the rings of continuous functions on a compact Hausdorff space. A manuscript of the paper was made available to mathematicians in Moscow by early 1936 and quickly stimulated some notes by Kolmogorov and Gelfand. A little later and at about the same time, Gelfand and I gave solutions of the characterization problem, he in the case of complex functions with his famous paper on Banach algebras and I in the case of real functions. Similar results were obtained for lattices or vector lattices of various types by a number of mathematicians, including Kakutani, the Kreins, von Neumann, and myself. Indeed, after reading a paper of Freudenthal, I had obtained an unpublished result along these lines as early as the summer of 1938. These different theorems had as a common feature the fact that they all furnished representations of certain algebraic structures by sets of functions subject to appropriate operations. The general principle on which they were all based was formulated in explicit terms in notes written at roughly the same time by Garrett Birkhoff, Kakutani, and McCoy. The greater part of these results was published in 1939—1940. I have no doubt that, if it had not been for the war, research along these lines and along certain other lines suggested by my Boolean algebra paper would have continued at the same level of activity during the period 1940—1945. Speaking for myself, I can say that I did indeed carry on my investigations of some of the general principles which appeared to lie at the foundation of functional analysis and topology. I lectured on this subject at the University of Buenos Aires in 1943 with Alberto Calderón as one of my auditors; returned to the subject again in my unpublished retiring address as President of the American Mathematical Society; and finally lectured on it again at the University of Chicago, in a class which included Richard Kadison as one of its members. I never succeeded in bringing my results to a point which satisfied me, and I was eventually drawn away from further work upon them by problems closer to my major interests in analysis. One thing that displeased me was that I did not succeed in showing the relevance to Galois theory of what I hoped were general principles with at least that much scope. It is quite clear to me that if the war and its aftermath had not interrupted my systematic work in this field I would have been in a far better position to make substantial progress in it. I think that under other circumstances I would inevitably have been led to explore the significance of category theory as it then was for my own work. No doubt what I have said about my own experience must be true of the experiences of many other mathematicians. And it must always be remembered that a great many young people were prevented by the war from starting their mathematical careers when they were ready to begin and that they were therefore delayed in exercising their talents on the problems we have been discussing.

In conclusion I would like to insist again on the value of a broad mathematical background in the conduct of research into the fundamental structure of our subject and even into somewhat specialized aspects or branches of it. A sign of the essential unity of mathematics is given us by the sort of experience which has been recounted here. It points to the need for offering our future research methematicians a broad preparation that will enable them to cope successfully with the increasingly complex interconnections that bind mathematics into a single whole.